FUTURE DIMENSIONS OF
WORLD FOOD
AND
POPULATION

A Winrock International Study

Future Dimensions of
World Food and Population
edited by Richard G. Woods

Is it possible to feed those who now are hungry in
the world in addition to the billions of people who
will be born by the end of the century? Or are we
headed for an inevitable Malthusian catastrophe because
the task is impossible? What can developing countries
do to increase agricultural self-reliance? What population dynamics accompany the transition from high birth
and death rates in developing countries to low birth and
death rates? What research can aid the struggle to
provide food to the world's masses?

These and other questions are explored by an array
of experts who participated in the Congressional
Roundtable on World Food and Population during 1979-80.
They offer this collection of papers in the spirit of
optimism about the future and about the U.S. role in
international development.

Richard G. Woods is a graduate of the University
of Tulsa and the University of Minnesota, where he
served on the faculty and as codirector of the Office
for Applied Social Science and the Future. He also
served on active duty with the U.S. Air Force as a manpower management specialist and has held staff positions
with General Mills, Incorporated, and Whirlpool Corporation. Most recently he has been a legislative aide in
the U.S. Senate, specializing in such areas as science
and technology, nutrition, and world food and
population.

Published in cooperation with
Winrock International

FUTURE DIMENSIONS OF
WORLD FOOD
AND
POPULATION

edited by Richard G. Woods

Routledge
Taylor & Francis Group

LONDON AND NEW YORK

First published 1981 by Westview Press

Published 2018 by Routledge
52 Vanderbilt Avenue, New York, NY 10017
2 Park Square, Milton Park, Abingdon, Oxon OX14 4RN

Routledge is an imprint of the Taylor & Francis Group, an informa business

Library of Congress Catalog Card Number: 80-54240

ISBN 13: 978-0-367-01919-8 (hbk)
ISBN 13: 978-0-367-16906-0 (pbk)

CONTENTS

Foreword -- Henry Bellmon and Paul Simon ix

About the Authors xiii

Section 1. THE GLOBAL PROBLEM OF BALANCING
            POPULATION AND FOOD

    Introduction 3

1. The Nature of the World Food and Population
        Problem -- Maurice J. Williams 5

2. Recent Trends in World Food and Population --
        W. David Hopper 35

3. The Worldwide Loss of Cropland --
        Lester R. Brown 57

Section 2. THE POTENTIAL SOURCES OF FOOD

    Introduction 99

4. Using Plants to Meet World Food Needs --
        Norman E. Borlaug 101

5. The Contribution of Livestock to the World
        Protein Supply -- John A. Pino and
        Andres Martinez 183

6. Aquatic Animal Protein Food Resources --
   Actual and Potential --
   Carl J. Sindermann 239

Section 3. THE PROCESS OF AGRICULTURAL DEVELOPMENT

   Introduction 259

7. Agriculture in Transition -- Sayed A. Marei. . 261

8. Conditions for More Rapid Agricultural
   Development -- D. Gale Johnson 281

9. Strategies for Rapid Agricultural Growth --
   A. T. Mosher 299

Section 4. THE INFLUENCE OF TRADE AND INVESTMENT

   Introduction 321

10. International Trade and Investment Policies
    That Influence Agricultural and Economic
    Development -- Ray A. Goldberg 323

11. Implications of the Multilateral Trade
    Negotiations for Agricultural Trade --
    Alonzo L. McDonald 359

Section 5. THE CONSEQUENCES FOR AMERICA

   Introduction 369

12. The Impact of International Development
    on American Agriculture --
    Morton I. Sosland 371

13. Meeting Energy Requirements in the Food
    System -- Emery N. Castle 389

14. World Food and American National Security --
    Steven Muller 405

FOREWORD

The circumstances of human life around the world have been changing so rapidly during recent years that all of us, even those who make it their business to keep informed, have difficulty in getting a secure grasp on a reasonably accurate description of the status quo, to say nothing of the more difficult and more important vision of a likely, or even possible, future. We have more information than ever before, but our perceptions of problems, and therefore our needs for information, are dynamic. This situation makes the job of the public policymaker a frustrating one. No one likes to make decisions which will affect the future well-being of the country, and perhaps of the world, without a solid understanding of what is being accomplished. Legislators, among other decisionmakers, often have learned how to decide under conditions of uncertainty, but they privately wish for the time and opportunity to make thoughtful decisions in the expectation that they would be better decisions.

The Congress of the United States has talented persons to help in gathering information and providing analyses. The Congressional Research Service of the Library of Congress provides a cadre of thoroughly professional men and women who can function under time pressure to provide specialized information to Senators and Congressmen. A steady stream of evaluations of existing legislation and programs is supplied by the General Accounting Office in an effort to assist the Congress in its deliberations. The Office of Technology Assessment offers special studies requested by congressional committees with a particular emphasis upon the long-term consequences of various policy choices. Special studies of the fiscal impacts of alternative decisions can be obtained from the Congressional Budget Office. Yet all these dedicated staff persons can but assist the Congress; they cannot do the thinking which each member needs to do in preparation for particular decisions. Often the member wishes he had a broad, long-term perspective within which narrower decisions could be made, but there is no convenient mechanism for achieving that perspective. The daily, immediate pressures of the legislative calendar, crises, constituent needs, and party politics conspire to crowd out the time for careful reflection.

This need for careful consideration of the problems associated with world food and population prompted the Congressional Roundtable on World Food and Population during 1979 and early 1980. In late 1978, when the Conference of Southwest Foundations indicated its willingness to help with a project of interest to members of Congress, it was clear to some of us that there were problems of ensuring the world's food supply, but there was considerable uncertainty about the long-term implications of population increases upon the food supply and about the appropriate role for the United States in responding to world food needs. Members were familiar with existing programs of food assistance and developmental aid, but some were unclear about what objectives were sought, or ought to be sought, and others wondered if the nature and magnitude of the future challenge could be met by tinkering with existing programs and legislation.

We decided to try a series of dinner meetings for members of both the House and Senate at which recognized authorities would present facts and ideas dealing with special aspects of the world food and population problem. We believed it was important, also, for staff members who might have to deal with subsequent congressional action to be informed, so the series of monthly dinner meetings was augmented by a repeat performance by each expert at a breakfast meeting for staff on the following morning. The purpose of the series was to focus the talents of outside experts on the food and population problems and to involve members of Congress in a continued examination of these problems.

It was not anticipated that the experts would necessarily present policy alternatives or that members who attended would reach some consensus as the basis for specific legislative actions. We thought that members of the several committees of both houses of Congress responsible for legislation connected with world food problems, as well as other members who attended, would gain some perspective about the broader implications of their legislative responsibilities and about the future implications of congressional action.

This Congressional Roundtable was not an isolated inquiry into world food and population problems. At approximately the same time the Presidential Commission on World Hunger got underway, and the Independent Commission on International Development Issues (the Brandt Commission) was preparing its report. These latter efforts were more wholistic and policy-oriented than was the Congressional Roundtable, which was structured to conform more nearly with congressional interests than with any need to reach conclusions or achieve consistency.

Attendance at the Roundtable meetings was bipartisan and represented political orientations which were quite diverse. The setting in the Whittall Pavilion of the Library of Congress made it easy for members to engage speakers in colloquy. The relaxed, informal atmosphere of a dinner meeting was a welcome break from the pressures of the office and committee hearings.

When finished, the Roundtable had produced a series of papers which constituted a valuable addition to the information needed by those actively considering the ramifications of world food and population. There was a need to assemble these papers as a handy reference for members of Congress, to respond to requests from various sources for copies of the papers, and to provide new staff persons with a compilation of presentations given at the series. Private groups concerned about world hunger, college students studying international development, and citizens who simply wanted to be informed needed access to the results of the Roundtable. Accordingly, it was decided to collect and publish the papers in this book.

We offer this collection in the spirit of optimism about the future and the U.S. role in international development. We believe that strong U.S. participation, however constituted, is an inescapable part of this country's leadership in the world, and we are confident that better-informed legislators and citizens can make the necessary decisions to ensure that leadership.

Many persons contributed to the success of the Roundtable. The Cooperative Projects Program of the Conference of Southwest Foundations provided generous support and technical assistance. Dr. Richard Wheeler, President, and Edward L. Williams, Development Officer, both of the Winrock International Livestock Research and Training Center, Morrilton, Arkansas, were particularly helpful in coordinating foundation support and in facilitating many of the necessary arrangements.

Without their consistent encouragement and support, the entire series would have been immensely more difficult. Barbara Scott and Shirley Zimmerman of the Winrock staff performed yeoman work.

The Conference of Southwest Foundations and the donor foundations (the Samuel Roberts Noble Foundation Incorporated, the Kerr Foundation, Incorporated, the Ewing Halsell Foundation) assume no responsibility for any conclusions drawn by individual speakers in the Roundtable series, and have taken no position either in support of, or in opposition to, any of the recommendations made by individual experts.

Initial planning for the series was aided by a Steering Committee consisting of Dr. Garland Hadley of the Kerr Foundation, Dr. Sterling Wortman of the Rockefeller Foundation, John March of the Noble Foundation, Larry Minear of Church World Service, Dr. James McCullough from the Congressional Research Service of the Library of Congress and the two of us. Meeting arrangements were supervised by Beverly Bennett and Chad Wilson of the Congressional Research Service. Technical assistance was provided by Barry Carr, Leo Mayer, and Dennis Little of the Congressional Research Service. Our gratitude is extended to these persons whose contributions to the Congressional Roundtable on World Food and Population were of particular value.

Henry Bellmon, U.S. Senate

Paul Simon, U.S. House of Representatives

ABOUT THE AUTHORS

HENRY L. BELLMON is a Republican from Billings, Oklahoma, who has served two terms in the United States Senate, beginning in 1969. A graduate of Oklahoma State University, he has long been active in farming and related activities. After combat in the South Pacific as a marine during World War II, he returned to Oklahoma to serve one term in the Oklahoma House of Representatives. He was elected Oklahoma's first Republican governor in 1962 and served four years. He is a former chairman of the Interstate Oil Compact Commission and a former member of the executive committee of the National Governors' Association. As the senior Republican member of the Senate Budget Committee, a position he has held since the inception of the congressional budget process, he is known as a pragmatic fiscal conservative.

NORMAN E. BORLAUG is director of the International Wheat Research and Production Program of the International Maize and Wheat Improvement Center in Mexico City. His decades of work in plant pathology, agriculture, and plant breeding began with advanced degrees in plant pathology and employment with the U.S. Forest Service. In 1944 he joined the newly formed cooperative Mexican-Rockefeller Foundation Agricultural Program. His research in producing new high-yielding wheat varieties was largely responsible for moving Mexico from a wheat-importing nation to the largest wheat seed exporter in the world. He developed semi-dwarf wheat varieties which have revolutionized wheat growing in India, Pakistan, Turkey, and North Africa. Borlaug has worked in developing wheat research and production programs throughout the Near and Middle East and in Latin America, focusing recently on wheat production problems in Argentina (a potentially large food-exporting country), and in Pakistan and India. He is generally regarded as the "Father of the Green Revolution." His work has been widely recognized by a large number of awards and honorary degrees. In 1970 he was awarded the Nobel Peace Prize.

LESTER R. BROWN is president and senior researcher for the Worldwatch Institute of Washington, D.C. He was advisor to the Secretary of Agriculture on foreign agricultural policy during 1964 and served as administrator of the International Agricultural Development Service of the U.S. Department of Agriculture from 1966 to 1969 when he coordinated programs to increase food production in forty developing countries. Mr. Brown was Senior Fellow with the Overseas Development Council from 1969 to 1974.

EMERY N. CASTLE is president of Resources for the Future, Washington, D.C. A graduate of Kansas State and Iowa State Universities, Dr. Castle has been an agricultural economist for the Federal Reserve Bank of Kansas City as well as professor of agricultural economics at Oregon State University. From 1972 to 1976 he served as dean of the graduate school at Oregon State University. He was named Fellow of the American Agricultural Economic Association in 1976 and Fellow of the American Academy of Arts and Sciences in 1977. In 1976 he served as chairman of the Priorities Study Team for the World Food and Nutrition Study undertaken by the National Academy of Sciences. He was president of the American Agricultural Economics Association during 1972-1973.

RAY A. GOLDBERG prefaced twenty years of business and teaching experience with advanced degrees from Harvard and the University of Minnesota. He has acted in the capacity of officer or director of businesses and foundations across the spectrum -- from seed and grain companies to hospitals, from investment committees to foundations. He has taught agribusiness, agricultural markets, and in related fields from Harvard to Chicago, and he has conducted seminars and directed continuing education in these fields in Berlin, Acapulco, The Hague, and London. He has been consultant to government agencies in agricultural matters for over ten years and has served as editorial advisor to important publications such as the American Journal of Agricultural Economics.

W. DAVID HOPPER is vice president for the South Asia Region of the World Bank. Canadian by birth, David Hopper served as agricultural economist and associate field director for the Rockefeller Foundation's Indian Agricultural Program in New Delhi. He was consultant to the Asian Development Bank and deputy team leader of the ADB Asian Agricultural Development Survey Mission. He also was a consultant to the government of Kenya on agricultural marketing and price policy. Prior to joining the World Bank in 1978, he was president of the International Development Research Centre, Ottawa, Canada.

D. GALE JOHNSON has extensive experience as a consultant and economic advisor in agencies such as the U.S. Department of State, in the RAND Corporation, on the President's Task Force on Foreign Economic Assistance and other presidential commissions, as a special representative for trade negotiations, and on numerous other national advisory commissions and boards. Since 1944 he has been at the University of Chicago where he has been active in research and teaching as well as in departmental and university administration. Currently, he is provost and Eliakim Hastings Moore Distinguished Service Professor in the Department of Economics.

SAYED A. MAREI now serves as assistant to the President of the Arab Republic of Egypt. A graduate of the Faculty of Fu'ad I (now Cairo) University, he directed the Agrarian Reform Committee which brought a reform-minded approach to Egypt's agricultural problems. Except for a 1961-67 period as chairman of the Egyptian State Bank, Mr. Marei spent the period from 1952 to 1972 in high-level posts related to agriculture, including two terms as Minister of Agriculture and one as Deputy Prime Minister for Agriculture and Irrigation. More recently he has been First Secretary of the Arab Socialist Union and an adviser to the President. He was president of the World Food Council from 1975 to 1977, and he served as speaker of the People's Assembly (Egypt's only elected body) from 1974 to 1978. Sayed Marei was appointed assistant to the President in October of 1978.

ANDRES MARTINEZ is an animal scientist with Winrock International Livestock Research and Training Center in Arkansas. His professional career in animal agriculture development includes management of an animal research facility and feedstuffs analysis laboratory in California, technical group leader of

livestock development programs in Iran, regional marketing manager for FMC International AG, and foreign
assignments in more than twenty-five developing countries. Dr. Martinez has authored over thirty scientific publications, technical articles, and reports.

ALONZO L. McDONALD graduated from college in 1948
and spent some initial years in journalism. Following
an advanced degree from Harvard, he has had a distinguished career of management in the business sector.
Seventeen years of that career were with McKinsey and
Company, Incorporated, an international firm of management consultants, where he had key posts as managing
director and chief executive officer for the firm's
activities in such worldwide offices as London, Paris,
and Zurich. In 1973 he became managing director of
that firm in the United States. Long active in professional and civic affairs, he was a trustee of the
Committee for Economic Development and a member of the
Council on Foreign Relations, among many others.
President Carter appointed McDonald ambassador to head
the United States delegation to the Tokyo Round in July
of 1977. Subsequently, he was named White House Staff
Director.

ARTHUR T. MOSHER holds a doctorate in international trade and economic geography from the University
of Chicago. A professional career spanning two decades
at the Allahabad Agricultural Institute, India, culminated in tenure as Principal from 1948 to 1953. He was
visiting (research) professor of economics and cultural
change at the University of Chicago where he assessed
technical assistance programs in Latin America from
1953 to 1955. Dr. Mosher has been with the Agricultural Development Council since 1957 as executive
director, as president, and as associate in Sri Lanka.
His book Creating a Progressive Rural Structure is one
of many publications.

STEVEN MULLER is president of the Johns Hopkins
University and president of the Johns Hopkins Hospital,
Baltimore, Maryland. A native of Germany, Dr. Muller
became a naturalized citizen of the United States in
1949. He is a Rhodes Scholar and holds advanced degrees from Oxford and Cornell. He began his career in
teaching as a professor of political science and government, and he remains a specialist in comparative
government and international relations, particularly
concerned with political developments in Europe. From
1961 to 1971 he was director of the Cornell University

Center for International Studies and then vice-president for public affairs for Cornell University. Dr. Muller serves on the board of directors of numerous business and financial institutions and is a trustee of the Whitney Museum of American Art in New York City. A monthly column by Dr. Muller appears in the London Times higher education supplement. Recently, he served as vice-chairman of the Presidential Commission on World Hunger.

JOHN A. PINO is director for agricultural sciences of the Rockefeller Foundation, a position he has held since 1970. A former professor at Rutgers University, he has held various agricultural program positions abroad under the sponsorship of the Rockefeller Foundation. Among his many appointments he has served on the board of trustees of the International Center of Tropical Agriculture, board of directors of Winrock International Livestock Research and Training Center, and the National Agricultural Research and Extension Users advisory board.

PAUL SIMON is a Democrat from Carbondale, Illinois, who has served in the U.S. House of Representatives since the beginning of the 94th Congress in 1975. At the age of nineteen he purchased the Troy Tribune of Troy, Illinois, becoming the youngest editor-publisher in the nation. After expanding this venture into a chain of fourteen weeklies, he sold his business in 1966 to devote full time to writing and public service. Elected to the Illinois House of Representatives at the age of twenty-five, he was reelected three times, then elected to the Illinois Senate and reelected once before being elected Illinois lieutenant governor in 1968. In 1973 he taught at the John F. Kennedy Institute of Politics at Harvard University. The author of several books, among them A Hungry World and The Politics of World Hunger (with his brother, the Reverend Arthur Simon), he has received the Best Legislator Award of the independent voters of Illinois seven times.

CARL J. SINDERMANN is director of the Sandy Hook Laboratory and assistant director of the Mid-Atlantic Laboratories, Northeast Fisheries Center, National Marine Fisheries Service, National Oceanic and Atmospheric Administration, U.S. Department of Commerce. A graduate of Harvard University and the University of Michigan, Dr. Sindermann is presently adjunct professor of veterinary microbiology at Cornell University. The

author or co-editor of recent books and more than one
hundred scientific papers on the biology and populations of marine fish and shellfish, he is president of
the World Mariculture Society and working group chairman of the International Council for the Exploration of
the Sea.

MORTON I. SOSLAND is editor and publisher of
Milling and Baking News, a weekly publication described
by the Wall Street Journal as "indispensable" to the
baking, milling, and grain industries. A native and
resident of Kansas City, Missouri, Mr. Sosland is a
graduate of Harvard University and currently serves as
a director of Trans World Corporation, H&R Block,
Incorporated, Kansas City Southern Industries, Incorporated, Hallmark Cards, Inc., Commerce Bancshares,
Incorporated, and ERC Corporation.

MAURICE J. WILLIAMS is executive director of the
World Food Council, Rome. A native of Canada, Mr.
Williams studied in international economics at the
University of Manchester (United Kingdom), Northwestern
University, and the University of Chicago. His long
career in the United States government culminated in
his appointment as deputy administrator of the United
States Agency for International Development. Mr.
Williams served as the United States coordinator for
International Food Relief and Rehabilitation during the
1970-71 emergency in India and Bangladesh and the
1972-74 drought in Africa. In March of 1974 he was
named chairman of the Development Assistance Committee
of the Organization for Economic Cooperation and Development (O.E.C.D.), and in 1978 he was named to his
present post by United Nations Secretary-General Kurt
Waldheim.

RICHARD G. WOODS is a graduate of the University
of Tulsa and the University of Minnesota where he
served on the faculty and as codirector of the Office
for Applied Social Science and the Future. He served
on active duty with the U.S. Air Force as a manpower
management specialist and has held staff positions with
General Mills, Incorporated, and Whirlpool Corporation.
Most recently he has been a legislative aide in the
United States Senate, specializing in such areas as
science and technology, nutrition, and world food and
population.

Section 1

THE GLOBAL PROBLEM OF
BALANCING POPULATION AND FOOD

INTRODUCTION

The idea that feeding people adequately constitutes a goal of almost universal acceptability is
central to the institutions and programs intended to
alleviate world hunger and stimulate food self-reliance. Certainly the more difficult problems of international peace-keeping, economic progress, and general
cooperation would seem to be beyond reach, as Maurice
Williams points out, if we cannot find ways to cooper-
ate in preventing hunger and ensuring an adequate food
supply for the people of the world. Civilization
itself rests upon reasonable assurances that such basic
necessities as food will be available for human groups.

However, the problem of balancing population needs
with food production is not well-understood, even among
those whose humanitarian impulses urge them to actively
contribute to the solution of world food difficulties.
There are no "quick fixes" standing ready to be employed at the last minute to ensure the continuity of
food supply to hungry people. While emergency food
programs are essential measures to prevent starvation
and massive malnutrition, they do not alter the food
production and distribution structure in ways which
will ultimately prevent crises. Persons seeking to
build a better world food system must first understand
the nature of the problem. Of necessity we turn for
the requisite insights to those with skill and experience in international development.

Maurice Williams begins his observations with "the
great humanitarian vision of a world without hunger" as
reflected in the recommendations of the 1974 World Food
Conference held in Rome. Williams gives us the details
of progress and prospects since 1974 as a means of
introducing the international dimensions of the world
food and population problem. The cause of rising food
deficits in the developing countries, he indicates, is

the rate of increase in the demand for food, which arises from a high rate of population growth coupled with a high-income elasticity of demand for food. In searching for supply-and-demand solutions, Williams describes the nature of the population dimension, the actions which developing countries can take to increase food production, the problem of malnutrition as it is related to the effective distribution of food, and the need for increased investment. Recommendations for the international community and for the United States conclude his comments.

David Hopper sketches the outline of the world food and population situation in numerical terms, paying special attention to the nature of the demand for food, the course of population growth, and the prospects for increasing food production to meet world demand. His description of the dynamics of population change is quite helpful in understanding the potential influence of population and development policies. He finds recent population trends encouraging. Like Maurice Williams, Hopper is optimistic about the world's food-producing capabilities. In the peasant farmers of the underdeveloped country, he finds willing learners and producers awaiting the right mix of incentives, institutional support, and technical help to increase productivity. Furthermore, he finds hope for the future in the experience of the past thirty-five years during which period no major famine was experienced due to man's ability to distribute surplus food to those in need. The problem of expanding world food production, he believes, lies in the social, economic, and political institutions of man, not in the absence of technical capacity.

Lester Brown stresses the difficulties of expanding food supplies when cropland throughout the world is decreasing. His message is that agricultural land can no longer be treated as an inexhaustible source of land for industry, urbanization, and the energy sector. Brown believes that existing cropland is at the limits of productivity now, and therefore the prospects for increasing world food production in the foreseeable future are not favorable. He argues for the adoption of land-use planning and management, but acknowledges that this solution most often is incompatible with existing social structures and therefore difficult for governments to establish.

Common to the papers in this section is the expression of concern that human institutions, particularly political institutions, need strengthening to meet the challenges inherent in feeding the world.

1

THE NATURE OF THE WORLD FOOD
AND POPULATION PROBLEM

Maurice J. Williams

VISION OF A WORLD WITHOUT HUNGER --
THE RESPONSE OF THE WORLD FOOD CONFERENCE

The problem of hunger and malnutrition has long plagued the human race with recurring famines during periods of political upheaval and food shortages induced by natural disasters. Concern that population explosion might outpace the production of food was seriously considered as early as 1625 by Francis Bacon. This concern was further developed by Malthus in the early 19th century. Whether a balance can be maintained between food production and the rising demand for food has become one of the major questions of our time.

Several recent writers have feared a major catastrophe for the human race in what they see as a finite world with a given stock of resources being depleted by ever larger numbers of people. Others, however, regard the great scientific advances of this century and the enhanced economic capability to produce, process, and distribute food -- as well as the effects of improved living standards on reducing birthrate -- as opening the prospect that hunger and malnutrition can largely be banished from the earth.

It is pointed out that the world now produces more than enough food to feed adequately all of its people, and that in some of the developed areas of the world food production is restrained as a matter of policy to avoid the disincentive effects of oversupply. Already, it can be said that the large-scale, devastating famines of earlier times can be greatly alleviated, if not avoided, because of worldwide communication and transport systems which permit the mobilization of food from surplus areas to meet the emergency needs of populations afflicted by drought and other disasters.

The Irish potato famine in the last century which sent millions to the New World and caused terrible suffering from starvation could not happen today. Further, recent studies have confirmed that there is a large technical and economic potential for increasing food production in the developing regions of the world, and that the lines of required development for the realization of this potential are reasonably well understood.

This great humanitarian vision of a world without hunger, as a central objective of our civilization in the last quarter of the 20th century, was embodied in the Universal Declaration on the Eradication of Hunger and Malnutrition, adopted by the World Food Conference meeting in Rome in 1974. The call for international cooperation to free the world from hunger and malnutrition is one that understandably engages the solidarity of the United Nations. At the Sixth Special Session of the General Assembly in New York, also in 1974, a great social and economic revolution was launched, a revolution which declared that the poor of the world had a right to share more equitably in the bounty of the world's resources and the fruits of development. The hope of the poor of the world for a new international economic order has since inflamed and challenged the deliberations of all nations, a challenge comparable in many ways to the earlier call in the political life of the United States for a "new deal for all Americans."

One is forced to conclude that if the community of nations cannot work cooperatively together on a problem so universally well-understood as food and the need to eradicate hunger as a central element of a new international order for all people, then there is little hope that the nations of the world can deal effectively with other common problems for the maintenance of world peace and economic advancement.

As often happens in human affairs, agreement to strive for a world without hunger was born of crisis, the great food crisis of 1972-73 when adverse weather brought combined crop failures in India, China, and the Soviet Union which, when combined with the drought in the Sahel of Africa, all contributed to two successive years of food production shortfalls. World food-grain reserves were dramatically reduced to less than a month's supply, and there was a worldwide scramble for food which was partly to meet immediate needs and partly speculative. The price of food increased over threefold in international markets. Many food producing countries limited access to their markets. Food aid for poor countries dwindled from earlier high levels, even as the need for food grew more acute. In this panic situation there was a near breakdown in international cooperation as each nation looked to its own advantage, and the weak and the poor were left to shift for themselves. The consequences were tragic. High food prices can be seen by those who have the means as an effective way for allocating scarce supplies, but high prices act as a regressive and grim tax on the poor. The results were that many of the world's people encountered deprivation, health damage, and death from the shortage of food in the first half of this decade.

Those most at risk in times of major crop shortages are an estimated one billion people in some forty low-income, food-deficit countries -- countries which have become increasingly dependent on external supplies for their principal food of cereals. Most of these people live on the margin of subsistence, spending over half of their incomes on food. For them dramatic shifts in the supply and price of food create immediate problems of hunger and malnutrition. Even when food supplies overall are adequate, it is estimated that over 400 million people in the low-income countries are chronically hungry and malnourished. A large number of the hungry are the children of families too poor to purchase the food they need. These children suffer substantial physical and mental damage from malnourishment and are unable to lead full and productive lives.

The 1974 World Food Conference, meeting under the pall of dire predictions that a crowded world would be unable to feed its growing population, adopted objectives for cooperative global action which constitute a comprehensive world food policy. Specifically, the Conference adopted resolutions for:

-- safeguarding populations affected by drought and disaster from the fearful consequences of inadequate food supplies;

-- increasing food production in countries where it is most needed;

-- broadening the effective distribution of food through measures for improving trade, consumption, and nutrition; and

-- building a better system of world food security which can avoid the disruptively wide swings in food prices, such as happened so dramatically in 1972-74.

The Conference called on developing countries to give higher priority to rural development and to the role of small farmers for meeting the food needs of poorer people. Successful rural development was linked, in effect, to the levelling off of high population growth. The Conference endorsed programs "to achieve a desirable balance between population and food supply" and emphasized the role of women as partners in food production, improved nutrition, and in decisions on the size of families.

The World Food Conference also called on developed countries to increase their assistance to low-income countries not only by more effective financial and technical aid but also by a redirection of economic policies which in the past had tended to depress development in economically weaker countries. These policies included economically unjustified restrictions against the exports of less-developed countries and policies for disposal of food surpluses which had the effect of depressing indigenous food production. In effect, the developed countries were asked to use their superior economic and technical strength to underpin the food security of poorer countries and to help them realize their potentials for stepped-up food production and development.

Thus, the World Food Conference dealt with the fundamentals of a multidimensional world food policy and produced a fair degree of agreement on its major elements. However, most of the specifics of how these objectives were to be achieved were left to future negotiations and to a ministerial-level World Food Council as a new institutional means for mobilizing support, for monitoring policies and programs, and generally for ensuring the coherence of overall efforts.(1)

The essential decisions for alleviating hunger and malnutrition in the world are primarily political, and only secondarily related to natural resource and technical factors. This is because, as most authorities have agreed, there are no major physical or technological limits to the expansion of the world food supply to meet the likely growth of population over the next three to four decades. Studies by the Food and Agriculture Organization (FAO) and others have pointed out that land resources are more than adequate and that the necessary investments can be made available to assure inputs of such essentials as water, fertilizers, research, and related extension services. However, it cannot be assumed that, without special efforts on the part of governments, the rate of increase in the production of food will realize these physical potentials and be adequate to world demand, nor that development will take place in regions where the food needs are greatest. The most difficult constraints to be over-come concern the adaptability of sociopolitical struc-tures -- within and among countries -- to realize the economic and technical potentials for eradicating world hunger and malnutrition. It is for these reasons that the work of the World Food Council is seen as being primarily oriented to political action and coordination within the framework of agreed world food objectives.

Progress and Prospects Since 1974

The progress of the last four years has been mixed. Good harvests for three successive years have contributed to the rebuilding of food stocks and a presently improved world food situation.

There has been increased investment in food pro-duction in developing countries, and governments have begun to give greater attention to the longer-term food needs of their peoples, but the efforts to date are still far less than adequate to the needs. A new International Fund for Agricultural Development, as recommended by the World Food Conference, has recently begun operations. The flow of development aid to agriculture has doubled in real terms from about $2 billion in 1973 to about $4 billion in 1977. Most of this increase has been through the programs of the international organizations, particularly the World Bank, and far too little has been contributed by na-tional bilateral aid programs which still represent 70 percent of total development assistance. The level of development assistance for agriculture needs to be substantially increased in the next few years, re-sponding to stepped-up efforts in investment alloca-tions by the food-deficit developing countries them-selves.

Mechanisms within the United Nations for emergency food relief have been strengthened. Contributions in 1978 nearly met a United Nations-administered 500,000 ton Emergency Grain Reserve. The recommendation of the World Food Council that this Emergency Reserve should be annually replenished has received general endorse-ment. Total commitments of food aid for fiscal year 1979 were slightly in excess of ten million tons. These are important contributions to world food securi-ty.

However, too much of the progress in the world food situation to date is the result of good weather.

It is bumper crops that have pulled the world back from
the danger of widespread famine, and governments have
been slow to deal with the fundamental issues of world
food security. In particular, we are concerned with
the slow progress in negotiating the new International
Wheat Agreement. The world still lacks an adequate
grain reserve system to underpin food security and
ensure that in time of food shortage the essential
needs of the poor as well as the rich will be met. And
the international Food Aid Convention is limited to the
four million tons agreed in 1971, as against the World
Food Conference target of ten million tons -- which, as
we have previously observed, is close to the present
level of food-aid commitment and minimal in relation to
current needs. The time to deal with important elements of world food security, such as internationally
agreed reserves and guaranteed adequate levels of food
aid, is the present moment when there have been several
years of surging food-grain production.

While a reasonable beginning has been made to deal
with the world food problem, much more needs to be
done. Future trends of food production and consumption
project an increasingly precarious situation in many
parts of the world. A principal concern is that, with
the rebuilding of world grain stocks due principally to
good weather, governments are being lulled into complacency and will fail to complete the work only recently
begun on a viable system of world food security, and
that far too little effort will be made to realize the
potentials for expanded production of food in the
developing countries. We must keep before us the fact
that millions of people are simply not getting enough
to eat to lead fully active and healthy lives, and that
the situation will continue to worsen unless large-
scale and sustained new efforts are undertaken.

FAO studies show that the world imbalance in food
production and consumption between developed and developing countries continues to worsen. Per capita
food production in all Third World regions except Asia
is less than it was two years ago. Even in Asia it is
difficult to tell how much of the increase in food
production should be attributed to good weather over
the past three years. The situation in Africa is
chilling in its implications. Not only was there no
increase in production in 1977, but per capita food
production was 10 percent below the level of a decade
ago.

These aggregate figures mask national disparities
and the seriousness of the problem for low-income
groups in the food-priority countries. Even when

allowance has been made for increased food imports and food aid, it appears likely that national-average food consumption for the lowest income groups in most developing countries has deteriorated markedly over the past twenty-five years. The result is due to a skewed distribution of income, in part resulting from policies biased in favor of urban development with adverse effects on the food purchasing power of the poor in rural areas.

Developing country imports of grain have risen considerably, from about ten million tons annually in 1960-61 to fifty-two million tons in 1977, of which food aid accounted for nine million tons. For some countries, like the OPEC group and certain East Asian nations, the increase in food-grain imports results from a conscious policy of emphasizing growth in non-agricultural sectors where they hold clear comparative economic advantage. However, this is not the case with most low-income developing countries whose increased food import requirements can be taken as a measure of the longer term lag in production behind domestic demand.

On present trends, aggregate import needs appear likely to reach 145 million tons by 1990, of which eighty million tons may be in the low-income countries of Africa and Asia. Growing trade restrictions will limit the ability of developing countries to earn more foreign exchange for such large food imports. Poorer countries will simply not be able to pay commercially for their mounting food needs, and aid on such a massive scale appears unlikely. Given a continuation of present trends and programs, the food outlook in low-income areas of the world, already precarious, is almost certain to deteriorate further, with alarming effects on the food situation of large numbers of people.

It is against this background of the deteriorating food situation for poor countries and peoples that the World Food Council at its recent meeting in Mexico proposed a program of urgent consultations among various groups -- developing countries, developed countries, aid donors and the major international development agencies -- to identify the major obstacles to increased food production and to work out specific agreements to overcome them. The first round of consultations has been initiated. Representatives of various groups are invited to participate with the World Food Council in developing action proposals which will have the commitment of all those involved and which should lead to more effective efforts to give priority to food production and reverse the deteriorating trend of increasing hunger and malnutrition in low income regions. Specifically, we have arranged that:

--   the Development Assistance Committee of the Organization for Economic Cooperation and Development (OECD) will discuss the issues in Paris at the end of January 1979;

--   the secretariat of the OPEC Special Fund will coordinate the views of the OPEC countries at a February 1979 meeting in the Sudan;

--   the World Bank in late February 1979 will seek the views of the major development financing institutions, the Food and Agriculture Organization and the United Nations Development Program;

--   China and the Socialist members of the World Food Council -- the Soviet Union, the German Democratic Republic, Poland, Yugoslavia and Cuba -- will have an opportunity to give their views and recommendations on the world food problem;

--   meetings of developing countries over the next few months will take place in four regional groupings -- Africa, East and West Asia, and Latin America -- in cooperation with the development banks and United Nations commissions for each region.

What we seek to achieve from these consultations is to break out of the present impasse of misunderstandings which impede food production and nutrition improvement in the developing countries. Each group presently appears to have differing explanations for the relative lack of progress, and there is a tendency to place responsibility on others. For example, developing countries draw attention to inadequate increases in development assistance, overly rigid aid procedures, and reluctance by developed countries to consider policy adjustments in such areas as trade. Developed countries, for their part, claim that governments of developing countries have given inadequate priority to food and agriculture and are slow to consider the necessary internal policy adjustments. International agencies have been inclined to assert that they are already doing everything possible within the limits of the resources available, and that, in any case, they are responding to the demands and priorities of member countries. These implied misunderstandings, in which everyone feels they are doing reasonably well and no one assumes responsibility for the overall inadequate results, are a major problem impeding adoption of more adequate efforts to alleviate the world food problem.

Our objective must be to achieve a framework of understanding which will promote mutually reinforcing national and international action in a global compact to overcome hunger and malnutrition.

Major Elements of the Food and Population Problem

Although the world food problem has international dimensions which affect almost all countries, in the first instance it is a problem which primarily concerns most directly some thirty-six to forty countries. The World Bank and the International Food Policy Research Institute have listed thirty-six countries as having major food deficits. The World Food Council has identified forty-three "food priority countries." Effective action must take place at the country level. It is for the government of each developing country to make decisions as to how it will address this problem and the precise mix of policies and practical measures which are necessary for success in the circumstances of its country. Without the right decisions fully and vigorously applied, success is impossible; but on their own, many food priority countries lack the resources to undertake programs adequate to the urgency and dimensions of the problem. The large food deficits foreseen by FAO, if present trends continue, can be eliminated only by the right combination of internal policies and programs and stepped-up international action.

Notwithstanding the traumatic setback caused by the world food crisis of 1972-74, the developing countries have made progress in their production efforts. The expansion of food and agricultural products over the last quarter century, 1950-75, occurred at a relatively high rate in both the developed and developing countries. Total agricultural production in the developed market economies doubled in this period, while the increase for the developing market economies was greater, or about 130 percent. This must be seen as a significant achievement, although much of the increased production in the Third World was based more on the expansion of area under cultivation than on increased productivity. Production increases in the future must increasingly be sought through modernization and accelerated rural development.

Essentially what is causing the food problem in developing countries has been the rate of increase in the demand for food -- which is a combination of a high

rate of population growth, averaging 2.5 percent annually, and a high-income elasticity of demand for food.
The combined effect was to raise food demand by 3.3
percent annually -- while food production in developing
countries has been increasing at 2.6 percent annually -- over the period 1950-75. This, then, is the
reason for the rising food deficits, particularly in
the last fifteen years.

The food problem lends itself to solution on both
the demand and the production of supply sides of the
equation. On the demand side, it would seem to many
observers that if developing countries cannot grow more
food the least they can do is produce fewer children.
Perversely, population is growing fastest where food
production increases are slowest. However, as we will
see, there are definite limitations to approaching the
food problem from the demand side.

The Population Side of the Equation

Historians studying our age will almost certainly
note the prescient scheduling of two closely related
United Nations Conferences in 1974, with the World
Population Conference preceding the World Food Conference by several months. The great food crisis of
1972-74 focused attention on the balance between food
and the burgeoning world population.

Recent unprecedented population expansion is a
phenomenon resulting in a doubling of population numbers within shorter and shorter periods as a result of
exponential growth. Triggering this demographic event
has been the rapid fall in mortality made possible by
inexpensive means of preventing death even for people
at very low levels of income while fertility has tended
to decline at a much slower rate, if at all. It is
noteworthy that, for the first time in three centuries,
the rate of population increase is now reported to be
declining. However, the built-in momentum of exponential growth is so great that the world's population
will almost certainly double again to eight billion in
thirty years.

Since the world is demonstrably failing to properly feed large numbers of people today, and with the
projected doubling of the population threatening to
vastly compound the problem, it seems humane and commensurate with the dignity of man to encourage voluntary measures for mitigating food demand pressures
through a rapid reduction in population growth. Most
of the world's leaders agree with this position. Three-
quarters of the developing world live in countries

where the reduction in population growth rates is an official policy, and a further 15 percent where family planning activities are supported for health reasons. Those charged with population programs in the most populous countries of Asia and the Middle East are seized with, and sobered by, the urgency of the situation -- understanding that every decade of delay in reaching replacement-level fertility results in a 15 percent higher utlimate level at which their population numbers will stabilize. They see development gains and the possibility of improving the quality of life of their people being nullified by the quantitative needs of the increasing numbers.

The dynamics of achieving the cooperation of parents to limit family size are, however, complex and deeply rooted in the experience of why families want children. Consider, for example, several constraints to family-size limitation which relate to food itself. In some areas of the world there is still very high infant mortality in the weaning period. Babies are taken directly from the breast to an adult diet, precipitating malnutrition which increases their susceptibility to infections and diseases such as measles. Death is common in these circumstances, and mothers, not unnaturally, seek to replace infants who die. Education for village mothers on proper weaning practices will save many of these babies -- and they must be saved -- but it will take a generation for parents to perceive and believe that most of their progeny will survive to adulthood and thus be willing to limit family size. This transition period of high fertility and low infant mortality -- which is difficult to side-step -- is part of the population equation.

Rural people view each additional child as a valuable extra hand in the field and an economic asset. Ironically, labor-intensive agriculture -- a favored approach for increasing rural employment and slowing the exodus to the cities -- gives an advantage to the larger family with more fieldhands. Thus efforts to increase agriculture yields through multicropping and other labor-utilizing methods to increase employment opportunities may result, at least initially, in larger family sizes in rural areas.

These several examples illustrate the complexity of achieving change in the reproductive behavior of people; there are no single, simple solutions to achieving replacement-level fertility. What seems to be required is a multifaceted, sustained effort which integrates family-planning supplies and services into readily available primary health care, coupled with

population education in every formal and nonformal
learning situation so that families are aware of their
options. These, plus a promising development situation
for the rural and urban poor which gives them hope for
an improved life as well as increased literacy and
enhanced status of women, will eventually lead to a
fertility transition which is only now beginning in
earnest throughout the developing world.

Countries which have substantially lowered birthrates, such as Korea, Taiwan and Singapore, appear to
be those effectively meeting the basic human needs of
the broad mass of their populations for food, health
care (including family-planning services), employment,
and improved roles for women. More food and other
improvements in the quality of life generally lead to
declining birthrates. However, we must conclude that a
dramatic decline in the demographic trends appears
unlikely in the next ten to fifteen years and that, in
any case, increasing incomes in themselves will bring a
rising demand for more and better food. Consequently,
the overall demand for food in the next two decades by
the developing countries will continue to rise as
rapidly, if not faster, than it has in the past.
Clearly this increasing need in developing countries
cannot be satisfied by a continuation of past trends of
performance in the production of food.

How Developing Countries Can Increase Food Production

The most promising option open to low-income
developing countries faced with growing food deficits
is to produce much more of their own food. They cannot
hope to meet the massive future deficits implicit in
current trends of consumption and production by drawing
on food supplies from developed countries; the scope
for increased commercial imports is constrained by
balance-of-payments difficulties while the possibilities of developed countries providing food aid on the
scale needed must be seen as politically unrealistic.

As we have previously noted, there is a large
technical capability among developing countries to
increase food production commensurate with economic
efficiency. Stated differently, the value of stepped-
up food production in developing countries would more
than meet the costs of its development given appropriate policies and adoption of cost-reducing technologies -- although for the poorest regions the calcula-
tion of economic and social returns would have to be
projected over a somewhat extended time period. That
most developing countries have not fully exploited in
the past their potential for food production is only

partially explained by the relative neglect of local food production by the former colonial powers in their dependent territories, and only partially attributable to the inherent difficulties of advancing tropical agriculture. More significant has been the policy inadequacy of developing countries in coming to grips with the food problem.

The priority given to food and agriculture, reflected in the volume and ratios of public investment, needs to be raised in keeping with the importance of agriculture in the economic life of the developing countries. In most of them, despite marked difference in situation and outlook, food requirements are far and away the most important development consideration. Agricultural development is the first essential for better nutrition and for generating increased income and employment as a solid basis for higher rates of economic growth.

The main thrust of investment for developing countries is seen as directed toward raising the productivity of existing cultivated land through increased yields and cropping intensities, with particular emphasis on improved seed varieties, fertilizer use, and better management of water resources. Irrigation investment, in particular, provides opportunities for multiple cropping. Opportunities also exist for raising output of rainfed crops on which most of the developing countries are dependent, but it will be difficult to achieve a quantum jump in yields from rainfed areas within the next decade. In low-income regions, rainfed agriculture is beset with several structural problems related to a need for major investments in institutional and physical infrastructure. One of the problems is that, under existing development criteria by most financing institutions, the return on investment in less developed areas is below the opportunity cost of capital. Consequently, investment capital is attracted to the more developed areas while the neediest rural areas tend to remain neglected.

Possibilities exist for the development of new irrigated lands, mainly in Asia and the Middle East in the next ten to fifteen years, as well as for the expansion into new rainfed areas, especially in Africa. Although there is a large potential for new African land development, its realization will prove neither easy nor inexpensive. What appears to be required is a much broader development approach (than in the past) in which investments are simultaneously directed toward improvement in infrastructure, elimination of disease, resettlement of people, protection of the environment through reforestation and reinforcement of research and technology delivery systems.

The development and diffusion to farmers of a continuing flow of new and improved agricultural technology is central to the production of the additional food required to reduce the projected food gaps. The principal contribution to increased productivity is likely to come mostly through larger and more efficient investment in applied research and extension services in the developing countries themselves. It is for this reason that the World Food Conference, in its resolution IV, placed strong emphasis on the need to reinforce international and national research institutions, to expand and raise the quality of extension services, and to train the additional personnel needed at all levels to staff those services, and to transfer new knowledge to farmers at the grass roots. It suggested a "several-fold" increase in funds for research and related training over the next decade. This priority has been under intensive review by the Consultative Group on International Agricultural Research (CGIAR), which has expanded the 1971 funding level from $12 million to more than $85 million in 1978 in support of an expanded program of nine international research institutes. Even this has not proved adequate to meet the diverse needs of the low-income countries for new technology, largely because the national institutions which need to adapt the research output of international centers to their local needs and to tackle other problems are often inadequately financed, understaffed, and cannot cope with the demands placed on them. The training requirements for helping developing countries achieve more effective agricultural programs are particularly large and will call for a much greater effort by both developed and developing countries, efforts which are foreseen in estimates made for the United Nations World Plan of Action for the Application of Science and Technology to Development.

It is important to underscore that, in addition to increasing investment and improving the application of technology, developing countries will need to adopt a number of critical policy changes. These range from a redirection of investment priorities toward agriculture as a core growth sector of their economies to the large-scale training and institutional changes which will vary with the past experience and level of development attained by individual countries.

A number of developing countries seeking to deal with their alarming food prospects are faced with difficult policy choices which have immediate implications for them and for their people.

Food deficit countries with economic structures essentially based on traditional agriculture can expand the production of food most rapidly to meet the needs of their growing urban populations by concentrating on the most modern agricultural sectors of their economies. This, however, means that the mass of the rural population would be neglected, their diet would remain at the same inadequate level, and they would continue to be a drag on the development of the economy as a whole. Alternatively, low-income countries can seek to raise their productivity and gradually modernize the preponderant subsistence agricultural sector thereby improving the food situation of the great poor majority of the rural population. During a transitional period, while this rural transformation is taking place, the food needs of the urban centers would have to be met by mounting food imports, increasing even further the size of external payments deficits.

A number of specialists believe it is possible to solve this dilemma between the needs of increasing food production and those of its wider and more equitable distribution by pursuing a balanced development strategy. Such a strategy would relate progress in the advanced economic sector to a more directly stimulating effect on the traditional sector in a broad advance of the economy as a whole. The in-vogue phrases of today's development planners are "appropriate economic linkages" and "appropriate technologies."

One of the political realities which limits the development choices of leaders in developing countries is their assessment of the prospects for assured external supplies of food and external assistance. For example, if social and economic inequalities in developing countries are to be reduced more rapidly than in the past, in line with the aspirations of the people, this almost immediately means an increase in the effective demand for food. Such increased demand calls for a combination of additional food imports and the stepped-up growth of agricultural production, both of which would be difficult to achieve without sustained and assured increases in external capital.

The "trade and aid" prospects of developing countries are critical to the realization of policies for increasing the production of food for their peoples. Common action by groups of developing countries, regionally and interregionally, can help in facilitating trade earnings and the transfer of appropriate technologies, but to a large extent more effective and accelerated programs for meeting food needs and related development on the part of low-income developing countries are dependent on their "trade and aid" prospects with the advanced industrial countries.

Problems of Malnutrition and the Effective Distribution of Food

Getting additional food into the diets of those who need it most involves much more than increasing production, difficult as that objective is. Greater production is essential, but it does not assure that the increased availability will reach the large number of hungry and malnourished people. Opinions vary among experts as to what constitutes an adequate standard for food requirements and the minimum level of energy intake below which essential functions cannot be maintained. The criterion of adequacy in calorie intake determines one's appraisal of the number of hungry and malnourished people in the world, with estimates ranging from 1.3 billion by the International Food Policy Research Institute to FAO's 450 million. The latter figure was derived by FAO using a standard which represents a minimum-intake level for survival, with no allowance for physical activity other than that incurred in performing essential personal functions. That over 400 million people do not have sufficient food to meet this very minimal standard is dramatic. Clearly, differences in calorie nutrition standards cannot materially affect the broad dimensions of the problem of malnutrition, nor its major policy implications.

The consequences of malnutrition in terms of poor health, low productivity, and general debasement of human dignity are incalculable. Inadequate diets are linked to high infant mortality, decreased human growth, greater susceptibility to disease, impaired mental development, and to low productivity in adults. Severe malnutrition associated with decreased growth also results in decreased brain size and altered brain chemistry. Such conditions retard development which requires intensive application of human energy and creativity.

The special vulnerability of infants and small children, as well as pregnant and lactating mothers, is well known, and effectively implemented nutrition and health programs have reduced child mortality and promoted physical growth for these vulnerable groups. National donor agencies, voluntary agencies and international organizations have expended considerable effort in developing programs, which in 1977 are estimated to have reached perhaps some twenty-eight million mothers and children. There are preschool children suffering from severe to moderate malnutrition.

Because of generally high administrative costs and personnel constraints, there are major obstacles to expanding directly administered programs for vulnerable groups. They depend heavily for their success on the existence of adequate social infrastructure, particularly in health and auxiliary services. The amount of food handled through vulnerable group programs is small, averaging in 1977 about 7 percent of total food aid to developing countries.

As useful as they are, selective programs for nutrition intervention cannot answer the problem of large malnourished populations afflicted with calorie insufficiency. Increasingly, it is becoming accepted that calories are the primary constraint in human nutrition in most circumstances. According to this view proteins are usually present in adequate amounts -- at least for most adults -- when calories are in adequate supply.

Basically, the problem of meeting the nutrition needs of large populations is one of poverty or lack of "effective demand." By "effective demand" is meant the ability of consumers to pay for food, and conversely of farmers to market food at prices which cover adequately the increased costs of sustaining higher levels of production. Consequently, a long-term solution must be sought by development policies which increases employment for the rural landless and the urban poor, and which increases production by small subsistence farmers.

These objectives can be best achieved through more integrated development of previously neglected rural areas. What is needed are basic social and economic changes to enable production of more food, its more equitable distribution and consumption, and an improved quality of life generally. These changes must be integrated in the sense that they are forwarded by concerted programs to improve medical and educational services, upgrade the role of women, enhance job opportunities and broaden land ownership. The World Food Conference concern with this approach to rural development was reemphasized by FAO in the 1979 Conference on Agrarian Reform and Rural Development.

While economic growth and steps to increase per capita incomes of the poor, ranging from increased employment to reduced family size, will have a fundamental effect on a country's nutritional status, another vitally important factor is the relative price of stable foods. This depends primarily on the growth of per capita food supply in relation to per capita incomes. For example, rising high prices of staple foods and slow income growth for the poor mean that the number of malnourished people in the world would steadily increase, while high income growth and stable food prices would mean a sharp reduction in the number of people undernourished and in the extent of their deprivation. Clearly both the supply of food at reasonable prices for the poor and the prospects of improving their incomes are major determinants of the extent of hunger and malnutrition among large populations.

It must be pointed out that while income relates closely to adequacy of nutrition for low-income groups, higher income and greater food availability are not always sufficient conditions for good nutrition. We all know that bad or uninformed food habits can lead to poor nutritional levels. It is here that nutrition education can complement such other forms of improved nutrition interventions as the promotion of home gardens and the home processing and preserving of food.

Programs for increasing the incomes of poor malnourished people in low-income countries are a long-term solution and far too slow; hunger is itself a drag on the development process. Hence, getting increased supplies of food to people most in need on a regular basis may involve special efforts by governments and potentially expensive programs.

The World Food Council encourages food-deficit countries to develop food policies which take into account food production and nutrition actions designed to meet more adequately the food needs of their people. The idea of a more consumer-oriented development plan is relatively new and as yet not well formulated. Countries need to examine means for increasing the supply of low-cost sources of calories for malnourished people, for reducing where possible the prices they pay for staple foods, and for increasing their ability to buy food. Taken together, these elements would constitute a nutrition-oriented food policy.

Within the context of such nutrition-oriented planning, food aid can play a more important role in helping to meet the needs of malnourished populations than it has in the past through subsidies and food-stamp-type rationing of food. These interventions on behalf of hungry people are at best interim measures to be applied where nutrition needs of large populations are urgent and until more permanent remedies can be implemented.

Many countries resist any idea of redistributing incomes; they are more willing to transfer large sums to subsidize the price of staple foods. Such programs are more likely to be supported because food is a political commodity, and meeting the food needs of malnourished people can be seen as having important humanitarian consequences. Also, benefits in food are widely perceived, by both the governments of developing countries and by aid donors, as more likely to be used for their intended purpose. Hence, programs for stabilizing the supply and price of staple foods are an important use of food aid while minimizing the negative impact on farmers' incentives.

Food subsidy programs directed to meeting the needs of malnourished people have not been attempted on a large scale. Most food price stabilization and subsidy programs, like those in India, Pakistan, and Bangladesh have had the objective of holding down the urban cost of living and urban unrest, rather than the more specific goal of alleviating malnutrition. However, the cost of food subsidies can be high. What a country can afford depends on its budgetary and food supply situation, as well as on the political importance attached to the task of implementing consumer and nutritionally oriented food policies.

In summary, if the number of malnourished people in the world and the severity of their malnutrition is to be reduced -- rather than to be increased -- in the coming decade, there is critical need to take the steps required for rapid economic growth in low-income regions with a growing distribution of income and benefits to the poor majority of people. Only through stepped-up, worldwide effort are the poor likely to gain the jobs and income they need to be able to purchase enough food, and governments to obtain the finances they need to carry out nutrition and other programs. In the interim, governments need to consider a range of food-policy issues which increase the production and effective distribution of low-cost staple food and ensure that it actually reaches their hungry and malnourished people.

The Need for Stepped-up Investment in Food

It is in the basic interest of the United States and of other advanced industrial countries to support stepped-up investment and a dramatic increase in food production in the developing countries. Such action would help these countries meet the nutritional needs of their growing populations -- a major factor affecting world political stability -- and hold down the soaring cost of food which has been an important factor in the current inflation in the United States and other OECD countries.

In the next fifteen years the demand for food grains in the developed countries will increase tremendously -- by an estimated 200 million tons, or nearly equivalent to the current production of grain in the United States. During this period, demand in the developing countries will increase by 350 million tons, and under current patterns of production well over 100 million tons of this increase in developing world demand would have to be met by increased exports from North America. The Washington-based Overseas Development Council has pointed out that the North American granary cannot provide another 100 million tons of exports for the developing countries while at the same time meeting growing needs in the United States and other developed countries, as well as the probability of large import demand by the USSR and China, without sharply increasing prices during the 1980s and beyond.

The United States and Canada cannot feed the world, nor should they attempt to do so. Already North America is on its way to becoming the breadbasket for the world, with exports increasing from thirty-four million tons in 1960 to 100 million tons in recent years. The inflationary implications of current trends in world demand and supply of food grains are considerable, not only because of the risk that world demand will surge ahead of world production, but also because of the almost certain increased costs of production in the United States resulting from the need to engage land which should not be put under intensive production and the use of scarcer water supplies, as well as the prospects of having to use even more intensive application of higher cost fertilizer and energy to gain increased yields.

The United States should not permit itself to be in the position of pushing levels of grain production to margins which will inevitably mean environmental deterioration of large land resources at higher food cost to American consumers, and implied obligations to provide even larger quantities of food aid to hungry masses of people overseas, when there are good capabilities for increasing food production in other countries where food needs are greatest.

For all these reasons, it is in the mutual interest of the people of developing countries and of the advanced industrial countries to step-up agricultural investment in developing countries. The humane and mutually beneficial response to the world food problem lies primarily in increasing production of food in the low-income developing countries, and to helping to ensure its more effective distribution.

If food production is to be increased in developing countries, and if agriculture is to be treated as a growth sector for generating increased income and employment, very large capital and recurrent investments will be needed in the next fifteen years over and above those currently being made by national and international agencies. The International Food Policy Research Institute has recently estimated that the identified capital and recurrent costs of raising the growth rate of food crop production in thirty-six low-income countries, which comprise over two-thirds of the developing world population, imply a cumulative capital investment exceeding $90 billion between 1975 and 1990 at constant 1975 prices, or $6.4 billion annually. An increment to investment in the range of $5 billion annually was estimated by a recent World Bank study to generate an additional forty-five million tons of food grains annually in the low-income countries by 1985. FAO and the World Food Council have estimated increased capital costs in the order of $8 billion annually for all developing countries, which includes livestock and fisheries development as well as cereals. Hence, it appears that these various estimates are reasonably consistent.

The extent to which it will be possible to mobilize the additional resources is uncertain. No firm figures are available on the developing countries' national expenditures in the agricultural sector; the World Bank has estimated that low-income countries were spending just under $5 billion a year on agriculture in 1976. Concessionary development assistance, mostly for low-income countries, was about $3 billion in 1977. Doubling of current concessional assistance from all donors for agricultural development in low-income countries would appear to be a reasonable goal -- paralleling an implied 60 percent increase in the developing countries' own efforts to expand capital and current expenditures on agriculture.

A substantially stepped-up investment effort on this scale appears essential if the world community is to address more effectively the problem of growing hunger and malnutrition. But we must stress that stepped-up financial resources must be accompanied by efforts on the part of developing countries to more effectively use these resources, giving emphasis to investment in rural development and to means for more equitable distribution of income. A population program, humanely conceived and well-supported at all levels of government, is also an essential effort in an all-out national effort to eliminate malnutrition. Advanced industrial countries, for their part, should support these efforts not only with increased technical and capital assistance but also with measures to improve the trade and world economic environment within which poor developing countries must function.

These then are the dimensions of the world food and population problems.

THE RESPONSIBILITIES OF THE INTERNATIONAL COMMUNITY AND THE UNITED STATES

The World Food Conference envisioned international cooperation in building a system of world food security which would assure the availability of essential food at least at current levels of consumption despite crop shortfalls and sharp increases in prices, and an advance over time to more adequate levels of consumption for poor countries. Food security is of special importance to developing countries, and there is much that they can do on the domestic front to improve the supply and distribution of food. However, their vulnerability to food insecurity is now so great that unless their efforts are complemented by appropriate international policies, further progress in reducing hunger and malnutrition for a large number of people is likely to be limited.

A better system of world security is important to all nations, and each must do its part. Since the great food crisis of 1972-73, governments and international agencies have refined their understanding of ways to mitigate the food problem, and new international instruments have been created or are being considered for implementing a better system. As a major producer, exporter, stockholder, and aid donor, the United States already plays a leading role. However, there is much more that can and must be done. Specifically, action should be directed toward the following four elements of a world food-security system: increased food production; building stability in international grain markets; systematized ways for responding to local food emergencies; and higher levels of food aid targeted for hungry people. Last, but most important, the problem of food security is one which can be reshaped only through better and more integrated food policies in individual countries.

Priority for Increased Food Production

Without sustained increases in food production there can be no long-term food security for the bulk of humanity. This involves actions by both developed and developing countries. To meet the increased demand in the next decade-and-a-half:

-- the United States and other food exporting countries must maintain existing agricultural production capacity and continue their search for ways of increasing productivity and holding down the cost of growing more food;

-- most developing countries with growing food deficits should direct increased energies and resources towards more self-reliant policies in favor of food and agricultural development;

-- stepped-up international efforts are essential to help the developing nations achieve sustained increases in food production through more effectively directed and higher levels of capital investment. For the poorer countries this will mean substantial increases in development assistance.

The United States and other donor countries should place emphasis on food production and the eradication of hunger as a more central focus of aid programs. In particular, the current U.S. aid program appears to be overly diffuse and in search of a rationale which will command renewed and broader support by the American people. Consequently, the United States does not now carry its fair share of the international-aid effort in helping the poorest countries and peoples. These issues are serious and should be squarely faced by the Administration, the Congress, and the American people.

A central focus on food for the hungry and malnourished, in all of its aspects, would provide an essential revitalization and discipline for on-going aid efforts, thereby solving the problem of a new rationale. The American people will understand and strongly support the need to help hungry people earn their daily bread. Also helpful would be a more coherent engagement of the private sector in U.S. assistance efforts, thereby placing the full U.S. experience in agriculture and food more directly into the struggle to solve the world food problem.

It also must be said that the United States has consistently committed more of its development assistance to population programs than any other country. It is to be commended for recognition of the development constraint imposed by high population growth, a recognition in which the Congress took a leading role. This U.S. effort should continue to receive the support it deserves, and other countries should find encouragement to provide fuller support of both bilateral and international population efforts. In doing so, they should take into account that high population growth is a stage in the development process, and that helping countries advance beyond that stage is the soundest way to solve the population problem.

Critical Importance of International Grain Market Stability

The second essential element of world food security is to enhance the capability of all nations to offset weather-induced fluctuations in production through building an internationally agreed system of reserve stocks and through enhancing international trade opportunities. A large number of nations are in a position to rely on trade and financial reserves to meet shortfalls in their food production, providing there is assurance of access to essential supplies in time of need. The reduction of trade barriers in a more open world market system strengthens world food security, but an adequate system of grain reserves also is essential to help avoid wide swings in prices. These fluctuations are disruptive to production incentives on the downswing and to the essential needs of consumers when prices are flagrantly high, at which time the per capita consumption levels of the poor are forced to adjust downward from already unacceptably low levels.

Recent multilateral trade negotiations in Geneva attempted to achieve a system of nationally held and internationally coordinated grain reserve stocks which would provide for the accumulation and release of stocks. This Wheat Trade Convention was an attempt to avoid extreme price fluctuations in the world wheat market primarily through the operation of national wheat revenues. While the negotiations did not yield the hoped-for new agreement, consensus was reached on major elements and tentative agreement on others. An indicator price mechanism would trigger the accumulation or release of reserves to keep wheat prices within a predetermined band. However, negotiations were unsuccessful concerning several issues: the size of the total reserve and individual country shares; the size of the price band; special provisions for the less-developed countries (e.g., a stock financing fund

to help the LDCs hold reserves and first option to buy
released reserves before developed countries); the
matter of limits on the use of export subsidies; and
supply assurances for importers. While the original
wheat agreement -- essentially an agreement to exchange
economic data -- had to be extended because of the
negotiations impasse, a valuable beginning has been
made in the process of accommodating the trade and
development needs of the less-developed countries.

The largest international grain market is in
wheat, but rice is the staple food for about half the
world's people. Means for introducing a greater element of stability in rice markets would have an important effect on food security. Since only a small
portion of the rice produced is traded internationally,
rice-consuming countries can depend on external trade
only to a limited extent. Reserves for them assume a
particular importance. The Association of South East
Asian Nations (ASEAN: Thailand, Indonesia, Philippines,
Malaysia and Singapore) have made a beginning in establishing a modest reserve scheme. This initiative needs
to be enlarged to extend stability in rice trade and
supplies to more rice-consuming and -producing countries.

More Effective Means for Meeting Local Food Emergencies

In the past, responsibility for responding to
local food emergencies, which occur in various countries every year, was left to voluntary actions by
individual governments and private agencies, with a
leading role having been taken by the United States.
The World Food Council's concern has been to regularize
and strengthen the emerging international system for
meeting emergency needs through a standing and regu-
larly replenished International Emergency Reserve
administered by the World Food Program. A new international food-funding facility also is being considered.

A food-financing facility would extend loans to
countries which suffer an unexpected shortfall in their
food production or a sharp rise in the costs of essential food imports. Many countries in the recent past
have undergone severe hardship because they could not
afford to buy the food they needed at the time they
needed it. The International Monetary Fund has opera-
ted a facility to help countries finance essential oil
imports, and its present Compensatory Financing is
available for loans to countries which suffer unexpected shortfalls in export earnings. One element of a
world food security system should be special lending

facilities available to help countries maintain their essential food supplies when there are seasonal disruptions either in local production or international prices.

For helping the poorest countries faced with immediate local food emergencies -- which are not likely to affect world prices -- there is no better means than strengthening the International Emergency Reserve now targeted at 500,000 tons of grain annually. This grain is available to the World Food Program in order that it can respond to emergencies as they occur. The World Food Program is steadily improving its response capabilities. The United States has been a leading contributor to the Emergency Reserve, with an allocation of 125,000 tons in each of the last two years. Sweden and Canada also are important contributors. The main problem is that many countries are lagging behind in tangible support of sensible emergency arrangements. The task must be to build a strengthened, central, and professional response capability to meet emergency food situations

It should be noted that the U.S. set-aside of six million tons of wheat provides a currently important emergency reserve capability.

The Growing Importance of Food Aid

Food aid has for some years provided a significant supplement, and it has for this reason been highly valued by recipients, although many have deplored its disincentive effects on local food production, on the one hand, and the uncertainty of supply on the other hand. Food aid as an outlet for excess production and market development is certain to continue, and the excess production of developed countries provides an important element of security for meeting emergency situations.

Since the World Food Conference, there has been an increasing effort to direct a larger portion of ongoing food aid programs to population groups outside the market economies of developing countries at assured levels of supply, thereby permitting improved consumption by hungry people, and at the same time clearly avoiding disincentive effects on local production. Also, food aid is seen as an essential though temporary adjunct to programs for assisting countries which seek to place a much higher priority on agricultural development. Developing countries need to be able to rely on external food aid until their own investment programs can more adequately provide for the food needs of their people. As the number of hungry people has continued to grow, the need for greater attention to food production in the developing countries, matched with increased external capital and food assistance, has become clearer. Hence, there is a strong rationale in favor of larger quantities of food aid, to the extent that it can be provided on a reasonably stable and continuing basis over several years.

The World Food Conference called on developed countries to undertake the substantial commitment of ten million tons in stable multiyear food aid programs for feeding hungry people, and related to agricultural development. The Food Aid Convention (FAC) of 1971 sought to guarantee that a minimum flow of 4.2 million tons would be available every year regardless of the market situation. This minimum has proved totally insufficient, and in most years, with the exception of the food crisis of 1972-74, food aid has been well above the FAC commitment level. In an effort to remedy this situation, a new Food Aid Convention is being negotiated together with the Wheat Trade Convention, with the objective of guaranteeing an annual minimum of ten million tons. Negotiations so far have achieved commitments up to 7.7 million tons. We are concerned that the agreed target of ten million tons, which is minimal to current needs, may not be reached. For this reason the World Food Council is appealing to OPEC and other donors to assume responsibility for financing a share of larger food-aid commitments on a sustained basis.

Given the magnitude of needs for increased food to alleviate hunger and caloric undernutrition, a fourteen- to fifteen-million ton target in sustained levels of food aid for multiyear agricultural-development and direct-feeding programs would not be excessive. The issue involved is whether the world is prepared to make food, agricultural development, and the feeding of hungry people a real priority or not. Such a commitment also would provide farmers in the United States and other grain exporting countries with an incentive to continue producing at capacity instead of taking land out of production.

Food Security Policies and Country Action Programs

With better understanding of the nature of the world food problem and renewed determination to eradicate hunger and malnutrition as a major objective of the development decade of the 1980s, it should be possible for the international community to achieve greater stability in international grain markets.

Countries in a position to provide assistance should be able to agree on sustained and higher levels of capital and food aid in support of developing country action programs to meet population food needs. It is at the individual country level that the specifics of the food problem must be tackled.

What is needed is a much more systematic approach in the formulation of individual country food and agricultural policies. These policies must take into account: current and projected food needs; production potentials and the investment and related programs for their realization; import and pricing policies to realize food consumption targets; special nutritional needs and direct feeding programs; and essential marketing, transport, and storage facilities to buttress local food security arrangements. Up to now in many countries such issues of integrated country policies have been largely neglected.

The World Bank and FAO are doing crucially important work, along with a number of other agencies, to improve food security at the country level. FAO has worked out, in collaboration with the governments concerned, food-security action plans and project proposals for a number of countries. This activity is worthy of increased financial support. Better assessments of on-going activities would help to evaluate overall progress, highlight the gaps, and focus more directly on the best specific programs of action.

The essential basis for a stepped-up program for eradicating hunger and malnutrition must be the responsibility of developing countries; they must work out their own goals and policies. As with the Marshall Plan, the initiative must rest with the countries seeking external assistance for the implementation of food policies -- especially the thirty-five to forty food-priority countries. Thereafter, governments and agencies in a position to provide assistance should engage their support. In this manner, increased technical and material resources would be available to help developing countries launch self-reliant programs to meet the food and nutritional needs of their people.

The issue facing the international community is whether the potential for increasing food production in low-income countries can be realized. The stakes are unmistakably high. Present trends and policies mean a continuing deterioration of food security and nutritional standards in many low-income countries, and especially among poorer millions in these countries. The growing number of hungry and malnourished people as well as the vicious inequality and relative neglect are a repudiation of the very idea of development. A sharp reversal of present trends is necessary. What is needed is a greater policy commitment by developing countries themselves and a global marshalling of the technical, organizational, and capital requirements for stepped-up investment in food production as part of national food-security programs of development.

A truly major effort to eradicate hunger, with its human degradation and despair, is a political imperative for building world cooperation and solidarity among all people and nations.

FOOTNOTE

(1) Established in 1975, the World Food Council is made up of thirty-six member states elected on a three-year rotating basis and representing the interests of countries from all continents, rich and poor, food exporting and food deficit, free market and centrally planned economies -- in brief, a food security council with universal membership and wide responsibility for all aspects of the world food problem. Its principal roles are those of advocate, catalyst, and coordinator in stimulating governments and international agencies to adopt mutually beneficial policies and programs for alleviating hunger and malnutrition.

The World Food Council's first president was Dr. Sayed Marei of Egypt who had presided over the World Food Conference in 1974. It was in large part due to his initiative that the idea of a high-level prestigious Council was realized. As advisor to President Sadat of Egypt, Dr. Marei understood well the importance of the political role of the World Food Council. Since 1977 the presidency of the World Food Council has been carried forward by Minister of Agriculture Arturo Tanco of the Philippines. Under his guidance the Council sharpened its focus on issues related to overcoming widespread malnutrition and hunger in a call to action known as the Manila Communique which was strongly endorsed by the United Nations and has contributed to a growing awareness of food problems. Following a further declaration on the need to eradicate hunger and malnutrition at its 1978 meeting in Mexico, the Council is presently

struggling to build support among governments and
international agencies on more specific operating
proposals, particularly in areas of food production and nutrition.

The initial work of the Council was under the
direction of Dr. John A. Hannah as its first full-
time executive director. As his successor, I am
determined that the Council retain its nonopera-
tional orientation and, with a small secretariat
of food specialists, function principally to
mobilize support for coherent policy action among
governments, working with the leading concerned
agencies -- the Food and Agriculture Organization
(FAO), the World Bank, the regional development
banks, United Nations Development Program and all
others who can contribute effectively.

2
RECENT TRENDS IN WORLD FOOD AND POPULATION

W. David Hopper

It seems probable that leaders of social groups have always been concerned with the problem of food and people. The 18th century observations of the Rev. Thomas Malthus articulated these concerns with numerical precision and, accordingly, endowed them with a particular foreboding. Although the opening of North America and the consequent export flows of food to Europe banished the Malthusian specter from Western civilization over the century-and-a-half that followed his writings, the underlying fears that prompted his warnings remain a source of unease today. As economic writers and journalists are not the sort who ignore any simple model of impending darkness, there has been in each century since Malthus a periodic revival of his warnings. While the projected crises have never materialized, at least not to the extent imagined, the specter of human numbers overwhelming available food supplies has remained good fare for pamphlets, the platform, and the press.

The basic fact is: Malthus saw population growth as exponential, that is, increasing as its power function of its base amount, while food output would grow arithmetically, that is, increase by a constant annual amount. He was incorrect about the latter. Over many centuries the long-term growth of food output in England and Europe was about one percent per year, a doubling every seventy years. This was a rate of growth about equal to that of population. It is true that food and population suffered periodic catastrophic declines as a result of adverse weather, crop pests or pathogens, or the outbreak of massive epidemic diseases such as the Black Death in the 14th century which almost destroyed European civilization, but these periodic reverses did not for long disturb the continued exponential growth in both numbers and sustenance.

In assessing the present state of world food and population, we are examining not a Malthusian fear but the fear that the exponential growth of each of the two blades of the "scissors" that could threaten mankind will close through a catastrophic mismatch between the supply of nourishment and the need for it.

This paper will touch briefly on three aspects of the match of food to population. First, on the nature of the demand for food; second, on the course of population growth as a component of global demand; and third, on the prospects for augmenting food production to meet world demand.

THE NATURE OF THE DEMAND FOR FOOD

Before one can assess the sufficiency of supply, it is necessary to say something about what consumers want. Estimates of present world population range around four billion people. The population clock on Washington's Connecticut Avenue puts our global numbers at 4.4 billion, and the increase at 172 per minute, or a growth of 90.4 million per year, at a rate this year of 2.05 percent, an implied doubling of numbers in thirty-four years. For want of a better estimate, this paper uses 4.4 billion as the base figure for 1979. If all the world's population were to consume food in the quantity and quality now consumed by the average American, that is, a consumption level of about 2,000 pounds of grain equivalent (1) per year, most of it as livestock products, world demand for food would be about 4.0 billion metric tons (2) of grain or its equivalent. That demand would be roughly 2.67 times present world food output of 1.5 billion tons of grain equivalent. The per capita food consumption in the poorest developing countries is about 400 pounds per year, or roughly 1,750 calories per day from grain sources. At this level of intake for all people, global grain consumption would be about 900 million metric tons per year, or 60 percent of present world production. In fact, this production evenly distributed among the global population would permit a diet of 750 pounds per capita per year or about 3,300 calories per day.

The imbalances in consumption among rich and poor nations make interesting statistics. The U.S. and Canada, with 240 million people, consume about 220 million metric tons of grain equivalent. Western Europe, which began the period after World War II at about 1,200 pounds per capita per year, now consumes around 1,700 pounds, about where the U.S. was in pre-war days. In the past thirty years, Eastern Europe

has reached the pre-war level of Western Europe at
about 1,300 pounds, and the Soviet Union is now about
where Eastern Europe was pre-war at around 1,000 pounds
per capita per year. Indeed, the great grain "steal"
of 1972 was an effort on the part of the USSR to maintain the livestock consumption levels, so recently
attained by its population, through the purchase of
grain from abroad. These purchases enabled the government to avoid reducing diet quality through belt-
tightening techniques that would have cut back livestock and poultry output in the time-honored Soviet
response of increasing the direct use of grain by its
citizens.

In all, the people of North America, Western
Europe, Japan, Australia, and Eastern Bloc countries
eat about 700 million metric tons of grain equivalent
annually, roughly half of present world production.
Together, these nations hold about one quarter of world
population, approximately 1.05 billion people, whose
average food consumption is approximately 1,470 pounds
of grain equivalent per capita per year, or about 6,400
calories per day per person. The rest of the world's
3.35 billion people average roughly 530 pounds per
person per year, or about 2,300 calories per day.

It is fashionable in some circles to call for a
better distribution of food among the world's population. For example, a cut back in North America from
close to a ton per person per year would save roughly
thirty-five million tons of grain equivalent, or about
5 percent of what is consumed by the people in low-
income nations. This saving would be enough to add
twenty-three pounds per year to each person's consumption in the poor countries. Unfortunately, such a
saving would be only enough to feed a two-year increment in world population.

The growth of world population, at an estimated 2
percent per year, or, to read from the Connecticut
Avenue population clock, at 172 persons per minute, is
the greatest single factor in the dynamics of food
demand. The growth is mainly in the developing countries. The Western industrialized nations, Eastern
Europe, Japan, and Russia account for about 10 percent
of the growth in world numbers; the remaining nine-
tenths (about eighty million people per year) are
increments to the population of the low-income, developing nations. At present food consumption levels, an
additional nineteen million tons of grain equivalent
per year are needed to meet this 2.3 percent expansion
of population in the poor nations. The additional
requirement, just to hold per capita levels constant,

takes an annual increase in total world food output of approximately 1.3 percent per year. To this required increase must be added the population growth in the richer nations; at their levels of consumption this growth necessitates another six million tons or an increment of 0.4 percent to present levels of world grain output. In other words, an annual growth in food output of 1.7 percent, distributed as it is now, could accommodate present levels of population expansion.

Population growth is not the only variable affecting food demand. In both high- and low-income countries, additions to per capita income arising from general economic growth result in higher consumer food expenditures. The income elasticities of food demand are not easily derived. They are complicated by cultural idiosyncrasies of taste and dietary preference and by such factors as the general and specific movements of prices for food and nonfood commodities. Estimates of the income elasticities for cereals vary according to each country. In the very poorest, such as Bangladesh, with a per capita income in 1976 of US$110, and India ($150), the elasticity is approximately 0.5, that is, a one percent increase in income will generate a 0.5 percent increase in cereal consumption, all prices remaining constant.

In slightly higher income countries such as the Philippines ($300) and Egypt ($280), the cereal elasticity falls to between 0.15 and 0.25. In middle income countries such as Brazil ($1,140) and Mexico ($1,090), the elasticity is around 0.1. In most countries, the income elasticity for meat consumption is around 1.0, again varying from country to country, a little higher in the poorer nations, lower in those with high incomes, and again dependent on dietary preferences as well as the ease with which consumers can secure meat and livestock products.

As the world's total economic product increases at about 3.4 percent per year, and per capita income at the rate of about 2.5 percent (weighing growth by income level on a global basis), there will be a significant opportunity for consumers to expand food consumption through an increase in personal purchasing power. In low-income countries, that is, those with a per capita gross national product of $250 or less, the annual growth of national product is about one percent per year. These countries will add about 0.5 percent per year to the demand growth for food which is generated by a population expansion of approximately 2.4 percent, therefore giving an annual increase in cereal requirements of roughly 2.9 percent. In the middle-

and low-income countries, where per capita incomes vary between $250 and $1,000 per capita, individual income is growing at a rate of 2.4 percent per year adding about 0.55 percent to the population growth rate of 2.6 percent for a needed annual food increase of 3.15 percent. In the countries from $1,000 to $3,000 per capita income, the economic growth rate is approximately 4.8 percent per year, and the demand growth for food due to this increasing affluence will be approximately 0.6 to 0.7 percent. This growth rate, combined with increments to population, would necessitate an expansion in food availability of 2.9 to 3.0 percent annually.

In the high income countries, the impact of the income elasticities is focused mainly on an expansion of quality and diversity of diet. These factors have a significant impact on grain needs, but a full analysis of them is properly beyond the scope of this paper except to say that, as demand for livestock and livestock products (including poultry) grows (as it has in the past years in Europe and Japan), there is a marked growth in the off-take of coarse grains, oil and fish meals, and other feed products. For the richer nations, that is, those with per capita income of about $3,000, the growth in grain demand will likely follow present trends that range from about 1.25 to 2.25 percent per year, representing both population and income growth requirements.

In all, the demand for food on a global basis can be expected to increase at close to 3.0 percent per year, approximately four-fifths of which is population expansion with the remainder due to income growth. A growth of 3.0 percent per year implies a doubling of food requirements in twenty-four years. Holding the present pattern of food distribution constant, and assuming a continuation of present population and economic expansion growth rates, there will be a world food need of approximately 3.0 billion metric tons of grain by the turn of the century. (3)

THE RATE OF POPULATION GROWTH

In an era when the credibility of all forecasters has been deservedly lost, the willingness of people to quote population projections with abandon is quite amazing. Indeed, if one looks at the scholarly literature on population of twenty years ago and extrapolates the errors of slower growth made then for a comparison with the facts as we believe them now, it must give us all great concern about our capacities to predict for the next twenty years to that convenient cut-off of the century's end. In fact, no one can say with the confidence of a probability estimate what world population will be two decades from now, let alone number the people likely to be in any country, particularly any developing nation.

If presently estimated trends persist, world population will likely be somewhere between 6.28 billion and 6.74 billion by the year 2000.(4)

There is agreement, however, that present growth trends are not likely to persist. While there is no such agreement on the likely future rates or the rate over the next twenty years, most scholars argue that the rate will be lower, perhaps significantly lower than at present, providing the current advances in economics, education, and health in the developing nations are sustained or accelerated and are accompanied by increasingly vigorous programs directed to family planning and population limitation.

Therein lies the uncertain story of the earth's future members. It is bound within the conundrum of how quickly the less economically developed societies will pass through a stage of change known as the "demographic transition," the transition from high birth, high death rates through high birth, low death rates to low birth, low death rates. From the data at hand, all countries have started the passage. The diffusion and application of modern medical technologies (antibiotics and epidemic control) plus the early conditions of general economic advance have lengthened life expectancy and lowered infant and child mortality, thereby reducing the annual rate of death per thousand population in every part of the world. In traditional societies, annual birth and death rates attain a rough equilibrium at somewhere between forty and fifty per thousand people. The death rate increases sharply with famine, pestilence, and war and drops with the bounties of peace and good harvests. Modernization brings both opportunities for better health care and for the modalities of coping with economic setback. It is striking that since the end of World War II there has not been a major famine anywhere in the world that has resulted in deaths numbered in the hundreds of thousands or millions. Never in world history has there been a thirty-five year period so free from the grim harvest of famine. It is an incredible record. It is equally striking that smallpox is no longer a world scourge, perhaps entirely eliminated, and that malaria and yellow fever are no longer a major threat to life and productivity in vast regions of the tropics that harbored these afflictions since recorded time. It is not surprising, then, that death rates in all the world's nations are now below twenty-five per thousand and the average for even the poorest societies is about twenty per thousand. It should not be unexpected that, as the per capita income of a nation increases, its death rate declines still further to a stable level of eight to twelve per thousand, a level now found in the richest of the industrialized countries.

The death rate is relatively easily influenced. Investments in health care, in the development of food distribution systems, in improvements in sanitation and potable water systems, and so on, all contribute to extending life expectancy and lowering the high mortality of traditional societies. In contrast, human reproduction is not easily influenced. In the past fifteen years, death rates in the developing low-income countries have fallen by 20 to 30 percent; on the other hand, birthrates have fallen less than 10 percent and virtually not at all in most of the nations with the lowest economic base. Notable exceptions are the world's two most populated countries, India and China. While statistics from each must be treated with caution as both nations have similar problems in providing reliable data on births and deaths, each has made substantial progress in lowering birth rates over the past fifteen years by roughly 17 percent to an estimated thirty-six per thousand in India and twenty-six per thousand in China. The death rate estimate for India is sixteen per thousand and for China, nine.

There is strong evidence from the work of social demographers that birthrates in all societies will decline as modernization proceeds. While the time pattern of the attenuation is not yet ascertainable, projections based on present trends of a decelerating decline in death rates and an accelerating reduction in birthrates in many developing nations suggest that world population will be roughly 5.9 billion in the year 2000, an increase of 34 percent over today's 4.4 billion, an implicit average annual rate of growth of roughly 1.5 percent.

Most of the projected increase will be in the poorer countries of the world. The present lowest income nations will have a 64 percent increase in their numbers, a growth not appreciably different from today's 2.4 percent per year. Population in the middle-income countries will grow by about 52 percent, or at a rate of 2.2 percent per year. On the basis of these projections, the population of the industrial nations, including Russia and Eastern Europe, will grow

by only 21 percent, an average rate of less than one
percent per year. Taken together, the nations now
classed as low- and middle-income countries will hold
four-fifths of all mankind by 2000, compared to approximately 70 percent today.

Obviously, knowledge about the transition from
high birth and death rates to low birth and death rates
is the key to an improved understanding and assessment
of population trends. While many observers rather
accurately predicted the decline in death rates, the
record on birthrate estimation is not good. Only in
recent years have national observations provided solid
evidence that human fertility is declining and declining at a higher short-term rate than previously expected. The observed facts are causing a reassessment of
all population projections by demographic scholars.
This paper will not seek to add to this reassessment,
but the fact that a reassessment is taking place must
give the layman interested in food and population
questions careful pause.

We know too little about the elements underlying
the dynamics of change in human numbers. It is clear
that human reproduction does not occur as a random
phenomenon in any society. In every society the numbers of children born to couples is a reflection of
deep cultural values and carefully patterned aspects of
personal and family behavior. This is not to say that
all births are planned; it is to say that the vast
majority of the world's children are wanted and are
given birth within the prevailing social mores of the
shared culture and sanctions of the group to which
their parents belong, and in which they, as newborn
infants, will be nurtured. In any society, a new bride
will respond to the question of how many children she
wants to have on the basis of a mixture of cultural
upbringing and personal wishes for the future of her-
self, her new husband, and her future offspring.
Cultural values are not easily changed. Sanctioned
expectations of the proper behavior of each member of a
society give stability to that society and are its main
source of intergenerational integrity. In such a
social environment, changes will occur slowly. Indeed,
if they occur too quickly, they can be a disruptive
threat to the structural strength and permanence of the
society itself. It is not surprising, then, to find
that in a traditional social setting young men and
women establishing new families reflect the time-
honored outlook of their kith and conform to the pressures of their kin for proper behavior. In fact, in
most developing countries there is a close correspondence between the desires of a new bride and the

average number of children she and other mothers of her society will bear during their period of reproduction. It seems that in low-income countries the average number of births realized exceeds the average number wanted by a small percentage; in the advanced countries the error is the other way. There is good reason for the direction of these errors. In the industrial societies undergoing rapid advance, the cost of child-bearing also undergoes change; in traditional socie-ties, predominantly agrarian in nature and not yet experiencing great or rapid change, children are often a major family asset, contributing to family work and bringing small increments to family income from an early age, assisting in family affairs as they grow older, later assuming the full responsibility for family affairs and providing the warm environment in which aged parents may find security and succor for their final years. It is always better to err on the high side when creating assets, especially so when infant and child mortality risks are high and life is hazardous for any offspring.

It follows that, if human reproduction is indeed nonrandom, the underlying elements that determine it should be capable of discovery, of scholarly assess-ment, and perhaps of policy manipulation.

This was the basic message of the 1974 World Population Conference at Bucharest. A superficial reading of global population data suggests that human fertility declines as economic growth and personal well-being increase. If an increase in per capita economic product is the aim of development, the focus must be on both components of that ratio. Economic advance is the numerator; number of people the de-nominator. The delegates from the less-developed, poorer countries charged at Bucharest that the indus-trialized, richer nations gave greater emphasis to the denominator by striving to alter its rate of change than to the means of accelerating the growth of the numerator, for example, by expanding trade opportuni-ties for poor countries, providing larger aid trans-fers, encouraging and financing price stabilization plans for internationally traded raw material commodi-ties, and so on. It was argued that if such assistance were forthcoming, economic advance would be accompanied by an eventual population decline. The assertion was that population variables are dependent on economic development action, not independent of that action. The delegates from the industrial countries argued the importance of population control policies and programs in the developing nations and pledged assistance for the efforts of low-income societies to attack directly their rapidly growing numbers by means that would slow this growth. They argued that as personal income rose through economic advance and simultaneous population decline, the dual approach would be mutually reenforcing and multiplicative in its effect in promoting progress.

Obviously, a nation's development policy and actions must focus efforts on both numerator and denominator in the per capita product ratio. While it is true in many traditional societies that the protagonists of family planning are regarded with suspicion, even some derision,(5) there is general appreciation in most nations, even in the remote villages, of the problems that arise from the rapid growth in human numbers. After all, in all but a few nations, the land resource is filled and added numbers only create pressures on the functioning economic and social structures of rural and urban societies. These can be seen and felt by those still living who remember the way it was in their childhood before the drop in the death rate brought multitudes to share a limited quantity of social resources.

There seems to be little doubt that there is an unfilled demand among many women in developing societies for more knowledge about family planning and its practice. A more aggressive provision of family planning services to all couples is a desperate need in most developing countries. There is ample evidence to suggest that such services, if supported by national leadership and provided to all with a repeated call for family limitation, can be a significant factor in generating fertility reduction.

But the availability and the strength of a family-planning program is not the only factor fostering fertility decline. Population restraint programs interact with other aspects of national development activities. They reinforce and are reinforced by: (a) the schooling given to girls; (b) the role of women in contributing to family income and the status they are accorded by the society; (c) the likelihood of infant and child survival to the age when he or she can contribute to family welfare and provide future old-age security; (d) the age at which marriage takes place; (e) the socially acceptable age for the first pregnancy; (f) proper child spacing; (g) the cost of educating children and the length of the period when children are a family liability rather than an earning asset; (h) the degree of urbanization of the population; (i) the extent to which nontraditional occupations replace the customary tasks of an agrarian economy, and so on. Indeed, fertility seems sensitive to: (a) issues surrounding the social and economic role and status of women in the family and society; (b) the value of children to the family, that is, whether they are early contributors to the family's collective economic well-being or whether they are family economic liabilities for a substantial period of time as they grow and are educated; and (c) whether there are significant social, economic and political pressures, and opportunities for changes in traditional values and behavioral modes.

Many of the component elements that interplay with fertility are amenable to the interventions of public policy. To a greater or lesser extent, developing country governments are finding these levers and, in the course of pursuing economic advance, are consciously or unconsciously manipulating them. The degree of manipulation has yet to be measured satisfactorily and the likely outcome is still well beyond the predictive power of most scholars and demographic observers. However, it seems certain that the next two decades will hold as many surprises in population studies as the past two decades have held. Probably, although not certainly, these surprises will be pleasant; perhaps as much as the previous ones were unpleasant.

There is an added reason why they are likely to be pleasant. The convoluted numbers game that engages the attention of demographers disguises and ignores the fact that, while human numbers have increased in the years past, so has the quality of the population enumerated. Life expectancy has risen which gives to every society a better memory and an easier access to experience, perhaps even wisdom. Schooling has increased, and the number and proportion of boys and girls in primary and secondary schools has grown markedly. In India, between 1960 and 1975, primary school enrollments grew from 41 to 65 percent for boys and from 27 to 52 percent for girls; university enrollment went from 2 percent of the twenty to twenty-four age group to 5 percent, a huge increase in numbers given access to advanced training. In the group of the poorest of the developing countries, adult literacy rose from 10 percent in 1960 to 23 percent in 1974, a remarkable measure of progress. And in this progress will be found an effect on both the rate of national economic advance and the course of human numbers. Improved health and schooling are the essential elements of human productivity in a modern society. They give to people an enhanced ability to react, respond, and manage the physical and social environment in which they find their purpose. As health care expands, as

schooling is increased, the behavior of people will
alter and the future will be different from that
elicited by the simple projection of the past. The
observations of today will be all too easily mocked by
the facts of tomorrow.

It seems certain that even our most sophisticated
estimates of what the world's numbers will be by the
year 2000 will prove to be unduly pessimistic. As
traditional social and cultural structures change,
people will learn, and learn quickly, that they can
control the course of their own reproduction to their
own advantage. It will not be done without an expansion in the opportunity to enhance personal well-being
through an easy access to a wide range of economic,
educational, health, social, and political services
that include the services of population-limitation
programs and techniques, but if these are provided on
an ever larger scale throughout the world, the outlook
for an unpredicted slowing in population growth is one
of optimism, not of pessimism.

THE GROWTH OF FOOD AVAILABILITY

The fact that in the past thirty-five years the
world has not experienced a major famine on the scale
of the calamities of old is a credit to the achievement
of man's capacity to distribute food that is surplus to
the needs of one group to the needs of another. The
conditions that resulted in the famines of history have
occurred many times in the past three-and-a-half decades. India, Bangladesh, Pakistan, China, the Sahel,
parts of Latin American, the Soviet Union, and elsewhere, have, during this time, been haunted by the
stark fear of imminent tragedy. In each case the food
surplus nations have responded and moved reserve stocks
through a complex of international and national channels of distribution to bring help to those afflicted
and otherwise doomed. The capacity of the world response to the food needs of its most destitute citizens
has written a truly remarkable record. Whether this
capacity can continue in the light of population- and
income-generated-demand expansion is the final concern
of this paper.

The modern era in which science-derived technologies have been applied to meeting man's material needs
began in the latter part of the eighteenth century.
The use of machines powered by nonmuscle energy laid
the foundation for transforming traditional techniques
of production that previously relied on simple agricultural methods and the tools and skills of the craftsman

and artisan. For 10,000 years these techniques were the basis of man's well-being and their yield founded his civilizations. The essence of development is the substitution of modern production methods for those derived from the neolithic period; it is the process of shifting from pastoral to industrial systems of generating society's economic output.

It is not a simple transformation. It involves deep social and political changes as well as fundamental alterations in economic structures and relationships. It involves also a major investment by each nation in the formation of a capital infrastructure that encompasses both manufacturing and farming, for the technological transformation is as important to the future of the rural economy as it is to the establishment of the urban-industrial economy.

There are many indicators of the progress of development. The growth of per capita output of a nation's economy is one. Traditional agrarian societies using ancient farming and artisan techniques can produce not much more than about $100 per person per year. As modern technologies find their place in manufacturing and agriculture, output per person will rise and crudely measure the extent to which, and the rate at which, change is taking place. This mix of national output is another measure. In agrarian societies, the major part of the economic product, low though it is, is generated by agriculture. In such societies the products of farming, fishing, and forestry comprise 40 to 70 percent of output, and this proportion gradually declines as industrialization advances until, as it is in the U.S. today, primary products account for less than one-twentieth of total annual income. A third measure of the transformation is the change in the place where people live. In the older, static societies, people are largely rural with upwards of nine-tenths of the population living in farming villages, or following livestock herds, or applying their trade in small towns. Modernization of a society is characterized by a marked and steady shift of population from rural to urban residence.

By these measures, all but a few countries in Africa are now undergoing significant change. Among low-income nations as a whole, that is, the twenty-four countries listed by the World Bank and the U.N. as having incomes of $250 per capita or less, agricultural output declined as a percent of total output from 51 percent in 1960 to 45 percent in 1976; the proportion of people in the rural areas dropped from 92 percent to 87 percent in roughly the same period; and per capita real income grew from $130 to $150.

As might be expected, the countries with a higher per capita income grew faster. In the fifty-seven noncommunist nations with per capita incomes between $260 and $4,000 -- the latter being Israel -- urbanization went from 32 percent in 1960 to 43 percent in 1975; the proportion of agricultural product and total national output fell from 26 percent to 21 percent; and per capita real income rose from $480 to $750.

The changes resulting from development are taking place worldwide, and they are having a significant impact on food output. However, this impact has not yet been enough to lift from the world the potential threat of famine. In the lowest income countries the growth of agricultural output has actually lagged behind the growth of population. Farmers have expanded food production at the rate of 1.6 percent per year; reproduction has added numbers at the rate of 2.4 percent per year. The result has been a decline in food availability per person, but not in all countries. India, Sri Lanka, Indonesia, Tanzania, to name but a few of the most important, have increased per capita food availability from the mid-1960s to the mid-1970s by as much as 7 to 15 percent. However, for most of the poorest nations, agricultural output has not grown to match increases in population, nor met the growth in total food demand that results from an expansion in numbers and purchasing power.

In many of the middle-income countries, the picture is somewhat better. As a whole, these countries enjoyed a 3.2 percent growth in farm output, enough to accommodate population increases of 2.7 percent and just sufficient to cover the growth of total off-take. As a result, the world's food problem lies less in these countries (such as Angola, Ghana, Congo, Nigeria, Peru, and Mexico who are not doing well in agriculture) than in the lowest income societies.

Most of these lowest income countries lie within or are adjacent to the tropics. Most have the sunlight, warm temperatures, and water supplies necessary to permit cropping throughout the year; however, because of a pattern of seasonal rainfall, water storage and subsequent dry-period irrigation are generally necessary conditions for such an intensive use of land. The soils of the tropics are not always the best. But with research and the application of physical and fertile improvements and careful husbandry, even the poorest earth can be made very much more productive than it is now under time-honored methods of cultivation.

The combination of soil, climate, and water can yield tremendous crops per hectare per year. Present technologies have enabled agricultural scientists in the tropics to obtain yields of grain equivalent in excess of ten tons per acre per year, or more than a total of 440 bushels of corn equivalent per acre over a twelve-month cropping cycle. In contrast, traditional farming-method yields per acre per year on the best tropical lands are less than eighty bushels, and the overall average is below thirty.

The reasons for the low yields are not hard to find: low soil fertility (especially nitrogen levels), poor farming methods based on muscle power and tools of limited capability, inadequate means for controlling pests, few services to assist farmers, and so on. In fact, these reasons parallel the attributes of early farming in our own history.

This level of farm husbandry has developed and persisted logically. Traditionally used varieties of plants are suited to production at low levels of fertility. They have been selected by generations of cultivators for their adaptation to local conditions and for their resistance to the depredations of local pests and pathogens. They are usually of an early, vigorous growing habit so that they outpace the competition from weeds. They yield a grain or a tuber whose tastes and cooking characteristics are suited to local dietary preferences.

It has not always been easy to break into the cycle of traditional farming. The most notable success was of the so-called Green Revolution, which has greatly expanded wheat output from Mexico to Bangladesh and rice output in Asia and parts of Africa and Latin America. The key to this output expansion was the development in Mexico of high-yielding, fertilizer-responsive, dwarf varieties of wheat, and in the Philippines of similar varieties of rice. During the late 1960s these varieties were introduced to widespread use in the South Asian countries of India and Pakistan. Accompanied by fertilizer and assured irrigation, they gave farmers yields of two to three times the output of the indigenous varietal types cultivated under traditional systems. India more than doubled its wheat output over the ten years since the dwarf varieties were widely introduced in 1968. The use of these varieties continues to spread, and their yield is the foundation of India's current twenty-million-ton reserve stock of grain.

The rice story is not as spectacular as the story of wheat. Rice production is more complex than wheat, and rice varieties must be more carefully tailored to local environments than is necessary for wheat. Nevertheless, this tailoring is taking place and rice output throughout Asia is showing a marked acceleration from its older growth rate of less than 2 percent per year.

The high-yielding varieties of wheat and rice are being complemented now by improved varieties and cultivation practices for corn, milo, millets, various grain legumes, and oil seeds through the work of the worldwide network of international agricultural research institutes located in the tropics and strongly supported by the U.S. and other bilateral, multilateral, and private donors. In essence, these advances in agricultural research are providing the biological and agronomic methods that are needed to move grain production in the tropics from traditionally low yields to yield levels comparable to those attained under the intensive agricultural regimes of the developed temperate zone nations. However, in the tropics, the gains are potentially much larger because of the multiple cropping potential. In fact, it is now clear that the food-production potential of the tropics is enormous beyond belief. Properly harnessed and carefully farmed, present world output could be doubled or tripled or quadrupled in the decades ahead. This is surely a potential that is more than sufficient to meet foreseeable population growth and the expansion of demand that will result from increased income over a future of fifty or more years.

The issue of food for people born and yet to be born is only a minor technical problem. World resources can, if mobilized, insure an undreamed abundance for all mankind. An example of this potential is found in deeply impoverished, underfed South Asia. Crossed by some of the world's largest rivers, the Indus-Gangetic-Bhramaputra plain of Pakistan, India, and Bangladesh is home to more than 630 million people. At the present time it produces less than one metric ton of grain per year from each of its 150 million acres. Experiments at various centers on this huge alluvial expanse indicate that with irrigation, proper husbandry (including fertilization, pest and disease control), and carefully matched intercropping and cropping sequences, this land could yield from 900 million to one billion metric tons of grain equivalent per year, a more than sixfold increase over present total output. This is not a spectacular yield. It is equivalent to three 120 day crops yielding eighty-eight bushels of corn or seventy-three bushels of wheat per acre each, an easily attained yield on irrigated land. Indeed, with added research, yields should be significantly higher.

The great South Asia northern river basins, if developed, could add two-thirds to present world food output. Add to this the 500 million acres of Southern Sudan that are now unused grazing land and gigantic swamps of the White Nile; the prairie lands of Colombia, Brazil, and Venezuela that border the Amazon rain forest basin on the north and south; the vast tracts of the transhumant zones of Africa, and the prospects for future food production become almost too large to comprehend.

Why, then, is food a fearful concern? Indeed, why do peasants of Bangladesh eke out an existence of want while walking upon soil of such immense potential? The answers must lie in the institutions of men. Traditional societies are balanced and stagnant. The capacity of people is harnessed and channeled within narrow confines of action. Major innovation is something alien to most human cultures. The shared legacy of customary ways and conventional usage is not easily broken from within the confines of an ancient vision. The oft heard expression "How are you going to change them, they have been doing it for a thousand years" is a shorthand acknowledgement of the hold of tradition. Although I have never met a thousand-year-old farmer to test the hypothesis; the very much younger ones I have met, while skeptical, are alert, careful, acute observers anxious to be more productive. Within the framework of the knowledge and experience they have received from their elders in the teachings of their culture, there is little room for a manifest expression of their desire and willingness to change. But they do change, and change quickly, when convinced that new farming methods will yield personal gains because of technical soundness and a proper balance of risk and reward, of cost and return. This is the real lesson of the Green Revolution. It is a lesson from tens of millions of farmers, large and small, adopting new practices, learning new ways, experimenting with new perceptions, and gaining yields from their land and profit from their venturesomeness. They exhibit dynamism that surpasses the belief and experience of the oldest and wisest members of their society. Once caught up in rapid change, these farmers will never be successful carriers of static traditions, nor will their societies revert for long to the closed doors of the past.

But providing nontraditional opportunities costs money. To irrigate properly for intensive cultivation, the 150 million acres of South Asia's northern river basins would entail an investment on the order of one hundred billion dollars or more in physical infrastructure alone, plus a few tens of billions of dollars more for the roads for the market facilities to handle product and farm supplies, for the fertilizer plants required to meet the nutrient needs of intensive farm production, for the farm equipment factories (animal power and human muscle will not suffice to meet the demands of rapid land preparation required by a system of multiple cropping), for the research and extension facilities, indeed, for the huge array of appurtenances of a modern, high-technology agriculture.

The combined gross annual product of Pakistan, India, and Bangladesh is less than US$110 billion per year. Domestic saving in these countries is less than $20 billion. In fact, in Bangladesh, with twenty-five million acres of potentially productive land, domestic savings are below $210 million per year. Annual external aid of $1.6 billion, much of which is used to keep the economy running, accounts for about 80 percent of the nation's development investment of less than $1.0 billion per annum. The vision of Bangladesh producing the 150 million metric tons of foodgrain per year, which its agricultural acreage is capable of generating with modern technology, fades very quickly amid the immediate problems of feeding its eighty million citizens from today's output of 14.5 million tons of rice and wheat -- about 350 pounds of grain per person per year. The dream of what might be vanishes before the stark reality of national poverty -- the ten to thirty billion dollars needed to turn the nation's potential into edible production, a reality as distant as the moon was to our forefathers.

If we are serious about conquering hunger, the dream must be attained in the manner of the moon: through persistence, through an organized will to accomplish, through the mobilization of talents and hardware, through organization, and through the expenditure of treasure. This is not an easy prescription for a world of separate nations, but unless the task is begun, its difficulties will not be encountered and be overcome, and people will not realize the promise of abundance that lies so close at hand yet so far beyond their present reach. Malthus can be right or wrong depending on how men of good will respond to this challenge of opportunity. It is not a question of food and population; it is really just a question of how much food do we, the world's people, wish to provide.

CONCLUSION

Permit me to close on a personal note. This paper leaves much unsaid. The organized political will and its concomitant elements of political action have not been mobilized to focus on food production in either the developed or the developing societies. Responsibility for the sad tale of the unfulfilled promise of the now discarded Consultative Group for Food Production and Investment that was launched at the 1974 World Food Conference can be laid with equal weight upon the discordant rings of national self-interest pealed by the carillonneurs of rich and poor countries alike. The predispositions of the leaders of new nations to see development as factory smokestacks and superhighways and steel mills has generated hunger among their peoples. Their preoccupation with cheap food for the urban masses may have contributed to the longer-term political stability of their societies, but it has destroyed the incentives for their farmers and has left their nations bereft of a strong and dynamic rural economy. The predispositions on the part of the leaders of the rich countries to leave development as a hostage to foreign policy or to use it as a tool for the penetration of foreign markets for domestic, industrial, and agricultural goods has distorted priorities and engendered consequences counter-productive to the long-term purposes of both giver and receiver. There can be no true sense of security about man's food until the political leaders of most of the world's nations agree to set aside their differences and their short-term interests to forge the hard alliance which will, in time, transform the world's potential for sustenance from an image of what could be to the reality of what is.

This paper does not touch on the issues of social and economic and political reform in traditional agrarian, often feudal, societies; nor has it examined the issues of rural poverty and the set of questions of greater equity and social justice in the context of generating development and allocating its benefits. It does not grapple with the difficult issues of the trade-offs between an expansion in the output of food and the immediate alleviation of destitution.

Perhaps, most significantly, this paper is deficient in assessing the likelihood that a Malthusian specter will haunt our next decades. On this question I have no predictive power. I have no doubt that available food can keep pace with demand if world leaders decide it <u>should</u> keep pace. There is really no mystery about how to do it. The deficiency is in the

will to act together, to commit resources, and to persist. If this is not done, then it is possible, perhaps probable, that food and the demand for it will
slip slowly into a famine imbalance. But if sustained
determination to reach for plenty is the answered call,
then I believe the 1974 World Food Conference promise
that no child need go to bed hungry will be fulfilled.

I do know that the time to begin is now if we want
the assurance that calamity will be avoided. The
transformation of traditional agriculture is a process
neither quickly begun nor finished. To delay imperils
too many millions, and with the answers and opportunities so close at hand, delay can only be regarded as an
act of callous disregard or of inexcusable ignorance.
But to the question: Will men of good intentions put
their influence and energy to the task of insuring all
mankind its daily bread? I have no answer.

FOOTNOTES

(1) The term "grain equivalent" includes the conver-
     sion factors for grain to livestock and livestock
     products, some allowance for grazing lands, a
     caloric match for nongrain production of food
     staples such as root crops, legumes, bananas and
     plantains, and oil seeds. About 85 percent of
     world food production excluding fruits and vege-
     tables, livestock and livestock products, is in
     the form of grain, that is, as rice, wheat, corn,
     barley, millets, other coarse-grains and legume
     pulses such as chickpeas, beans, pigeon peas,
     lentils, etc. The latter class is equal to about
     5 percent of total world grain output. Only in
     Subsaharan Africa does nongrain production account
     for more than 20 percent of food supply; root
     crops and ground nuts total about 40 percent of
     the output of this region.

(2) A metric ton of 2,205 pounds is the standard
     measurement used in world food statistics. That
     measurement is used in this paper.

(3) It must be noted that this approach to demand
     analysis is essentially mechanical. Alterations
     in population growth rates and in income expansion
     rates are two obvious elements which could change
     the projected outcome, perhaps radically. Both
     these rates are explored in detail in the next

sections of the paper. There is another approach to demand analysis through the projection of nutritive needs, and while this is an important approach for any assessment of a strategy of development to ensure human basic needs, the nuances of its details are beyond the reach of this paper.

(4) The range depends mainly on whether present population is approximately 4.2 billion or 4.4 billion, and whether the present growth rate is 1.92 or 2.05 percent per annum. Various sets of estimates can be found in various publications, but these two sets seem to be in the middle of the range of optimistic and pessimistic projections.

(5) In the Punjab villages of India, for example, protagonists of family planning are called "those who do not want to see the smile on the face of a child."

3

THE WORLDWIDE LOSS OF CROPLAND

Lester R. Brown

Ever since the beginning of agriculture, the area and fertility of cropland has determined the availability of food. When land is scarce, people often go hungry. When soils are depleted and crops are poorly nourished, people are often undernourished as well.

The growth in world population since mid-century has intensified pressures on the world's croplands, raising doubts about long-term food security. Growing populations demand more land not only for food production but for other purposes as well. Even as the demand for cropland is expanding at a record rate, more and more cropland is being converted to nonagricultural uses. In addition, the soaring world demand for food since mid-century has led to excessive pressures on the more vulnerable soils. This in turn has led to soil degradation and cropland abandonment.

Cropland loss and deterioration is not a new problem. The Tigris-Euphrates Valley, once described as the Fertile Crescent, may have formerly supported more people than it does today. The food-deficit lands of North Africa were once the granary of the Roman Empire. What is new is the scale of cropland loss and soil deterioration, a problem that now affects all countries, rich and poor alike.

Direct evidence of the mounting pressures on the global cropland base is seen in accelerating soil erosion, the spread of deserts, and the loss of cropland to nonfarm uses. In the central Sudan and northwest India, sand dunes now occupy the land where villages once stood. In the midwestern United States, shopping centers occupy land that only a few years ago was planted in corn. Factories are being built in southern China on land that for generations yielded two rice harvests each year.

Indirectly, excessive pressures are reflected in falling crop yields, growing food deficits, and a greater instability in the world food economy. Crop failures are more frequent, as cultivation is extended onto land with less reliable rainfall. During the seventies, prices of wheat, rice, and soybeans have doubled in times of scarcity. After a quarter-century absence, famine returned in the seventies, claiming hundreds of thousands of lives in Bangladesh, Ethiopia, and the Sahelian zone of Africa. The food shortages that led to these deaths were initially blamed on poor weather, but weather was often the triggering event that brought into focus a more fundamental problem, the growing pressure on local land resources.

A valuable national resource, cropland is the foundation not only of agriculture but also of civilization itself. Yet monitoring the condition of the global cropland base has been handicapped by a lack of comprehensive data. Virtually all governments collect statistics on the land being brought into production through irrigation and drainage projects, or through landclearing, but relatively few systematically compile data on the land being taken out of production. Fewer still attempt to measure topsoil loss through erosion.

Obtaining a complete picture of the changing world cropland situation is not possible because many key pieces of the global mosaic are missing. Nonetheless, enough pieces are now available to bring the overall picture into focus. Some bits of the mosaic are described graphically in the project documents in the files of the Agency for International Development (AID), the World Bank, and the U.N. Development Program. Dimensions of the unfolding land shortage are evident in the publications of the Food and Agriculture Organization (FAO). Evidence of excessive pressures on cropland can be seen in the computer printouts on national grain-yield trends from the U.S. Department of Agriculture (USDA). The preparatory documents for the 1977 U.N. Conference on Desertification included a compilation of data on the spread of deserts, one of the major threats to the world's cropland.

Scientific journals are beginning to carry detailed analyses of local cropland loss and deterioration. In recent years, Science has published several landmark articles on soil erosion and cropland loss, principally in the United States. Translations from Soviet scientific journals reveal the extent of official concern with soil deterioration. Improved access to information on China indicates an extraordinary effort to preserve the cropland base in the face of both the food and nonfood claims on land of nearly 900 million people.

In 1976, Erik Eckholm's Losing Ground detailed the extent of soil erosion, the deterioration of mountain environments, deforestation, the firewood crisis in the Third World, desertification, the silting and salting of irrigation systems, and the other sources undermining the inherent productive capacity of the world's soils. The first global overview of the complex of forces affecting the cropland base, this book captured the grim dimensions of the threat to the world's croplands.

The information now available on the changing global cropland situation indicates a growing worldwide shortage of productive cropland, acute land hunger in many countries, and soaring prices for farmland almost everywhere. Despite the greater availability of information on soil erosion and deterioration, most governments take their cropland for granted. The loss of productive cropland is helping to drive food prices upward, thus contributing to inflation and creating troublesome political problems. Yet, many political leaders seem oblivious to these pressures, perhaps because the loss of cropland occurs in the countryside, while most political leaders live in the city.

THE HISTORICAL EXPANSION OF CROPLAND

The historical expansion of cultivated land has been closely related to the growth in human numbers. In response to population pressures, farmers moved from valley to valley and continent to continent, gradually extending the area under cultivation until, today, one-tenth of the earth's land surface is under the plow. By the mid-twentieth century, most frontiers had disappeared. Up until that time, increases in world food output had come almost entirely from expanding the area farmed. Increases in land productivity were scarcely perceptible within any given generation.

Historically, as the demand for food pressed against available supplies, farmers devised several techniques for extending agriculture onto land that was otherwise unproductive. Dominant among these techniques are irrigation, terracing, and fallowing. Irrigation enables farmers to cultivate land where rainfall is too low to support crops. In mountainous regions, the construction of terraces permits the extension of agriculture onto steeply sloping land that would otherwise be uncultivatable. Centuries of laborious effort are contained in elaborate systems of terraces in older settled countries such as Japan, China, Nepal, and Indonesia, and in the Andean regions once inhabitated by the Incas.

In vast semiarid regions -- such as Australia, the western Great Plains of North America, the Anatolian plateaus of Turkey, and the drylands of the Soviet Union -- where there is not enough rainfall and moisture to sustain continuous cultivation, alternate-year cropping has evolved. Under this system, land lies fallow in alternate years in order to accumulate moisture; all vegetative cover is destroyed during the fallow year, and the land is covered with a dust mulch that curbs the evaporation of water from the soil.

The crop produced the following year thus draws on two years of collected moisture. In some situations this practice would lead to serious wind erosion if strip-cropping were not practiced simultaneously: alternate strips planted to crops each year serve as windbreaks for the fallow strips.

In the tropics -- much of Africa south of the Sahara, Venezuela, parts of Brazil, and the outer islands of Indonesia -- where available nutrients are stored more in plant materials than in the soil, fallowing is used to restore the fertility of the soil. Stripped of the dense vegetative cover, soils in the humid tropics quickly lose their condition. Farmers clear and crop land for two, three, or possibly four years, and then systematically abandon it as crop yields decline; moving on to fresh terrain, they repeat the process. These cultivators return to their starting point after twenty to twenty-five years, by which time the soil has become fertile enough to support crop production for a few more years.

Over time, irrigation, terracing, strip-cropping, and shifting cultivation enabled farmers to produce food on land where conventional agriculture would not survive. Today, land farmed with these specialized techniques provides a large share of the world's food supply. Without them the earth's capacity to feed people would be far lower than it is. Although these practices have withstood the test of time, in some areas they are beginning to break down under the pressure of population growth.

Agronomists have been aware of the mounting pressures on the world's cropland, but their efforts to measure the impact have traditionally been handicapped by the absence of reliable data on overall cropland trends.

An alternative to using data on total cropland is
to use data for cereals only. This sacrifices some
comprehensiveness, but the greater availability and
precision of data more than compensate for this loss.
Cereals occupy some 70 percent of the world's cropland
and are the dietary staple in virtually every society.
Cereals are also a rather homogeneous commodity group,
making yield trends and comparisons relatively easy to
compile and analyze.

In addition, the USDA now has the most reliable
historical data on cereals available anywhere. The
product of several years of painstaking effort, the
USDA computer bank contains information on area, production, and yield for every cereal in all countries
from 1950 through 1978. An invaluable source, it
provides new insights into changes in both cropland
area and productivity.

Although rising cereal output per hectare has
accounted for most of the growth in the world food
supply since 1950, expansion of the area in cereals has
also made a significant contribution. Between 1950 and
1978, the area in cereals expanded by 140 million
hectares, or some 23 percent.(1) During this time there
are two periods of rapid expansion. From 1951-56, the
area in cereals expanded by 10 percent, or 2 percent
each year. Close to half of this increase was
accounted for by the extension of grain production onto
the "virgin lands" of the Soviet Union.

Between 1956 and 1972, the area in cereals increased only 4 percent. This was a period of excess
production capacity in the global food economy, a time
when the world's farmers could produce more grain than
the world's consumers could afford at prevailing
prices. During most of this time the United States,
the world's leading food producer, idled some twenty
million of its 140 million hectares of cropland.(2)

The second period of rapid expansion occurred from
1972-76 in response to a worldwide food shortage and
soaring food prices. The Soviet wheat purchase in
mid-1972 was the triggering event that brought the
tightening world food situation into sharp relief.
Over four years, some fifty million hectares were added
to the world's harvested area of cereals, increasing
the overall area nearly 8 percent. At least one-third
of this increase was due to the return to production of
U.S. cropland previously idled under government programs. A small share came from reduction of the land
in fallow in the United States and the Soviet Union.
The soaring prices encouraged expansion of the

cultivated area in several developing countries --
importantly, in Argentina, Brazil, and Nigeria.

As the food supply tightened in the early seventies, there was enough slack in the system to allow
this increase of several percent in the land planted to
cereals. Much of the increase, however, was due to
special circumstances that cannot be repeated, including the return to production of idled cropland and the
reduction in fallowed area.

Table 1. World Population and Area in Cereals, 1950
          and 1975

Year	Population (billions)	Area in Cereals (million hectares)	Area per Person (hectares)
1950	2.50	602	.241
1975	3.97	729	.184

Source: United Nations and U.S. Department of
Agriculture

During the third quarter of this century, world
population increased by 1.5 billion, or 59 percent,
while the area in crops expanded from 602 to 729 hectares, a gain of 21 percent. (See Table 1.) This was
almost certainly the first generation during which the
growth in human numbers greatly exceeded the growth in
cropland. It led to a precipitous drop in area per
person from .241 hectares to .184 hectares. Such a
drop did not pose any problems as long as the yield per
hectare was increasing rapidly. Once that began to
slow, however, as it did in 1972, a global food shortage developed.

CONVERSION TO NONAGRICULTURAL USES

As human population continues to expand, it generates needs for land for purposes other than the production of food. Principal among these are urbanization,
energy production, and transportation. Each of these
sectors is now claiming cropland in virtually every
country.

Whenever national data are available, they usually
show the growth of cities to be a leading source of
cropland loss. Within the United States, cities are
consuming cropland at a record rate. Land-use surveys

by the USDA in 1967 and 1975 indicated that some 2.48 million hectares of prime cropland "were converted to urban and build-up uses" during the eight-year span. A study of urban encroachment on agricultural land in Europe (grasslands as well as croplands) from 1960 to 1970 found that West Germany was losing 0.25 percent of its agricultural land yearly, or one percent every four years. For France and the United Kingdom, the comparable figure was 0.18 percent per year, or nearly 2 percent for the decade.(3)

While data are available on the loss of agricultural land to cities in the industrial countries, there is little information on the Third World, where the most rapid urbanization is occurring. Scores of cities in developing countries, such as Lima, Ankara, and Manila, are growing by 5 to 8 percent yearly. Inevitably, some of this growth comes at the expense of cropland.

An unknown amount of the world's cropland was claimed by cities during the third quarter of this century, and it is only possible to estimate how much cropland cities will cover by century's end. To get some idea of the land needs of expanding cities, information is needed on the expected growth in urban numbers and on the amount of land required per person. According to the U.N. demographic division, the urban share of world population, which was 29 percent in 1950 and 39 percent in 1975, is projected to reach 49 percent by 2000. (See Table 2.) This means the urban population would expand from 1,547 million in 1975 to 3,064 million just twenty-five years later. If these projections are reasonable, the absolute increase in urban dwellers during the final quarter of this century will be nearly double the increase from 1950 to 1975.

Table 2. The Urbanization of World Population, With Projections to 2000

Year	World Population (millions)	Urban Population (millions)	Urban Increase (millions)	Portion of World Urban (percent)
1950	2,501	725		29
1975	3,968	1,547	822	39
2000	6,254	3,064	1,517	49

Source: United Nations

To estimate the extent of urban encroachment on cropland, we need to know how much each person will use and the share of that which will be cropland. Within the United States, the average urban resident required .09 hectares in 1970, an increase from .08 hectares in 1950. In Canada, analysts assume that each person added to the urban population requires .09 hectares. And in projecting future land requirements of urbanization outside the United States, John McHale of the Center for Integrative Studies assumes 0.16 hectares is used by each city dweller.(4)

If it is assumed, perhaps conservatively, that the projected increase in world urban population of 1,517 million from 1975 to 2000 will need only .04 hectares per person, then the world's cities would occupy sixty-one million hectares of land by 2000. Assuming that 40 percent of this total is cropland, expanding cities will cover twenty-four million hectares of cropland between now and the end of the century.

Although this loss would amount to only 2 percent of the world cropland base, the share of world food output involved would be larger, for in most countries cities are usually situated on the most fertile soils. A study of changing land-use patterns in Canada reports that "half of the farmland lost to urban expansion is coming from the best one-twentieth of our farmland."(5) This appears to be the case in most countries, although it may not apply to the United States. But more importantly, at current levels of land productivity and food consumption, this loss of twenty-four million hectares represents the food supply of some eighty-four million people.

All additions to the world's population require living space, whether they live in the city or in the countryside. In the Third World, where land is in critically short supply, cropland is lost each year to village growth. In an analysis of village spread, using data over several decades for his native Bangladesh, Akef Quazi, a lecturer in science and technology at the University of Wales, concludes that growth in the number of families and growth in the area occupied by the village are closely related. One reason for this direct correlation is that homes are "made of locally available materials, such as bamboo, thatch, and corrugated iron sheets, and, as such, are never strong enough to hold an upper story." Quazi reports that "every new village homestead is being built on cropland."(6) While there are undoubtedly occasional exceptions, Quazi's general point is a sound one, since Bengali villages are usually surrounded by the fields on which they depend.

A similar situation exists in India, a country of 600,000 villages, nearly all of which are expanding. Living-space requirements of the fourteen million people added each year are at least partly met by homes built on cropland. In its report to the 1976 U.N. Conference on Habitat held in Vancouver, Canada, India estimated that between 1970 and 2000, nonagricultural land use would expand from 16.2 to twenty-six million hectares, an increase of 60 percent.(7)

Further east, China, the world's most populous country, is also losing cropland to cities and industry. USDA China specialist Alva Erisman reports that "water control projects, urban growth, and the appropriation of agricultural land for roads, railroads, airfields, industrial plants, and military uses have removed good farmland from cultivation." Dwight Perkins, Chinese scholar at Harvard, makes the same point, noting that the 10 percent growth in modern industry, with factories commonly sited on the edges of cities, has undoubtedly used up cropland. He believes Chinese planners are aware of this continuing loss of cropland, but that "there is no way around the fact that good farm land (flat, located near transport, et cetera) often makes an excellent factory site."(8)

Rivaling the urban sector as a claimant on cropland is the fast-growing, global-energy sector. During the final quarter of this century, as during the third, the consumption of energy is projected to increase more rapidly than that of food. Like the production of food, the production of energy requires land. Hydroelectric dams often inundate vast stretches of rich bottomland; electric generating plants can cover thousands of hectares. More often than not, oil refineries and storage tanks are built on prime farmland along rivers and coastal plains. Strip-mining of coal and the diversion of irrigation water for coal gasification both tend to reduce the cultivated area.

No one knows just how much of the world's cropland lies under the reservoirs behind hydroelectric dams, but it surely numbers in the millions of hectares. Nigerian ecologist Jimoh Omo-Fadaka outlines the losses in Africa: "The modern Kariba Dam in Africa has put 29,000 farmers out of business. Some of Ghana's rich farmland lies beneath the Volta Dam, and prosperous cocoa and coffee farms in Ivory Coast's Bandama Valley are dying as the waters rise behind the Kossou Dam, sealed in 1971."(9) Some hydroelectric dams, of course, have many purposes and may add to the irrigated area even while subtracting from the cropland base through inundation.

In the United States, the Senate Committee on Interior and Insular Affairs studied the land requirements associated with the development of various energy resources and facilities. Among other things, they analyzed President Ford's 1975 State of the Union address in which he asked for "200 major nuclear power plants, 250 major new coal mines, 150 major coal-fired power plants, thirty major new refineries, twenty major new synthetic fuel plants, and the drilling of many thousands of new oil wells" in the next ten years. The Federal Energy Administration estimated that the program would disturb land on some eighteen million hectares. The Senate Committee noted that "this amounts to a tripling of the amount of land currently devoted to energy thus far."(10)

Although the Senate Committee did not estimate the share of this land that would be cropland, some studies conducted at the local level have done so. One such study compiled data on the loss of cropland to the strip-mining of coal in Illinois, a state with some of the most productive land in the Corn Belt. It reported that "as of 1976, 202,422 acres [81,981 hectares] in forty Illinois counties have been affected by surface and deep mining. Surface mining accounted for over 94 percent (191,874 acres) of this affected acreage."(11) Although mining companies are now required by law to restore land to its original productive state, many people doubt that it will be possible in most situations. Given the rich, extensive resources of coal close to the surface in some fifty-one Illinois counties, the potential disruption is a matter of great concern.

Although there are no detailed surveys of the worldwide loss of cropland to strip-mining, it is safe to assume that the measured loss in Illinois is only a small share of the U.S. total and a minute percentage of the global total. While estimates of future cropland losses to urbanization are possible, those for losses to the energy sector are more difficult. The potentially large claims, however, emphatically underline the need for energy conservation programs. To conserve energy is also to conserve cropland.

All transport systems require land, but some systems use much more than others. Automobile-centered transport systems are voracious consumers of land. An enormous amount of U.S. cropland has been paved over for the automobile. Millions of hectares are required just to park the nation's 143 million licensed motor vehicles. But even this is rather small compared with the land covered by streets, highways, filling stations, and other service facilities. Moreover, the automobile has encouraged other inefficient uses of land, such as urban sprawl. Societies moving toward an automobile-centered transportation system should weigh carefully the sacrifice of cropland that is sure to be involved. Almost any other form of transportation requires less land. Societies with well-developed public transport systems are able to use land far more efficiently than those where most people rely on cars.

Just as rising income increases the per capita demand for cropland, so too it increases the land required for other purposes. In effect, high-income man is a space consumer. All the principal nonfarm uses of land are greater among high-income groups than they are among those with low incomes.

The amount of cropland that will be paved over, built on, strip-mined, or flooded by a dam by the end of this century is unknown. However, if world population projections are correct, 2.3 billion people will be added between 1975 and the year 2000, a far larger increase than the 1.5 billion added during the preceding twenty-five years.(12) Given these population projections and the projected gains in income, every nonfarm claimant on cropland -- urbanization, energy production, transportation -- will be greater during the last quarter of this century than during the third.

ABANDONED CROPLAND

Although the world demand for cropland is greater than ever before, the amount of cropland abandoned each year may also be at a record level. The reasons for cropland abandonment, usually the product of economic pressures interacting with ecological forces, include desertification, severe erosion, waterlogging and salinization of irrigated land, and the diversion of irrigation water to nonfarm uses.

Deserts are now expanding on every continent. Documents prepared for the U.N. Conference on Desertification detailed this expansion and the human creation of desertlike conditions. Some 630 million people, or one in every seven of the world's people, live in arid or semiarid areas. An estimated seventy-eight million people live on lands rendered useless by erosion, dune formation, changes in vegetation, and salt encrustation.(13) Desertification has deprived these people of their means of production, of their livelihood.

Agronomists, specializing in the management of arid and semiarid croplands, have been aware for some time of the mounting pressure on these fragile soils and of their progressive deterioration and loss of productivity. It was not, however, until the droughts of the late 1960s and early 1970s in the Sahelian zone of Africa that the social consequences of desert expansion became more evident.

Although public attention has focused on the southward spread of the Sahara, the desert is also marching relentlessly northward, squeezing the populations of North Africa against the Mediterranean. FAO agriculturist H. N. Le Houerou estimates that the North African tier of countries -- Morocco, Algeria, Tunisia, and Libya -- are losing 100,000 hectares of range and cropland each year. The ever-growing Sahara is expanding westward into Senegal and eastward into the Sudan. Readings taken of the Sudan in 1958 and 1975 indicate the desert expanded southward by some ninety to one hundred kilometers during the seventeen-year period.(14)

Fed by the human pressures on their fringes -- overgrazing, deforestation, and overplowing -- virtually all the world's major deserts are expanding. As human and livestock numbers multiply, the creation of deserts or desertlike conditions is accelerating throughout the Middle East and in Iran, Afghanistan, Pakistan, and northwestern India. Brazilian ecologist J. Vasconcelos Sobrinho reports that the semiarid tip of Brazil's northeast is being desertified; similar conditions are developing in the Argentinean states of La Rioja, San Luis, and La Pampa.(15)

Soil erosion is an even more widespread problem. Soil scientists classify erosion into main categories -- sheet erosion and rill or gully formation. Continuous erosion is usually found on sloping lands that are farmed without soil conservation techniques such as terracing and strip-cropping. This leads to productivity declines in the short run, and to land abandonment over the long-term. The difference between rills and gullies is one of scale: the depth of rills can be measured in inches, but that of gullies is usually measured in feet. Gullies usually appear where subsoils are also erodible.

Some countries have studied extensively the process by which their croplands are being degraded by gully information. Soil scientists at the soil erosion laboratory at Moscow University have analyzed gully formation in the southcentral Soviet Union --

Kazakhstan, Central Asia, and West Siberia. They note that although in only 2 percent of the area is there now strong gullying, there could be medium and strong gullying on as much as 50 percent of the land as efforts to intensify agriculture proceed.(16)

A parallel Soviet study analyzing the present gully network in the Steppe and Forest Steppe regions of the European part of the Soviet Union, concluded: "One factor that more than any other has fostered the spread of gullies has been the plowing up of the land. The annihilation of the natural vegetation cover, the systematic loosening of the soil, the network of dirt roads, plow furrow, etc., produce conditions that give rise to a dense network of gullies, especially on steep slopes." In examining the history of the process, the authors report that gully formation has probably accelerated as "good land reserves became exhausted and sloping land began to be plowed."(17)

Although severe gullying, leading to land abandonment, is now common in many parts of the world, it has received the least official attention in those countries where the process is most advanced. A U.N. report on soil deterioration and cropland loss in Latin America notes its serious dimensions in the Andean regions.(18) Throughout the Andean countries gullies are advancing through the steeply sloping countryside like the tentacles of a giant malignancy. As these gullies eat their way across fields, farmers who are already desperate for land continue to till what is left, right up to the very edge of the gully. In so doing, they feed the gully, accelerating its progress across the countryside.

A report by the Organization for Economic Cooperation and Development (OECD) describes the extensive abandonment of farmland in Italy, noting that "it is generally agreed that in Italy two million hectares have been abandoned in the last ten years....the farming methods used on this marginal land have led to deterioration of the soil so that the land was consumed in the literal sense of the term."(19) Similarly, some of the decline in the harvested area of cereals in Yugoslavia and East Germany over the past two decades reflects abandonment of eroded, worn-out soils in farm areas with the more rugged terrain.

A report from the U.S. Embassy in Jakarta indicates that "soil erosion is creating an ecological emergency in Java. A result of overpopulation, which has led to deforestation and misuse of hillside areas by land-hungry farmers, erosion is laying waste the land at an alarming rate, much faster than present reclamation programs can restore it." Similar pressures are building in the rainfed agricultural regions of Pakistan. An AID officer in the Punjab area reports the abandonment of some 4,800 to 12,000 hectares of cropland yearly because of severe erosion and degradation.(20)

The irrigated lands that provide a disproportionately large share of the world's food are threatened both by ecological forces -- waterlogging and salinity -- and economic forces that divert water to competing uses. Waterlogging and salinity develop wherever surface water from rivers or streams is diverted to irrigate land that has inadequate underground drainage. This net addition to the natural underground water supply gradually raises the water table. Once the water table is within a few feet of the surface, the growth of deep-rooted crops is impaired and the early symptoms of waterlogging appear.

As the water table rises further, water begins to evaporate through the remaining few inches of soil, thereby concentrating the minerals and salt near the surface. Eventually the salt concentrations reach a level that prohibits plant growth. Glistening white expanses of heavily salted, abandoned cropland are visible from the air in countries traditionally dependent on irrigation, such as Iraq and Pakistan.

The problem is as old as irrigation itself. Indeed, the decline of some of the early civilizations in the Middle East is now traced to the waterlogging and salinity caused by their irrigation systems. Although the designers of the earliest irrigation systems in the Tigris-Euphrates Valley did not understand the subterranean hydrology well enough to prescribe corrective action, modern irrigation engineers do. A U.N. report, however, estimates salvage costs at $650 per hectare.(21)

Worldwide data compiled in a 1977 U.N. report indicate some twenty-one million hectares of irrigated land were waterlogged -- one-tenth of the total area irrigated. The productivity on this land had fallen by 20 percent. A similar area, often overlapping with the waterlogged area, of some twenty million hectares was affected by salinization, with a reduction in productivity estimated at roughly the same level as that of waterlogging.(22)

There are signs of waterlogging and salinity in many countries. Although fully half of the world's irrigation capacity has been developed since 1950, waterlogging and salinity are already impairing the effectiveness of many systems. The agronomic litera- ture of China is liberally sprinkled with references to the problem. Heavy investments in agricultural land reclamation and irrigation in the Soviet Union during the 1970s have been aimed at improving drainage systems for the reclamation of irrigated land that has become waterlogged and salty. The more intensive irrigation in the Nile River Valley made possible by the Aswan Dam has upset the long-standing water balance in some areas, leading to the waterlogging and salinity of Egyptian soils that have historically been free from this problem.(23)

Governments in some Middle Eastern countries have been trying for some time to identify or develop crops that are salt-tolerant. A case history of the Greater Murrayeb Project in the Euphrates plain of Iraq reports that 50 percent of the soils in the region are now moderately to severely saline, with 16 percent severely saline. The consequences are "lowered yields, loss of irrigable land, and abandonment of some holdings."(24)

A U.N. survey found that, in Argentina, "some two million hectares of irrigated land have declined in productivity due to salinization and alkaliniza- tion."(25) In Peru, the survey reported that of 800,000 hectares of irrigated agricultural land in the coastal desert, 300,000 hectares are affected by poor drainage and salinization.

Salinization and waterlogging also plague irri- gated soils in India. The progressive loss of pro- ductive land to salinization is a particularly acute problem in northwestern India. In one irrigated area in Rajasthan, twenty-four square kilometers of produc- tive land have been lost through salinization, and a U.N. study predicts that "if the process continues, it may extend to another 40.3 square kilometers."(26)

In Pakistan, virtually all the irrigated land in some districts is plagued with waterlogging and salini- ty. In the Mona reclamation area, a project that began in 1901, about 90 percent of the soils were waterlogged by the early 1960s.(27) Since that time, the recla- mation effort has led to some improvement, but the rehabilitation of waterlogged soils is a time-consum- ing, capital-intensive process.

Salinization has also become a major problem in the southwestern United States and northern Mexico. American farmers have designed drainage systems that

remove the salt from their irrigated fields; unfortunately, some of the salt ends up in the river waters
that eventually irrigate Mexican fields. The result
has been salt concentrations so high that many Mexican
farmers have helplessly watched the productivity of
their land fall.(28) This unintentional salting has
posed a sticky diplomatic issue for the two governments.

In some regions crop production is threatened not
by too much irrigation but by too little. The diversion of irrigation waters to other purposes is particularly a problem in arid areas with irrigated agriculture and with rapid urbanization and industrialization,
such as the southcentral regions of the Soviet Union
and the southwestern United States.

Soviet efforts to regain food self-sufficiency
rest heavily on plans to expand irrigation. Consequently, investment in irrigation has been the fastest
growing sector of the economy over the past fifteen
years. But irrigation plans and capital commitments
may not be enough. Thane Gustafson, Soviet scholar at
Harvard, reports that "the entire irrigation program is
threatened because the southern half of the country is
running short of water."(29) Barring the massive diversion to the south of northward flowing rivers for
irrigation purposes (a project that is now officially
questioned on environmental grounds), severe water
shortages appear to be inevitable.

In the United States the problem has been carefully detailed for some of Arizona's leading agricultural counties. The irrigated area in Maricopa County
is shrinking steadily as water is being diverted to the
rapidly expanding Phoenix metropolitan area. Between
the late 1950s and the early 1970s, the irrigated
area fell from 224,000 to 177,900 hectares. But even
at this reduced level of water withdrawal, groundwater
levels have been dropping by some ten to twenty feet
per year. In Pinal County, another leading agricultural center in Arizona, the land planted in crops fell
from 138,000 hectares to 93,000 hectares during the
same span.(30)

A similar situation exists in California, though
on a much larger scale. As thirsty Los Angeles requires more and more water, the demand can often be
satisfied only by diverting water from irrigation
canals. Irrigated valleys that were once lush green
have now turned a dusty brown. Fields that were once
among the most productive in the world have now been
abandoned.(31)

In the western Great Plains -- Nebraska, western Kansas, Colorado, and Wyoming -- the withdrawal of underground water now often exceeds the rate of natural recharge. As water tables fall, pumping costs rise, until eventually farmers can no longer afford to irrigate. Efforts to develop the extensive coal resources of the northern Great Plains and the oil shale resources of western Wyoming and Colorado will divert more water from agriculture.(32) In Montana, farmers are battling energy firms in the courts in an effort to retain water for their fields. And Denver's expanding water needs are being met by the diversion of water from central Colorado farms.

While waterlogging and salinity have long been claiming irrigated land, the problems associated with the wholesale diversion of water from agriculture are relatively new. During the final quarter of this century, the diversion of water to pressing municipal and industrial needs will continue. In a Science article assessing the world food prospect, Neal Jensen of Cornell University predicts that "western U.S. irrigation agriculture faces gradual elimination as the pressures for higher priority water needs become evident."(33)

Responding effectively to the threats to cropland associated with mounting food demands poses a dilemma for farmers and government planners alike. Both economic pressures and political instincts encourage a short-term focus, a desire to expand current output to satisfy immediate needs. But this pressure to wring more food out of the land can have devastating consequences for the soil.

THE THINNING LAYER OF TOPSOIL

Over much of the earth's surface, the thin layer of topsoil is only inches deep, usually less than a foot. Any loss from that layer can reduce the soil's innate fertility. Forty years ago, G. V. Jacks and R. O. Whyte graphically described the essential role of topsoil: "Below that thin layer comprising the delicate organism known as the soil is a planet as lifeless as the moon."(34)

The erosion of soil is an integral part of the natural system. It occurs even when land is in grass or forests. But when land is cleared and planted to crops, the process invariably accelerates. Whenever erosion begins to exceed the natural rate of soil formation, the layer of topsoil becomes thinner. Once enough topsoil is lost, the cropland is abandoned. Eventually the topsoil will disappear entirely, leaving only subsoil or bare rock. But the gradual loss of topsoil and the slow decline in inherent fertility that precedes abandonment may take years, decades, or even centuries.

As population pressures mount, cultivation is both intensified and extended onto marginal soils. Some of the techniques designed to raise land productivity in the near term lead to excessive soil loss. In the American Midwest, the intensification of farming has led to the continuous cropping of corn, thereby eliminating the rotations that traditionally included grass-legume mixtures along with the corn. The transition to continuous cropping has been abetted by cheap nitrogen fertilizers that replaced nitrogen-fixing legumes. While these chemical fertilizers can replace nutrients lost through crop removal, they cannot make up for the loss of topsoil needed to maintain a healthy soil and structure.

Although the strong world demand for U.S. grain has made the continuous cropping of corn profitable, it has also led to an unacceptable level of soil erosion in many parts of the Corn Belt. A 1977 report from Iowa State University Experiment Station notes that "the often unobserved thinning of productive topsoil by erosion gradually removes the basic resource that supports agricultural production in an area." The report, from one of the world's most productive farming areas, adds: "The 200 million tons of soil lost from Iowa's cropland each year, for example, simply cannot be replaced within our lifetimes or those of our children. The eroded soil is gone, depleting the fertility of the land."(35)

In other parts of the world, the doubling of demand for food over the past generation has forced farmers onto land that is either too dry or too steep to cultivate and is thus highly vulnerable to erosion. Explosive Third World population growth has forced farmers onto mountainous soils without sufficient time to construct terraces. Once the natural cover is removed from unterraced, mountainous land, the topsoil quickly washes down into the valley below.

In the Andean countries of Latin America, skewed land-ownership patterns aggravate this problem. Wealthy ranchers use the relatively level valley floors for cattle grazing, forcing small landholders onto steeply sloping fields for the production of subsistence crops. (36) This peculiar inversion in land-use patterns is leading to severe soil erosion on the slopes, impairing the productive capacity of both the mountainside and the valleys.

In dryland wheat areas of the world, pressures to reduce the area in fallow could lead to widespread drying up of soils. This proved catastrophic in the U.S. Great Plains during the Dust Bowl of the 1930s and in the virgin lands of the Soviet Union during the 1960s. Except where land can be irrigated, the basic natural constraints on cultivation under low-rainfall conditions cannot be altered substantially.

Soil scientists have established tolerable rates of erosion for various types and conditions of soil. This tolerance level -- or T factor, as it is commonly referred to -- equals the rate of new soil formation. As long as the soil loss does not exceed the T factor, the innate productivity and fertility of the soil can be maintained indefinitely. In a 1977 survey of Wisconsin soils, 70 percent were found to have a soils loss greater than the tolerable level. On these soils with a T factor of 3.6 tons, the actual loss was 8.4 tons, more than double the tolerable rate.(37)

The USDA Soil Conservation Service reports that farmers are not managing highly erodible soils as well today as they were a generation ago. The adoption of conservation measures is relatively easy when there is excess capacity in the system; when commodity prices are high and food is scarce, the temptation to forego these essential measures is strong. A nationwide survey by the Soil Conservation Service indicated that "in 1975, soil losses on cropland amounted to almost three billion tons or an average of about twenty-two tons per hectare. Although this was excessive, it was far less than the estimated four billion tons of topsoil that would have been lost in 1975 if farmers had followed no conservation practices at all.(38) The report noted that if U.S. crop production was to be sustained, soil loss would have to be reduced to 1.5 billion tons annually.

The Council for Agricultural Science and Technology, supported by a consortium of midwestern universities, reported in 1975 that "a third of all U.S. cropland was suffering soil losses too great to be sustained without a gradual but ultimately disastrous decline in productivity." A summary document prepared for the 1977 U.N. Conference on Desertification reported that just under one-fifth of the world's cropland is now experiencing a rate of degradation that is intolerable over the long run. The U.N. report estimated that productivity on this land has now been reduced by an average of 25 percent.(39)

These broad global estimates become more meaningful when examined at the national level. In the Soviet Union, attempts to regain food self-sufficiency through heavy investments in agriculture are stymied by soils that have lost some of their inherent productivity. In an analysis of the government's agricultural plans, Thane Gustafson of Harvard notes that Soviet efforts must reckon with "fifty years of neglect [that] have left a legacy of badly damaged soils." The Nepalese Government estimates that the country's rivers now annually carry 240 million cubic meters of soil to India. This loss has been described as Nepal's "most precious export."(40)

In Ethiopia, the deterioration of soils was brought into focus by a drought that culminated in famine in 1974. A foreign ambassador in Addis Ababa described the origin of the problem in graphic terms: "Ethiopia is quite literally going down the river." More recently, the U.S. AID mission reports that "there is an environmental nightmare unfolding before our eyes...It is the result of the acts of millions of Ethiopians struggling for survival: scratching the surface of eroded land and eroding it further; cutting down the trees for warmth and fuel and leaving the country denuded...Over one billion -- one billion -- tons of topsoil flow from Ethiopia's highlands each year."(41)

According to U.N. estimates, erosion robs Colombia of 426 million tons of fertile topsoil each year.(42) Colombia's topsoil is relatively thin; given the poorly nourished condition of many of its people, the country can ill afford such a massive loss of this life-sustaining resource. In Pakistan, the progressive deforestation of recent decades has led to widespread soil erosion. At some point these losses in Ethiopia, Colombia, Pakistan, and other severely affected countries may begin to sabotage farmers' efforts to expand output.

Few efforts have been made to precisely measure the relationship between the cumulative loss of topsoil and the inherent fertility of cropland. Luther Carter writes in Science that "even where the loss of topsoil has begun to reduce the land's natural fertility and productivity, the effect is often masked by the positive response to the heavy application of fertilizers and pesticides which keep crop yields relatively high." The implications are, nonetheless, disturbing. David

Pimentel of Cornell University cites three independent studies, all undertaken in the United States, indicating that, other things being equal, corn yields decline by an average of "four bushels per acre for each inch of topsoil lost from a base of twelve inches of topsoil or less."(43)

In a world facing an acute shortage of productive cropland, any loss of topsoil should cause concern, for such a loss is essentially irreversible in the short term. Even under normal agricultural conditions -- including the heavy use of fertilizers and tillage practices that gradually mix subsoil into topsoil and that incorporate organic matter -- creating an inch of new topsoil can take one hundred years; if left to nature, it may take many centuries.

National political leaders and ministers of agriculture are faced with the task of adequately feeding an ever-expanding population without irreparably damaging one of the world's most essential resources -- the soil. Mounting demand pressures, whether the subsistence survival efforts of Peruvian peasants or the responses of American farmers to market forces, can have the same effect. In a great many countries, efforts to produce more food are now leading to a slow but gradual decline in the inherent fertility of soils.

LAND PRODUCTIVITY TRENDS

The ancients calculated yield as the ratio of grain produced to the seed that was planted. For them the constraining factor was the supply of seed-grain. But as agriculture spread, and as seed became more abundant and land more scarce, the focus shifted from the productivity of seed to that of land.

From the beginning of agriculture until World War II, land productivity increased very slowly. For long periods of time it did not increase at all. Rice yields in Japan during the nineteenth century were only marginally higher than those during the fourteenth century. Corn yields in the United States during the 1930s were the same as those during the 1860s, the first decade for which reliable yield estimates are available.(44)

Following World War II, however, crop yields began to rise rapidly in virtually every industrial country. During the 1960s, introduction of the fertilizer-responsive varieties of wheat and rice enabled yields in many Third World countries to also rise. From the end of World War II until the early 1970s, the steady rise in cereal yield per hectare was one of the most predictable trends in the world economy. Between 1961 and 1971, there were particularly impressive gains.

Since 1971, however, the increase has slowed; yields have become much more erratic and less predictable. From 1950 to 1971, the cereal yield per hectare climbed from 1.14 metric tons to 1.88 tons per hectare, or 2.4 percent per year. Between 1971 and 1977, the yield increased at only one-fourth that rate, or 0.6 percent per year. This marked slow-down was due to a combination of pressures, including the addition of marginal land to the world's cropland base, higher energy prices, a reduction in the land in fallow in the dryland wheat-growing regions, a shortening of the fallow cycle in areas of shifting cultivation, and the cumulative loss of topsoil in major food-producing regions.

Within the United States, the yield turned abruptly downward in 1973. While the yield of all cereals combined increased by a rather spectacular 4 percent per year from 1950-72, it has actually declined since then. The 1977 yield was 6 percent below that in 1972. Per hectare yields of wheat, barley, oats, rye, and rice peaked in 1971; the yield of corn peaked in 1972. As of 1978, none of these cereal yields had regained its earlier high level. The return to production of most of the twenty million hectares of idled U.S. cropland (nearly one-seventh of the cropland base) helps explain the initial fall in yields in 1973. Combined with uncommonly adverse weather, it contributed to the further decline in 1974. But it does not adequately explain the leveling off since then.

Another agriculturally advanced country that has experienced a yield downturn is France. Closely paralleling the U.S. experience, the cereal yield in France nearly tripled between 1950 and 1973, climbing from 1.6 metric tons to 4.35 metric tons per hectare; but after this impressive climb, it turned downward. During the four years since, it has fluctuated between 3.4 and 4.2 metric tons per hectare.

Egypt and China, two Third World countries with rather intensive agricultural systems, have also experienced a leveling off in per hectare cereal yield during the 1970s. Egypt's yield nearly doubled between 1950 and 1971, but it has not increased since. China's yield, more modest overall, has not increased at all since 1974. Except for the sharp falloff in the aftermath of the 1958 "Great Leap Forward," China's cereal yield had been edging upward throughout most of the 1950 to 1974 period.

The reasons for the interruption in national yield trends are not precisely the same for all countries. The main contributor to the downturn in U.S. yield per hectare was undoubtedly the return to production of cropland previously idled under government programs. When participating in agricultural limitation programs, farmers invariably retire their poorest, least productive land.(45) Since the United States is the world's leading food producer, bringing this poorer quality land back into production obviously had an effect on the world trend as well.

The high cost of energy has also influenced agricultural yields. The fivefold increase in the price of petroleum during the 1970s slowed the growth of energy use in agriculture. This influenced fuel use not only in tractors and irrigation pumps but also, indirectly, in the production of chemical fertilizer, an energy-intensive commodity. Fertilizer prices during the mid-1970s, influenced both by cyclical trends in the industry and by rising energy costs, soared to record levels.(46)

A third influence on the productivity of cropland is a shrinkage in the fallowed area in dryland farming regions. As world wheat prices rose during the 1970s, U.S. fallow land dropped from seventeen million hectares in 1969 to thirteen million hectares in 1974. This decline led Kenneth Grant, head of the USDA Soil Conservation Service, to warn farmers that severe wind erosion and dust bowl conditions could result. He cautioned farmers against the lure of record wheat prices and short-term gains that would sacrifice the long-term productivity of their land.(47)

Stresses are evident in other dryland farming regions as well. The U.S. agricultural attaché in Moscow reports there has been a reduction in fallow land in the Soviet Union. During the late 1960s and early 1970s, the Soviets were consistently fallowing seventeen to eighteen million hectares in the dryland regions; but after the massive crop shortfall and heavy imports of 1972, the fallowed area was reduced by one-third to eleven to twelve million hectares, where it has remained.(48)

In tropical and subtropical regions, where fallowing has evolved as a method of restoring fertility, mounting population pressures are forcing shifting cultivators to shorten the rotation cycles. This in

turn acts to reduce crop yields. Students of African
agriculture refer increasingly to that continent's
"cycle of land degeneration." In Losing Ground, Erik
Eckholm noted that "where traditional technologies persist -- due either to the lack of proven alternatives
or to the failure to reach farmers with new methods --
deterioration of the land begins as soon as human
numbers in a local area surpass the level that shifting
cultivation can support."(49) As cycles are shortened,
land productivity falls.

In Nigeria, where both the addition of marginal
land to the cropland base and a shortening of the
fallow cycle are lowering cropland fertility, cereal
yields have been falling since the early 1960s. Since
1950, the harvested area in Nigeria has multiplied
two-and-one-half times, but the quality of the added
land has been such that the average yield has fallen by
some 15 percent over the past decade and a half. The
fall in average productivity of land has overridden the
combined contribution of all advances in agricultural
technology, including chemical fertilizers, improved
varieties, and the expansion of irrigation. A World
Bank study reports that "fallow periods under shifting
cultivation have become too short to restore fertility
in some areas."(50) In some locales the original
cropping cycle of ten to fifteen years has already been
reduced to five years.

Similar pressures on the land are evident in Latin
America. According to FAO researchers: "There is
abundant evidence in certain regions of Venezuela that,
with growing population pressure, the fallow period is
becoming increasingly shorter so that soil fertility is
not restored before cropping. This leads to a fall in
the organic content and the water-holding capacity of
the soil. Soil structure deteriorates and compaction
becomes more common....In other words, with the population of modern times, formerly stable, shifting cultivation systems are now in a state of breakdown."(51)

Pressures associated with the growing demand for
food affect more than just the humid tropical and
subtropical regions where the fallowing systems of
shifting cultivation are breaking down. Although the
process of soil erosion is not always highly visible,
its cumulative effect on land productivity can be
graphic. A U.N. study of Latin American agriculture
ties the falling potato harvests to soil erosion in the
Peruvian Andes, the region that gave the world the
potato.(52)

German geographer Rober Schmid describes a situation in Nepal that is similar to Peru's. "During the last century man in search of more arable land had greatly extended the terraced fields...Increasingly, the farmers had to be self-sufficient, and therefore marginal land was taken into cultivation..." The farming of increasingly marginal land, Schmid goes on, combines with progressive deforestation, as trees are felled for wood and branches are lopped from remaining trees for use as animal fodder, until a cycle of soil erosion and deterioration is set in motion. "On the permanently farmed fields the soil showed signs of exhaustion and the yields declined slowly."(53)

Schmid's account resembles those of others who have witnessed the deterioration of hill agriculture in Iran, Pakistan, northern India, and Ethiopia. The experience of the last several years suggests there is relatively little fertile new land that can readily be brought under the plow. There are indications in several major countries, including the United States, Canada, Nigeria, China, and Brazil, that the falloff in the quality of cropland is rather steep.

In Canada, the inherent productivity of the cropland is being reduced by the continual substitution of marginal land for prime land. Land being lost to urbanization includes some of the most productive soils in the country, whereas that being added is far less productive. The Canada Land Inventory uses seven classes of land with classes one to three being considered arable. Almost all of the class one, two, and three land is now being farmed. An estimated 233 hectares of class four land, described as marginally productive, is required to replace 100 hectares of class one land.(54) The net effect of substituting lower quality land for higher quality land lost to nonfarm uses is a reduction in the average fertility.

An even sharper falloff in land quality is reported in China, where population pressures are far greater. Leslie T. C. Kuo, a specialist on Chinese agriculture, reports that "the use of one mou of cultivated land for construction purposes must be offset by the reclamation of several mou of wasteland."(55) Kuo goes even further, indicating that a sizable portion of the present cultivated land is so poor as not to be profitable for crop production. This suggests there are few opportunities for profitably expanding cropland area.

A somewhat similar, yet slightly less dramatic, situation exists in Brazil, the largest country in

Latin America. Brazil has more than doubled its harvested area since 1950, but, despite heavy investments in fertilizer and other modern techniques, crop yield per hectare has not increased. The declining quality of successive additions of land appears to have offset advances in agricultural technology and the increased use of fertilizer.

Apart from assessing changes in the innate fertility of soils, there is also the question of the biological potential for continuing to profitably raise crop yield per hectare. Despite its importance, this issue has received little attention from the research community. Among the few to consider the issue are Louis Thompson, professor of agronomy at Iowa State University, and Neal Jensen, a plant breeder and professor of genetics at Cornell University. From 1950 to 1972, Thompson has plotted the average cornfields recorded on several experimental stations in Iowa as well as those achieved by farmers in the state. In the early 1950s, there was a wide gap between average yields on the experimental plots and those on Iowa farms; but by the early 1970s, this gap had virtually disappeared.(56) In effect, the backlog of technology available to farmers in Iowa had been largely used up.

Neal Jensen analyzed the long-term historical rise of wheat yields in New York State and reached a conclusion even more sobering. Applying his findings to agriculture in general, he predicts that "the rate of productivity increase will become slower and will eventually become level for several reasons."(57) He believes that plant breeders and agronomists in the more advanced agricultural countries have already raised yields about as far as they can.

In summary, the outlook for raising cereal yields appears less hopeful now than it did at mid-century. A number of soil surveys and studies indicate that the inherent fertility of one-fifth or more of the world's cropland is declining. At the same time, the backlog of agricultural technology waiting to be applied in the agriculturally advanced countries appears to be dwindling. Together, these two factors cast a shadow over the prospect of a rapid, sustained rise in yield per hectare during the final quarter of this century to compare with that of the third quarter.

THE CROPLAND PROSPECT

There is no simple way of projecting the cropland area for the remainder of the century, much less assessing its adequacy. One way of gaining some perspective is to compare increases in population growth, harvested area, and yields for the third quarter of the century with those projected for the fourth quarter. Between 1950 and 1975, world population increased by 59 percent, or 1.5 billion. The harvested area of cereals expanded by 22 percent and average yield per hectare rose by 63 percent. Together the area and yield gains meant an increase in world cereal output of 99 percent, a near doubling.

In 1950 there were still substantial opportunities for adding to the world's cropland base. The potential for expanding the area under irrigation was impressive, particularly in the developing countries. And farmers had scarcely begun to exploit the potential for raising per hectare yields, even in the agriculturally advanced countries.

During the final quarter of this century, population is projected to increase by 57 percent; although slightly smaller in percentage terms, the expected addition of 2.3 billion people would be half again as large as the 1.5 billion added during the third quarter. As the final quarter of this century began, cropland was being lost to nonfarm purposes at a record rate. The abandonment of agricultural land because of severe soil erosion, degradation, and desertification was at an all-time high. The chances for substantial net additions to the world's cropland base were not good.

A review of the future potential for expanding the world's cropland by geographic regions does not provide much ground for optimism. In North America the area farmed has been shrinking for three decades. The OECD's Agricultural Committee reports that the opportunities for new land reclamation in Western Europe are negligible.(59) The countries of Eastern Europe have been hard pressed to maintain their cultivated land over the last fifteen years. In the Soviet Union, where farming has already extended into highly marginal rainfall areas, the prospect of further expanding the cropland base is not good.

In densely populated Asia the prospect is not very bright either. There is little new land to bring under the plow in China; given the already extensive development of irrigation, the opportunities for further expanding the area multiple-cropped are not good. In India, the region's other leading food producer, the cropland area is expected to increase little, if at all. But there is room to expand the area harvested through multiple-cropping, since much of the irrigation potential, particularly on the Gangetic Plain, is still to be developed. The one area in Asia that could sustain an increase in both cropland area and the intensity of cropping is the Mekong Valley and Delta, assuming that local political conditions permit these agricultural developments.

Although there are a few areas in Africa yet to be developed, overall there is limited opportunity for greatly adding to the cropland of the continent. In the vast northern section, the Sahara is reaching in every direction, claiming land in nearly a score of countries. The one country with potential for markedly expanding the area harvested is Sudan since it has developed only part of its share of the Nile waters reserved to it by treaty with Egypt. To the south, the principal hope for more farmland lies in opening the tsetse fly belt to cultivation -- if a way can be found to eradicate the tsetse fly.

In Latin America the cropland area can be expanded by plowing up some grasslands, as Argentina has been doing during the 1970s. This would, of course, reduce the scope for grazing livestock. The other major potential, often overrated, lies in opening new lands in the interior of the continent, principally in the Amazon basin. The principal constraint here is the inherent poor fertility of the soil and its limited capacity to sustain cultivation over an extended period of time.

Wherever recent data are available, they indicate a sharp falloff in the productivity of new additions to the cropland base. This has been documented for countries as different as Canada and China. In effect, it confirms what some analysts have suspected -- that the world's farmers have most of the productive cropland under the plow.

Although there will be numerous efforts to expand the cultivated area both through governmental efforts and the initiative of individual farmers, there will be offsetting factors. Offsetting some of the additions to the global cropland base will be the abandonment of millions of hectares of marginal land brought under the plow during the great push since mid-century to expand world food output. This land is on the sides of mountains in the Andean countries of South America, in the highlands of East Africa and the foothills of the Himalayas, and in the rugged valleys of Java. Well before the end of the century, the topsoil will be gone from many of these plots and they will have been abandoned.

On balance, it is difficult to see how the world as a whole can achieve much more than a 10 percent increase in the cropland area during the final quarter of this century. This increase of 10 percent, representing a substantially smaller increase than occurred during the third quarter of this century, indicates a continuation of the declining rate of expansion of the area under cultivation. Barring a dramatic rise in food prices, one that would make the cultivation of marginal land profitable, it may represent the realistic upper limit.

If this 10 percent increase is combined with the projected growth in population during the final quarter of this century, it will lead to a decline in cropland area per person from .184 hectares to .128 hectares. (See Table 3.) This is an impressive shrinkage, paralleling that of the third quarter century. If this cropland assessment is a reasonable one, it suggests some serious difficulty achieving the growing food output projected for the final quarter of this century.

Table 3. World Population and Area in Cereals, 1950 and 1975 With Projections to 2000.

Year	Population (billions)	Area in Cereals (million hectares)	Area per Person (hectares)
1950	2.50	602	.241
1975	3.97	729	.184
2000	6.29	802	.128

Source: United Nations and U.S. Department of Agriculture

Projections of world food demand by the FAO and the International Food Policy Research Institute indicate that the world food demand will roughly double between 1975 and 2000. This being the case, the required increase in yield per hectare during the final quarter of this century would have to be substantially larger than that achieved during the third quarter.

Recent trends indicate the potential for a continuing rapid rise in crop yield per hectare may be much less than has been assumed in all official projections of world food supply. The postwar trend of rising yield per hectare has been arrested or reversed in the United States, France, and China, each the leading cereal producer on its respective continent. Aside from the biological constraints on raising land productivity, a combination of pressures to extract even more food from the land plus poor land management is leading to the slow but progressive deterioration of one-fifth to one-third of the world's cropland. For most countries the mounting demand pressures are of domestic origin; for food exporters such as the United States and Canada, it is the growth in demand from abroad that is pressing the land beyond its limits.

The evidence now available raises doubts as to whether it will be possible to get a combination of cropland expansion and yield increases that will satisfy the growth in world food demand projected for the remainder of this century. Evidence of the difficulty can be seen in the reduced rate of growth in world food supplies. Between 1950 and 1971 the modest expansion in world cropland area and the unprecedented rise in crop yield per hectare raised world grain production per person from 276 kilograms to 360 kilograms. Since 1971 per capita grain production has actually declined, averaging only 354 kilograms over the following six years.(60)

This review of the period from 1950-78 indicates a clear loss of momentum during the 1970s in the efforts to raise food consumption. Since 1971 output has failed to keep pace with population growth, much less with the growth in demand generated by rising incomes. The result has been rising food prices. If the cropland assessment is reasonable, and if the projected increase in world population materializes, then steep rises in food prices in the years ahead may be inevitable. The dimensions of the problem will come into focus when two or more of the major food producing countries have poor crops. Just as years of bad weather tend to bring a deteriorating situation into focus, so years of good weather tend to obscure the long-term deterioration.

PUBLIC POLICY IMPLICATIONS

Few things will affect future human well-being more directly than the balance between people and cropland. If recent population and cropland trends continue, that balance will almost certainly be upset, leading to economic uncertainty and political instability. Avoiding large, politically destablizing rises in food prices between now and the end of the century may not be possible without a mammoth effort to protect cropland from nonfarm uses, to improve the management of soils, and, most importantly, to quickly reduce the rate of population growth.

The lack of well-developed, land-use policies is a legacy of an earlier era when land was relatively abundant and there were still frontiers to be pushed back, or when there was still a large unrealized potential for raising per hectare yields. Now the frontiers are gone and the loss of cropland is proceeding at a pace that threatens to undermine the balance between food and people.

Agricultural land can no longer be treated as a reservoir, an inexhaustible source of land for industry, urbanization, and the energy sector. Cropland is becoming scarce. In a world of continuously growing demand for food, it must be viewed as an irreplaceable resource, one that is paved over or otherwise taken out of production only under the most pressing circumstances and as a result of conscious public policy.

In some situations adjustments are needed in land use within agriculture as well as between agriculture and other sectors of the economy. The prevailing social structure in several countries in Latin America has led to grazing on valley floors and the plowing of mountainsides. Intelligent land-use policies would reverse this situation with a thorough land-reform program. In effect, existing social structures in some countries are simply not compatible with the wise management of land resources.

Historically, land in most countries was allocated to various uses through the marketplace, but unfortunately the market does not shield cropland from competing interests. As cropland becomes more scarce, it can be protected from competing nonfarm demands only through some form of land-use planning. Such planning can occur at the national level, the local level, or both. It can rely primarily on land-use restrictions in the form of legislation or government decrees, or it can rely on incentives such as differential tax rates.

Each society will need to employ approaches suited to its own circumstances.

The OECD reports that Japan is the only country with comprehensive zoning nationwide. In 1968 the entire country was divided into three land-use zones -- urban, agricultural, and other. In 1974 the plan was further refined to include specific areas for forests, natural parks, and nature reserves.(61) Given the acute pressures on its land resources, Japan has faced the issue first and in so doing has developed a model for other countries to follow.

Within the United States, national land-use planning is still at a rather rudimentary level, confined largely to setting aside national parks, forests, and wildlife reserves. At the local level there are often restrictions on land that can be used for commercial purposes. A number of states, including California, Massachusetts, New Jersey, and New York, are concerned about the need to protect agricultural land, but to date only a few have effectively done so.

During the 1960s, several European countries -- Belgium, France, Germany, and the Netherlands -- passed legislation establishing land-use guidelines, with planning to be done at the local level. In addition to protecting agricultural land as such, their laws also addressed such issues as the control of urban sprawl, the need for parks, and the establishment of green belts around cities.(62)

Rivaling the loss of cropland to nonfarm uses is the process of gradual soil degradation through erosion. If the loss of topsoil and associated soil degradation are permitted to continue, they will lead to ever higher food-production costs. But the effort needed to halt the deterioration of the soils on which humanity depends is staggering, one fraught with complicating social factors.

The one source of cropland degradation and loss officially dealt with at the international level is desertification. The World Plan of Action to stop the spread of deserts, agreed to at the Nairobi conference, concluded that money spent now to halt and reverse desertification and the loss of cropland promised high returns, ones that were competitive with other forms of investment. The plan specified the levels of investment by national governments, the World Bank, the U.N. Development Program, and the Inter-American Development Bank that would bring the process to a halt within the next decade and a half. They estimated that "a net zero desert growth can be achieved within the next ten to fifteen years, provided that the measures are started promptly and are effectively and comprehensively carried out."(63) Achievement of a net zero desert growth will be possible only with prompt, concerted action by national governments that are, unfortunately, not usually noted for such efforts.

The effort required to stabilize soils in many countries will require a strong national political commitment and a detailed plan of action. Such a program has been outlined for the United States by the Soil Conservation Service. In order to stabilize U.S. soils and bring the annual loss of topsoil down to a level that does not exceed the tolerance factor, the plan calls both for changes in cropping practices and for heavy investments in land improvements. The principal recommendations of the plan include an increase in the terraced land where farmers leave crop residues on the surface from 4.8 million hectares to 17.6 million hectares, an increase in the amount of strip-cropped land with minimum tillage from about 0.4 million to fourteen million hectares, and an increase in the strip-cropped land with crop residues left on the surface from 3.2 million to 7.2 million hectares. For land where farmers rely on a combination of contour farming and minimum tillage to keep their soil in place, they recommend a twenty-onefold increase -- from 0.8 million to 16.8 million hectares. The largest suggested increase of all is for land where contour farming and crop residues on the surface are used together -- from 3.2 million to 22.4 million hectares.(64)

These prescribed changes in farming practices are monumental in scale, involving half of all U.S. cropland. Adoption of these soil-saving measures would often run counter to the immediate economic interests of farmers and consumers, since they would lead to a 5 to 8 percent increase in food production costs.(65) Some efforts to conserve soil and protect long-term food production might reduce food output in the near term as well.

While soil scientists can chart a national plan of action in detail, they cannot generate the political support needed to fund and administer such a plan. Public support on the scale needed will not be forthcoming without a broader understanding of the costs to society of failing to act. R. A. Brink and his colleagues, writing in _Science_, observe that "in our predominantly urban society, it may be difficult to gain the public support needed for funding an adequate soil conservation program." But they predict that "public opinion will shift with the worsening of the world food crisis."(66)

Brink wonders whether "as a result of mounting pressures on the land, the need for soil-saving measures is outrunning the capacity of conservation agencies, as now financed, to assist farmers in meeting it." If this observation applies to the United States, one can only wonder about the future adequacy of the cropland base in those countries that have far less to work with, both institutionally and financially. The alternative facing governments is to not respond to the deterioration, and then be confronted with long-term food scarcity as the cropland base progressively deteriorates.

The future balance between people and cropland is likely to be affected more by population policy than by any other single factor. Growth in human numbers not only generates a demand for more cropland, but it also simultaneously generates pressures to convert cropland to nonfarm uses. Thus far, policymakers have assumed that projected increases in population over the final quarter of this century will materialize. But the social costs of remaining on the projected demographic path may become too great. A fundamental rethinking of population policies at the national level may well be unavoidable in many countries.

If so, pronatalist policies will have to be abandoned, and governments will need to actively encourage small families. Other governments may join those of China and Singapore in advocating a maximum of two children per couple and in making family planning universally available. With future rates of population growth bearing so heavily on the prospect of eliminating hunger, the fact that one-half of the world's couples do not yet have access to family-planning services is not only socially irresponsible, but politically inexcusable as well.(67) If population growth is to be reduced to a socially acceptable rate, unprecedented changes in government policies and in individual attitudes will be required. Such changes are not likely to occur without a better understanding of the precarious balance between people and cropland.

The convulsive changes in the world food economy during the 1970s reflect in part the growing pressures on the cropland base. They should be regarded not as an aberration but rather as a signal that pressures on the cropland are in some ways becoming unbearable. They may in fact be advance tremors warning of the quake to come.

Threats to the world cropland base from the non-farm sector are very real. Evidence that the soils on which we depend for our food are deteriorating is overwhelming. What is not nearly as evident is how political leaders will respond. But the time left in which to respond is measured in years, not decades.

The issue is not whether the equilibrium between people and land eventually will be reestablished. It will be. If the deterioration is not arrested by man, then nature will ultimately intervene with its own checks. The times call for a new land ethic, a new reverence for land, and for a better understanding of our dependence on a resource that is too often taken for granted.

FOOTNOTES

(1) All statistics on grain area in this paper are from Economic Research Service, "26 Years of World Cereal Statistics by Country and Region," U.S. Department of Agriculture, Washington, D.C., July 1976; recent statistics are unpublished estimates by the Economics, Statistics, and Cooperative Service, U.S. Department of Agriculture, Washington, D.C., 1978.

(2) Agricultural Statistics (Washington, D.C.: U.S. Department of Agriculture, annual).

(3) Linda Lee, "A Perspective on Cropland Availability," U.S. Department of Agriculture, Washington, D.C., 1978; European figures from Organization for Economic Cooperation and Development, Land Use Policies and Agriculture (Paris, 1976).

(4) John McHale, World Facts and Trends (New York: Macmillan Co., 1972).

(5) Science Council of Canada, "Population, Technology and Resources," (Ottawa, Ontario, July 1976).

(6) Akef Quazi, "Village Overspill," Mazingira, No. 6, 1978.

(7) Country Report of the Indian Government to the U.N. Conference on Human Settlements (Vancouver, B.C., Canada, June 1976).

(8) Alva Erisman, "China: Agriculture in the Seventies," and Dwight Perkins, "Constraints Influ-
encing China's Agricultural Performance," in U.S.
Congress, Joint Economic Committee, China: A
Reassessment of the Economy, Committee Print, July
10, 1975.

(9) Jimoh-Omo Fadaka, "Superdams: The Dreams that
Failed," PHP International, August 1978.

(10) U.S. Senate, Committee on Interior and Insular
Affairs, Land Use and Energy: A Study of Inter-
relationships, Committee Print, January 1976.

(11) Janet M. Smith, David Ostendorf, and Mike Schect-
man, "Who's Mining the Farm," Illinois South
Project, Hernin, Ill., Summer 1978.

(12) Population Division, United Nations Secretariat,
Population by Sex and Age.

(13) U.N. Conference on Desertification, Desertification: An Overview, (Nairobi, Kenya, August
29-September 9, 1977).

(14) H. N. Le Houerou, "North Africa: Past, Present,
Future," in Harold Dregne, ed., Arid Lands in
Transition (Washington, D.C.: American Association
for the Advancement of Science, 1970); H. F.
Lamprey, "Report on the Desert Encroachment Recon-
naissance in Northern Sudan, 21 October to 10
November 1975," Nairobi, undated.

(15) J. Vasconcelos Sobrinho, "O Deserto Brasileiro,"
Universidad Federal Rural de Pernambuco, Recife,
1974; J. Vasconcelos Sobrinho, "Problematica
Ecologia do Rio Sao Francisco," Universidad Federal Rural de Pernambuco, Recife, 1971; Cesar F.
Vergelin, "Water Erosion in the Carcarana Watershed: An Economic Study," Dissertation, University
of Wisconsin, Madison, 1971.

(16) B. F. Kosov et al., "The Gullying Hazard in the
Midland Region of the USSR in Conjunction with
Economic Development," Soviet Geography, March
1977.

(17) Ye. F. Zorina, B. F. Kosov, and S. D. Prokhorova,
"The Role of the Human Factor in the Development
of Gullying in the Steppe and Wooded Steppe of the
European USSR," Soviet Geography, January 1977.

(18) United Nations, Economic Commission for Latin America, El Medio Ambiente en America Latina (Santiago, Chile, May 1976).

(19) "Should Agricultural Land Be Protected?" OECD Observer, September/October 1976.

(20) Report from United States Embassy, Jakarta, March 1976; reference to cropland abandonment in the Punjab is in U.S. Agency for International Development, "Fiscal Year 1980 Budget Proposal for Pakistan," Washington, D.C., 1978.

(21) U.N. Conference on Desertification, "Economic and Financial Aspects."

(22) Ibid.

(23) Thane Gustafson, "Transforming Soviet Agriculture: Brezhnev's Gamble on Land Improvement," Public Policy, Summer 1977; Omo-Fadaka, "Superdams."

(24) U.N. Conference on Desertification, "Synthesis of Case Studies of Desertification," (Nairobi, Kenya, August 29-September 9, 1977).

(25) U.N. Conference on Desertification, Background Document, "Transnational Project to Monitor Desertification Processes and Related National Resources in Arid and Semiarid Areas of Latin America," (Nairobi, Kenya, August 29-September 9, 1977).

(26) U.N. Conference on Desertification, "Synthesis of Case Studies."

(27) Ibid.

(28) "Proceedings of International Symposium on the Salinity of the Colorado River," Natural Resources Journal, (January 1975).

(29) Gustafson, "Transforming Soviet Agriculture."

(30) Arizona Water Commission, Inventory of Resources and Users, State of Arizona, July 1975.

(31) Joel Kotkin, "Los Angeles Is Draining Once Verdant Owens Valley," Washington Post, September 8, 1976.

(32) U. S. Senate, Land Use and Energy.

(33) Neal Jensen, "Limits to Growth in World Food Production," Science, July 28, 1978.

94

(34) G. V. Jacks and R. O. Whyte, The Rape of the Earth -- A World Survey of Soil Erosion (London: Faber, 1939).

(35) Iowa Experiment Station, "Our Thinning Soil," Research for a Better Iowa, February 1977.

(36) Frances M. Foland, "Changing the Agrarian World: Focus on Latin America," Common Ground, Summer 1978.

(37) R. A. Brink, J. W. Densmore, and G. A. Hill, "Soil Deterioration and the Growing World Demands on Food," Science, August 12, 1977.

(38) Soil Conservation Service, "Cropland Erosion," U.S. Department of Agriculture, Washington, D.C., June 1977.

(39) C.A.S.T. study cited in Luther Carter, "Soil Erosion: The Problem Persists Despite the Billions Spent on It," Science, April 22, 1977; U.N. Conference on Desertification, "Economic and Financial Aspects."

(40) Gustafson, "Transforming Soviet Agriculture," Master Plan for Power Development and Supply (Kathmandu, Nepal: His Majesty's Government, with Nippon Koei Company, 1970).

(41) Ambassador in Addis Ababa quoted in Jack Shepherd, The Politics of Starvation (Washington, D.C.: Carnegie Endowment for International Peace, 1975); U.S. Agency for International Development, "Fiscal Year 1980 Budget Proposal for Ethiopia," Washington, D.C., 1978.

(42) United Nations, Economic Commission, El Medio Ambiente.

(43) Carter, "Soil Erosion"; David Pimentel et al., "Land Degradation: Effects on Food and Energy Resources," Science, October 8, 1976.

(44) All statistics on grain production and yields in this paper are from Economic Research Service, "26 Years of World Cereal Statistics"; recent statistics are unpublished estimates by the Economics, Statistics, and Cooperative Service, U.S. Department of Agriculture, Washington, D.C., 1978.

(45) Economic Research Service, "Productivity of Diverted Cropland," U.S. Department of Agriculture,
Washington, D.C., April 1969; Pierre R. Crosson,
The World Food Situation (Washington, D.C.: Resources for the Future, 1977).

(46) Economic Research Service, "Fertilizer Situation
(annual), U.S. Department of Agriculture, Washington, D.C., various issues.

(47) Figure on U.S. fallow land is private communication, Economics, Statistics, and Cooperative
Service, U.S. Department of Agriculture, Washington, D.C., August 22, 1978; Kenneth Grant, "Erosion in 1973-74: The Record and the Challenge,"
Journal of Soil and Water Conservation, January/
February 1975.

(48) Figures through 1970 from Soviet Statistical
Handbook (Moscow: date unknown); more recent
figures are from Agriculture in the USSR, 1971-77
(Moscow: date unknown) and The National Economy of
the USSR (annual) (Moscow: various issues).

(49) Erik Eckholm, Losing Ground: Environmental Stress
and World Food Prospects (New York: W. W. Nortun &
Co., 1976).

(50) Wouter Tims, Nigeria: Options for Long-Term Development (Baltimore: Johns Hopkins University Press,
for the World Bank, 1974).

(51) R. F. Watters, Shifting Cultivation in Latin
America (Rome: Food and Agriculture Organization,
1971).

(52) Ibid.

(53) R. Schmid, "The Jiri Multipurpose Development
Project," in Mountain Environment and Development
(Kathmandu, Nepal: University Press, 1976).

(54) "A Land Use Policy for Canada," Agrologist, Autumn
1975.

(55) Leslie T. C. Kuo, The Technical Transformation of
Agriculture in Communist China (New York: Praeger,
1972).

(56) L. M. Thompson, "Weather Variability, Climatic
Change, and Grain Production," Science, May 9,
1975.

(57) Jensen, "Limits to Growth in World Food Production."

(58) U.N. World Food Conference, Assessment of the World Food Situation: Present and Future, Rome, November 5-16, 1974; International Food Policy Research Institute, Food Needs of Developing Countries: Projections of Production and Consumption to 1990 (Washington, D.C., December 1977).

(59) Organization for Economic Cooperation and Development, Land Use Policies.

(60) All statistics on per capita grain production are from Economic Research Service, "26 Years of World Cereal Statistics"; recent statistics are unpublished estimates by the Economics, Statistics, and Cooperative Service, U.S. Department of Agriculture, Washington, D.C., 1978.

(61) Organization for Economic Cooperation and Development, Land Use Policies.

(62) Ibid.

(63) U.N. Conference on Desertification, "Economic and Financial Aspects."

(64) Soil Conservation Service, "Cropland Erosion."

(65) Ibid.

(66) Brink, "Soil Deterioration."

(67) Bruce Stokes, Filling the Family Planning Gap (Washington, D.C.: Worldwatch Institute, May 1977).

Section 2

THE POTENTIAL SOURCES OF FOOD

INTRODUCTION

In order to meet the challenges of feeding the world's people in the coming years, men and their institutions will have to skillfully use their accumulated expertise with plants, animals and fish. It will be necessary to develop new species and methods for their husbandry. While strategies must be geared to local and regional needs, it is important to gain some sense of the respective potential contributions of different food sources. These potentialities, when integrated with the particular resource capabilities of a country or region, and when meshed with the talents and desires of indigenous people, offer considerable hope for long-range improvement in human nutrition.

With a brief paper and a longer appendix, Norman Borlaug outlines the prospects for plants. His contribution is wide in scope and rich in detail, and he does not hesitate to give his personal assessment of opportunities and problems. After commenting on the remarkable nature of American agriculture, as well as the reasons for its rise and decline over the years, Borlaug notes its influence on foreign agriculture. With plants preeminent in meeting world food needs, it is important to assess the maximum genetic yield potential of current crop species and to consider the value of adding new species. There are substantial research needs if we are to stay ahead of population pressures, but much can be done to improve food production through increasing yields and cropping intensities and through remedying the defects of traditional agriculture.

A great deal of Borlaug's statement is fundamental to understanding agricultural development in low-income countries. He views the extreme manifestations of the environmental movement as inimical to the task of increasing food production where it is needed most, and he clearly chooses to intervene judiciously in the balance of Nature. In Borlaug's perspective, it is not the technical requirements of food production which are frustrating; rather, it is the imposition of inappropriate political behavior, the influence of activist groups from affluent societies pursuing causes, and the attendant distortions of the truth which erode support for effective action. These barriers, when added to the considerable technical requirements of our present world food situation, seem to him to be negative, but perhaps determining, forces.

With a thorough treatment of the role of protein in human nutrition and the place of animals in food production, John Pino describes the contributions of livestock to the world supply of food. Considering all sources of protein, Pino finds that there is presently enough protein produced to satisfy the needs of the projected population in the year 2000, but that poor distribution can be expected to perpetuate local protein malnutrition unless local production capabilities can be increased. Pino demonstrates that food from animal products is very important to the protein supply, and he indicates that animal products will continue to be important in the future. He concludes with a series of suggestions for U.S. policy which could affect both national and international welfare.

A limited but important role for aquatic animal protein sources is sketched by Carl Sindermann. A plateau of world production of food from aquatic sources was reached during the 1970s, and Sindermann believes that future expansion must come from well-managed but limited stocks of "traditional" species and from increased usage of "non-traditional" species. More potential for expansion exists with aquaculture, particularly in estuarine and coastal areas. A major problem with the husbandry of this source of food lies in the lack of research and development support, which has never been sufficient. Consequently, understanding of the population dynamics of aquatic animals is so limited that more effective management is questionable; nor can the impact of human actions on aquatic environment be accurately predicted. Habitat degredation, via pollution, is a serious deterrent to aquaculture, and steps toward environmental management are needed if food production from certain waters is to be increased.

There is optimism in these essays about the possible future contributions of plants, animals, and fish to solving the problem of world hunger. However, each of the three writers reporting on potential food sources finds reason to be concerned about the necessary actions of the governments of the world.

4
USING PLANTS TO MEET
WORLD FOOD NEEDS

Norman E. Borlaug

The use of plants to meet the world's food needs
is vital to civilization's survival. To explain their
importance, I will touch briefly upon the history of
agriculture and give you my insight as to why agriculture in the U.S. has flourished while it has stagnated
and lagged behind in the food-deficit, developing
nations. Moreover, I will give my impressions on what
research is required to continue the improvement of
agriculture in the U.S. I also wish to cite some of
the forces and factors that are likely to restrict
agricultural production in the next several decades
unless they are overcome. Finally, I will attempt to
give you some insight into the world's land-food-
population problems as I perceive them over the next
forty years.

THE UNIQUENESS OF AMERICAN AGRICULTURE

At the outset I will risk your feeling that I am
being both sentimental and nostalgic by stating that I
believe one of the greatest achievements of the U.S.
has been the outstanding development of its agriculture. Some can rightfully say that I am biased in
favor of agriculture by having spent the past thirty-
five years struggling to assist food-deficit nations in
many parts of the world in their attempt to expand food
production. Nevertheless, others, such as Dr. Milton
Friedman, renowned economist of the University of
Chicago, agree. A few months ago Friedman, writing
under the title of "What is America?" made this statement:

If you look at the achievements of the United
States, I think the most dramatic example is its
agriculture. If you go back to the founding of
this country in 1776, something like 80 to 90

percent of the people were on farms. It took
eight to nine people working full time -- twelve
to sixteen hours a day -- to feed themselves and
their families and perhaps one other person.
Today fewer than 5 percent of the American people
are on farms. One person today can feed himself
and his family and nineteen others and their
families and also have a great deal left to export
abroad. That is really -- in a nutshell -- the
miracle of America.

A September 1978 General Accounting Office publication, entitled Changing Character and Structure of
American Agriculture: An Overview, clearly indicates
the outstanding accomplishments of agriculture as well
as its changing structure. It emphasizes the past and
present importance of agriculture to the general economy of the country.

Agricultural assets in 1977 were $671 billion --
equal to 75 percent of the capital assets of all manufacturing corporations in America. Agriculture is the
nation's largest industry, employing directly and
indirectly a total of seventeen to twenty million
people in the entire food chain, from production
through transportation, processing, storage, distribution, marketing, and serving in restaurants.

About 7.8 million people lived on farms in 1977 --
almost 3.6 percent of the total population. However,
only 4.3 million, or about 2.0 percent of the total
population, worked in agriculture directly. Agriculture, besides being the country's major employer, is
also its most efficient industry. At the farm level,
one worker now produces 300 percent more than he did
twenty years ago, while the manufacturing worker's
production has increased only 1.7 percent during the
same period.

The 1976 direct sales of agricultural commodities
were about $94 billion, representing only about 3.6
percent of the total U.S. gross national product. If
direct-agriculture-support industries are included,
this rises to about 25 percent.

Farmers are also big spenders, using 75 to 85
percent of every dollar of income on production costs,
thus injecting back into the economy $80 billion
through purchases of inputs.

Despite falling prices for most agricultural
export commodities during the past three years, exports
represent one of the few bright spots in American

foreign trade. In each of the past five years, exports of agricultural commodities have totalled $20 billion or more, leaving a net positive balance of $8 to $10 billion after deducting for the import of agricultural commodities not grown in the U.S. Bad as our balance of payments has been, largely because of the vast expenditures for imported petroleum, it would have been much worse except for agricultural exports.

Certainly the greatest contribution of American agriculture to the consumer is the relatively cheap price of food. At present the average American family spends about 17 percent of its take-home pay, after taxes, on food. There is no other country in the world that has a comparably low level of expenditure for this first and most basic necessity.

What Contributed to Making American Agriculture Unique?

The U.S. was blessed with having within its boundaries vast areas of land with topography, soils, and climate potentially suitable for the production of a wide range of different types of agriculture and forestry. It was also blessed in its early years with millions of industrious farmers who immigrated to America in search of a better life and with hopes of eventually owning their own farm. They and their descendants took advantage of the opportunities for free education. They struggled and most became, sooner or later, owners of the land they worked.

It was not until 1862, however, that government enacted the basic policies and established the organizational structures that laid the firm foundation upon which the agricultural revolution of the past four decades was based.

Within a two-month period in mid-1862 three key legislative acts were passed by Congress. The Department of Agriculture was established and was given a broad mandate for agricultural development. The Homestead Act distributed federal lands free to farmers who occupied and worked them. The Morrill Land Grant Act appropriated funds -- from the sale of public lands -- to support agricultural and mechanical colleges in every state. These institutions were to train young people and also were to develop means of solving agricultural production problems. After a slow, uncertain, and often frustrating early period of development, the agricultural and mechanical colleges evolved into the land grant universities. These, during the past sixty years, have expanded into many disciplines in both research and education which have had profound impact on all walks of American life.

It is interesting to note that the three legislative acts passed in 1862, which set a new course for agricultural development, were pushed through Congress by the joint efforts of politicians, farmers, and journalists. Scientists played a very minor role. About twenty-five years later, with research floundering because of a discipline-oriented approach, cumbersome administrative bureaucracy, and inadequate funding, scientists began to exert pressures for change. With growing interest in agricultural science in Congress, this effort culminated in the passage of the Hatch Act of 1887, which established the agricultural experiment stations, and which stimulated research and coordination of activities at federal and state levels. This legislation soon began to catalyze not only agricultural scientific research but also many other activities within the land grant universities. These events created the need for a vehicle to disseminate the new knowledge and led to the establishment of the extension service.

There is a long gestation period from the initiation of research to its fruition. Agricultural problems are very complex involving climate, soil, biologic, and economic factors. Knowledge and materials must therefore be amassed, through research, which have the potential to solve these problems. Finally, inputs such as seed of improved varieties, fertilizer, pesticides, and machinery must be made available so that the new technology can be applied on a large commercial acreage if it is to be reflected in large increases in production.

The initial fruits of the agricultural research programs initiated in the 1890s, gradually and nonuniformly, had their effect. Depending on the complexity of the different problems, the impact began to be felt in the second decade of the 1900s. By the 1930s government and university-sponsored research had developed information and materials with substantial potential for solving many of the production problems. Meanwhile, during the 1920s and 1930s, private sector agribusiness developed tractors and ancillary machinery that rapidly replaced horses. This development had a twofold effect. It reduced the area of land needed for animal feed. It also reduced the number of people required to operate a family farm and set the stage for an accelerated migration of rural people to higher paying jobs in industry.

Private sector agribusiness in the 1930s also aggressively developed the fertilizer, hybrid seed corn, fungicide, insecticide, and herbicide industries.

To summarize, by the mid-1930s, research, initiated in the last decade of the 1800s, had developed information, materials, and inputs with a substantial potential for solving many of the production problems. The economic depression of the 1930s with its low prices for all agricultural commodities, however, intervened and delayed the application of this technology until World War II.

American Agriculture at Its Zenith

The agriculture of the United States reached its apogee, from the standpoint of both American and world public opinion, during World War II and during the first decade of rehabilitation following the end of the war. With much of the agriculture in Europe, Asia, and parts of Africa in disarray as the result of the war, American agricultural production was rapidly increased to feed a vast portion of the population in the devastated areas. This was achieved by applying on a large scale the new technology that had been developed by research during the first four decades of this century, but whose application had largely been held in abeyance during the 1930s because of reduced demand and depressed prices for agricultural commodities as a consequence of the worldwide economic depression. Yields per acre and production increased spectacularly in the 1938-40 to 1968-70 period as is indicated in Table 1. Never before in the history of world agriculture had such a spectacular increase been obtained on such a diverse group of crops over such a short period of time. The yield of wheat, rice, barley, rye, tobacco, and cotton essentially doubled. Corn and peanuts increased approximately two-and-one-halffold and sorghum yields increased fourfold. (Table 1).

In 1947, shortly after the war, the average American family spent about 25 percent of its expendable budget on food. With food prices relatively constant and incomes rising, the proportion of the family budget spent for food in 1971 had fallen to about 16 percent of take-home pay. Higher family incomes, a shrinking portion of the budget needed for food, and more time for travel and recreation seemed to downgrade the importance of food and agriculture for most of the general public. As surplus stocks of agricultural commodities increased alarmingly during the 1950s and 1960s, because of the fantastic increase in yields, there was a growing disenchantment with agriculture and the government's agricultural policy. Agriculture's public image was further tarnished in the late 1960s and early 1970s when it was branded by elements of the environmentalist movement as a polluter that was en-

Table 1. Acres Saved Annually in U.S. for Optional Uses During the Period 1968-70 As a Result of Increases in Yield Resulting from Improved Technology Since the Period 1938-40

Crop	1968-70 production (in millions)	Millions of acres 1968-70	1968-70 yield (acre)	1938-40 yield (acre)	Millions of acres needed to produce 1968-70 crop at 1938-40 yields	Millions of acres saved in 1968-70 through yield increases
Hay	126 tons	62.6	2 tons	1.3 tons	96.9	34.3
Corn(grain)	4,362 bushels	55.9	78 bushels	28.4 bushels	153.6	97.7
Wheat	1,470 bushels	48.6	30.1 bushels	14.2 bushels	103.5	54.9
Soybeans	1,121 bushels	41.5	27 bushels	19.2 bushels	58.4	16.9
Oats	932 bushels	18.0	51.8 bushels	31.3 bushels	29.8	11.8
Sorghum(grain)	725 bushels	13.7	52.9 bushels	13 bushels	55.8	42.1
Cotton	10.3 bushels	10.7	0.96 bale	0.5 bale	20.6	9.9
Barley	419 bushels	9.6	43.5 bushels	23 bushels	18.2	8.6
Corn(silage)	94 tons	7.9	11.9 tons	7.5 tons	12.5	4.6
Flax	30.7 bushels	2.5	12.2 bushels	9.2 bushels	3.3	.8
Rice	92.5 hundred weight	2.1	44.2 hundred weight	22.7 hundred weight	4.1	2.0
Sugar beets	25.3 tons	1.4	18.1 tons	12.5 tons	2.0	.6
Beans(edible)	18.6 hundred weight	1.5	12.3 hundred weight	8.9 hundred weight	2.1	.6
Potatoes	310 hundred weight	1.4	221 hundred weight	75.7 hundred weight	4.1	2.7
Peanuts	26.8 hundred weight	1.4	18.6 hundred weight	7.5 hundred weight	3.6	2.2
Rye	31.1 bushels	1.3	24.1 bushels	12.1 bushels	2.5	1.2
Tobacco	1,806 pounds	.9	2,008 pounds	947 pounds	1.9	1.0
Total		281.0			572.9	291.9

Source: Barrons, K. C., "Environmental Benefits of Intensive Crop Production," Agricultural Science Review 9(1971), pp. 33-39.

dangering the survival of wildlife and endangering
human health by the widespread, careless use of pesticides. When world grain reserves fell below a critical
level in 1973, as a result of widespread droughts in
the USSR, Asia, and Africa, and food prices, especially
meat, began to soar, there was an outcry from the
nonrural consumers. Apparently they had become convinced that food should always be cheap, even though
one seldom heard the same complaints about the price
increases in automobiles, television, and other non-
essentials.

American Agriculture's Declining Star

The agriculture of Europe and parts of Asia made a
spectacular recovery in the mid-1950s. As their production increased, there was less need for large imports of American grain and other foods. Consequently,
large grain surpluses began to accumulate in the U.S.
This soon evolved into all the difficulties that are
inherent in any government policy that is trying to cut
production so as to bring it more in line with demand.
Various types of food programs were established to
supply grain and other foods to food-deficit developing
nations under the P.L. 480 type of concessional sales
or outright gifts. These programs helped to reduce
American grain-surplus stocks and, at the same time,
provided food to hungry nations. Unfortunately, the
grain that was received by the recipient nations sometimes was handled in such a way that it kept the price
of domestic grain at an unrealistically low price and
it became a disincentive to the development of domestic
agriculture.

Not only was the change in government policy
designed to reduce production, but also it was designed
to discourage applied research that might open new
avenues to increasing yield which would, in turn,
increase production. As a result, the political word
was passed to officials of the colleges of agriculture
of the land grant universities and to the U.S. Department of Agriculture research scientists that they were
to discontinue applied research and deemphasize extension as much as possible so as not to worsen the surplus problem. All were encouraged to shift emphasis to
more abstract, theoretical research. The deemphasis of
agriculture and shift in emphasis shortly led to a
twofold dilemma: (1) the growing shortage of funds for
agriculture and (2) the inability of administrators to
evaluate the productivity of agricultural scientists
under the new orientation. It soon begot the disastrous situation of "publish or perish," which in turn
gave rise to an avalanche of publications of varying

degrees of quality, certainly many of them of less than earthshaking significance. These events gave rise to frustrations among many agricultural scientists, who then found themselves on an isolated and seemingly sinking island. Many of the most agile soon learned the "grantsmanship" game in order to support esoteric research projects only remotely related to agriculture.

It appears to me that the support for agricultural research at the program level continues to wither and shrink. Similarly, the public's understanding and interest in agriculture continues to erode away. Drs. André and Jean Mayer have succinctly summarized the sad state of agriculture as follows: "The failure of our secondary schools and liberal arts colleges to teach even rudimentary courses on agriculture means that an enormous majority, even among well-educated Americans, are totally ignorant of an area of knowledge basic to their daily style of life, to their family economics, and indeed to their survival. It also means that our policies of agricultural trade and technical assistance, as important to our foreign relations as food production is to our domestic economy, are discussed in the absence of sound information, if indeed they are discussed at all."

The lack of instruction in agriculture today is even more incomprehensible when one considers the great emphasis that is currently given to ecology and environmental issues at all levels of education. It is ironic that early agricultural man, employing a "slash and burn" migrant agriculture, was effectively managing ecosystems to plant his crops thousands of years before the terms ecology and ecosystem were coined. They had learned the value of opening the forest canopy to permit sunlight to enter, and they had learned the use of fire to recycle plant nutrients and to control weeds. As long as the human population pressure was low and the time interval between slash and burn cycles was long, the system was not deleterious to the overall ecosystem. Yet today these pioneering ecological achievements are ignored by modern environmentalists and ecologists.

THE ROLE OF PLANTS IN MEETING THE WORLD'S FOOD NEEDS

Plants are the most important source of food for man. The 1973 FAO Production Yearbook data show that, in terms of gross tonnage, without correcting for different moisture contents, 98 percent of the total food harvested came from the land and only 2 percent from the ocean and inland waters. Of the total world food harvest, plant products directly contributed 81.8 percent of the tonnage and animal and marine products together only 16.8 percent.

As world population increases, plant products will play an increasingly greater direct role in feeding mankind. Today the average American or Canadian consumes about one metric ton of grain per year, whereas the same amount of grain will feed five Indians. The difference is attributable to the difference in diet. The diet of the American and Canadian is made up of a large amount of animal products, which in the case of monogastric animals (e.g., swine and poultry), is produced largely from grain, whereas the Indian's diet contains very little animal protein, and by contrast is made up almost entirely of cereal grain and pulses.

Throughout history man has utilized more than 3000 species of plants as food. Of these, at least 150 different species have been grown in sufficient quantities to have entered into world trade.

Over the centuries man has gradually reduced the number of species upon which he depends for his food. He apparently has tended to continue to depend on those species which yield him the greatest return for his land and labor, while he has gradually discontinued cultivating the rest.

Currently, excluding a considerable number of species of vegetables and fruits, man depends primarily upon twenty-three different crop species. These include five cereals: wheat, rice, corn (maize), sorghum, and barley; three root crops: potato, sweet potato, and cassava; two sugar crops: sugar cane and sugar beets; six grain legumes (pulses): dry beans, dry peas, chick-pea, broad bean, and the two dual-purpose pulse-oilseed species, groundnut and soybean; three oil seeds: cottonseed, sunflower and rapeseed; and four tree crops: banana, coconut, oranges, and apples.

Cereal grains, collectively, are the most important group of crops. They are sown on approximately 50 percent of the cultivated land area of the world. They directly contribute 52 percent of the calories to the human diet worldwide, and about 62 percent of the calories in the developing world. Moreover, they supply nearly 50 percent of the total protein intake. Indirectly, cereals also contribute greatly to both human protein and calorie intake because about 40 percent of world cereal production is fed to animals to produce meat, eggs, and milk.

Wheat is the cereal which is grown on the largest area and which also produces the greatest total tonnage of grain. The worldwide tonnage of corn production, however, has been increasing rapidly during the past twenty years and is closing the gap between itself and wheat and rice. The average grain yield of corn world-wide surpasses all of the other cereals. The 1969-71 worldwide average yield, expressed in kilograms per hectare, of the five major cereals was:

1969-1971 Average Yield	
Cereal	Kilograms per Hectare
Corn	2567
Rice (as paddy)	2320
Barley	1843
Wheat	1524
Sorghum	1108

Maximum Genetic Yield Potential of Each Crop Species

The anticipated food requirements for the next forty years are indicated in the Appendix. It is indicated that the 3.3 billion metric tons of all food that was produced in 1975 will need to be doubled to 6.6 billion metric tons by 2015, assuming that human population growth continues at the present rate. This production will be a difficult target to achieve under the best of circumstances, and much of the success or failure will depend on our ability to further increase the yields of the major cereal crops which are sown on about 50 percent of the cultivated area of the world.

There is no way of knowing now with certainty what the future maximum genetic grain yield potential will be for each of our important food crops. Grain yield per acre is an expression of the interaction between the genetic makeup of the variety and the environment in which it is grown. Maximum grain yield per acre is produced only when the variety with the highest genetic yield potential is grown in an environment that is optimum for the crop from the standpoint of sunlight, soil, moisture, plant nutrients, temperature, and pest free. Under farm conditions grain yields are often limited either by shortage of moisture, inadequate levels of soil fertility, or reduction in yield imposed by weeds, diseases, or insects.

In all probability, we are closer to approaching the maximum genetic yield potential in crops such as corn, wheat, rice, and sorghum, which have had the benefit of extensive and intensive breeding efforts over the past fifty years, than we are in the case of barley, oats, rye, millets, beans, chick-pea and pigeon pea.

During the 1920s and especially in the 1930s there was gradual improvement in the agronomic type and disease resistance of varieties of wheat, rice, oats, barley, flax, peanuts, and cotton. Nevertheless, there was little increase in the genetic yield potential per se in these crops. The first major increase in genetic yield potential of any important field crops was achieved with the development and introduction of hybrid-corn varieties in the second half of the 1930s and early 1940s. Within the last two decades a similar spectacular increase in yield was obtained when hybrid sorghum varieties were developed and widely grown.

I have spent most of my scientific career trying to increase the yield and production of spring wheat in many different countries of the world. I do not wish to imply, however, that my interests have been confined to wheat, which is generally grown as a cool season (winter) crop in the semitropical areas of the world. Rather, in areas where temperature and moisture are favorable for plant growth throughout the year, research scientists must develop methods and crop varieties which can produce the maximum tonnage of grain per acre per year. This can generally best be achieved by growing on the same land a cool weather crop, such as wheat, during the winter, and a hot weather crop, such as rice and corn, during the summer season.

For the first seventeen years of our work in breeding spring wheat varieties in Mexico we were unable to obtain yield of more than 4.5 metric tons per hectare (sixty-eight bushels per acre) even when the crop was grown under irrigation and heavily fertilized. This yield barrier was imposed by susceptibility to lodging when the crop was heavily fertilized. (Lodging is the collapsing of the plant stalk before harvest.) When the Japanese NORIN 10 dwarfing genes were introduced into our breeding program we were able to develop dwarf wheat varieties of 140 to 150 day maturity, which were resistant to lodging, with a yield potential of 8.5 to 9 metric tons per hectare or 127 to 137 bushels per acre. These newer types of wheats have a higher grain-to-straw ratio than the older, tall varieties. Consequently, they are also more efficient in converting moisture and fertilizer to grain, which is the portion of the plant that is valuable as human food. It was these broadly adopted, high-yielding, semidwarf

Mexican wheat varieties that revolutionized wheat production in India, Pakistan, Bangladesh, Tunisia, and Turkey in the past fifteen years. The impact of the semidwarf wheat varieties and new technology on Indian wheat production is shown in the Appendix.

In the same manner the semidwarf rice varieties that have been developed during the past fourteen years have succeeded in doubling the maximum genetic-yield potential of this crop.

Referring specifically to the future maximum genetic yield potential of wheat -- the crop with which I am most familiar -- one may wonder: Can the grain yield of 140- to 150-day-maturity varieties of spring wheat be doubled again from 125 bushels per acre to 250 bushels in the near future? I sincerely doubt this can be achieved! I will venture the risky guess that we have already achieved 75 to 85 percent of the maximum genetic yield potential in wheat -- excluding the utilization of F_1 heterosis (hybrid vigor) as is currently being exploited in hybrid varieties of corn and sorghum. My reason for making this statement is based on the fact that, despite a massive breeding effort during the past ten years which employed a wide range of germplasm, additional increases in grain yield beyond the 125 bushels per acre were at best marginal. In all probability, future increases in genetic yield potential will be of smaller and smaller increments, and they will be more and more difficult to achieve. To continue to make progress in increasing yield will require a greater research effort.

Of course, this does not imply that there is currently little or no benefit being derived from the wheat-breeding programs. Although it has become apparent that a genetic yield plateau or barrier has been reached that is difficult to break through, progress in varietal improvement in a number of other important respects continues to be made. The newer varieties now being released are superior to those of ten years ago in a number of different characteristics such as: better adaptation to certain problem soil types (e.g., acid soils), better straw strength, better grain test weight, improved milling and baking quality, and broader resistance to diseases. The latter consideration is exceedingly important in crops such as wheat and rice because their major disease causing pathogens mutate and change. As a result, the varieties that were bred and released to farmers as being resistant to a given disease sooner or later will become susceptible to newly-evolved races of the same pathogen. The only protection against severe crop

losses that may result from such changes in the pathogens is a dynamic, aggressive breeding program manned
jointly by a team of top-flight plant breeders, plant
pathologists and cereal technologists. Obviously, such
research programs must have continuity in both funding
and scientific staffing. Failure to provide them
invites disaster.

Although I have used the case of wheat to illustrate that our varietal breeding program appears to
have reached a genetic yield plateau, the same seems to
be true of corn, sorghum, and rice.

I will briefly mention the situation on corn where
there has been only a modest increase in the maximum
genetic yield of corn hybrids in the past twenty-five
years. Of course, there have been improvements in
hybrids in other respects such as: (1) better adaptation to many soil and climatic areas where the crop was
formerly poorly adapted; (2) better adaptation to
mechanical harvesting; (3) better resistance to stalk
breakage; (4) improvements in insect and disease resistance; and (5) although not yet available commercially,
the outlook is for the development soon of hard endo-
sperm varieties with improved nutritional value that
will be competitive in terms of yield with conventional
dent-type hybrids.

With the development of high-yielding varieties of
shorter stature combined with the development of better
weed-control practices and more appropriate use of
fertilizer, corn is likely to become much more important as a crop in tropical areas of the world during
the next twenty years. Moreover, within the next
twenty to thirty years corn in all probability will by
necessity, if the world is to be fed, play a much more
important direct role as a human food than it does at
present. At present, perhaps no more than about 15
percent of the total world annual production of 325 to
350 million metric tons is used as food, principally in
Mexico, Central America, the Andean countries, Brazil,
and a number of Central African countries. The remainder is used almost entirely as feed for livestock,
with small amounts used for industrial purposes.

Although I have emphasized that we appear to have
reached a maximum genetic yield barrier that is difficult to break through in corn, wheat, rice, and sorghum
in the past fifteen or twenty years, this does not mean
that world production of these crops cannot be increased greatly. World production can be increased
with the more widespread and intensive use of the
improved production technology now available such as:

(a) the more widespread use of high-yielding varieties, (b) proper rates and dates of sowing, (c) proper land preparation, (d) proper conservation and utilization of moisture, (e) proper use of the right kind and amount of fertilizer, (f) control of weeds, insects, and diseases, and (g) adoption of government financial policies that will stimulate production of the new technology.

If the world is to be fed and disaster averted during the next twenty to thirty years, much of the increase in production must be achieved in the developing nations where crop yields are now low compared to those of the U.S. The first prerequisite toward achieving this goal is a continuing adequate expansion in world fertilizer production. Today, there is great enthusiasm among some microbiologists who feel that the efficiency and levels of biological fixation of nitrogen can be greatly enhanced through research. Some claim that this can lead to a large reduction in the fossil fuel needed for nitrogenous fertilizer production. Although I favor expanding the basic research expenditures in biological nitrogen fixation, I am strongly opposed to those who advocate cutting back on the use of fossil fuels for the production of nitrogenous fertilizer. I contend, as indicated in the Appendix, that such a policy makes no sense while the world continues to ignore the tremendous wastage of fossil fuels in many other sectors of the economy where savings of energy and substitution of other energy sources can be more easily accomplished.

Soybeans -- valuable addition to American agriculture. The development of soybeans as a highly important commercial crop in the U.S. during the past forty years is one of the outstanding accomplishments of research. The genetic materials upon which this program was based were introduced from China in the 1920s and 1930s. Today the U.S. is the largest producer and exporter of soybeans in the world. Among foreign exchange earnings from agricultural commodities, soybeans rank first. Moreover, they play an important role in rotation with corn in the Corn Belt, and at the same time, being a legume, they fix considerable nitrogen.

However, there has been only a very modest increase in maximum genetic yield potential of this important crop species in the past two decades.

The spectacular success of soybeans as a crop necessitates examination of the possibility of introducing and/or developing other new crops in the decades ahead.

Triticale -- a potentially valuable new crop species for the decades ahead. Although many of the benefits of the so-called Green Revolution in the developing countries are still to come, we are already asking ourselves where and how we can add another dimension to the present revolution in crop production. Although it is obvious that we must continue to expend most of our research effort and budget on the crops that feed the world today, I feel we must also explore in a modest way other new approaches.

All of our major present-day food crops are the result of the guiding evolutionary hand of Mother Nature exerting her influence over millions of years. About 10,000 to 12,000 years ago, Neolithic women in different parts of the world domesticated mixed wild populations of virtually all of the important crop plants upon which we continue to depend for our food supply. Over the thousands of years since the discovery of agriculture, perhaps 500 to 600 generations of farmers exerted, first unconsciously and later consciously, selection pressures under different environments on these early mixed populations and, as a consequence, progressively developed many locally adapted, improved cultivars (a reproducible genetic strain). It has only been during the past 100 years, and especially during the past fifty, that scientists -- geneticists, plant breeders, agronomists, plant pathologists, entomologists, and cereal technologists -- have joined hands and worked together across disciplines. This led to the development of the present dwarf, high-yielding, disease- and insect-resistant wheat and rice varieties and to the high-yielding corn and sorghum hybrids with their desirable quality and agronomic characteristics.

Benefits from the genetic improvement of wheat, corn, rice, sorghum, peanuts, and soybeans in the past five decades have been enormous. However, each increment of improvement becomes more and more difficult to achieve as we approach the upper limits of the genetic potential of our present food crops. How can we hope to achieve the increased food production needed to keep pace with the increases in population when the areas of productive land are limited and shrinking? One major possibility lies in the creation of new food crops tailor-made to produce more food on soil already in production, and to utilize unproductive soil for food production.

A major step in this direction is being made by the development of the new man-made crop triticale. The new species was created by producing a "sterile mule" plant by hybridizing wheat and rye and manipulating the chromosomes with chemicals to induce fertility. The new form possesses all of the genetic mechanisms of two species, wheat and rye, in a single species. The genetic potential of this species in increasing food production and improving nutritional quality is indeed promising.

Although the possibilities of the wheat-rye hybridization were reported by a Scotsman named Wilson a century ago, nothing came from his effort. Moreover, although Dr. Anne Muntzing at the University of Lund in Sweden and Dr. S. Sanchez-Monge at the University of Madrid in Spain, two outstanding scientists, had spent most of their professional lifetimes trying to develop this hybrid as a new crop species that would compete with other grains, they were unsuccessful. Our interest in triticale at the International Maize and Wheat Improvement Center dates back to 1958 when my late colleague Dr. Joseph A. Rupert and I attended the First International Wheat Genetics Symposium in Winnipeg, Canada. As part of the "living demonstrations" associated with the symposium, Dr. L. Shebeski and Dr. C. Jenkins of the University of Manitoba had plots of their triticale breeding materials growing as well as a considerable collection of our Mexican dwarf wheats, which were then still in early stages of development. Both Dr. Rupert and I were greatly impressed with the vigor of the triticale and with the size of the spikes, despite the fact that they were highly sterile, the grain shrivelled, and the plants too tall. Nonetheless we decided to initiate a small "bootleg" breeding program on triticale and were given triticale material by Drs. Shebeski and Jenkins with which to initiate our effort. After three years the Rockefeller Foundation, at our request, awarded a modest grant to the University of Manitoba, which permitted them to expand their breeding effort and to collaborate with our modest Mexican breeding effort. Our Mexican triticale program was officially funded for the first time in 1966.

At that time the best triticales yielded only half as much grain per acre as the best Mexican dwarf wheats. The best varieties, despite the impressive size of the spike, had more than half of the spikelets sterile (didn't set seed), and the seed which did set was very shrivelled. In 1968 the first really significant step to correct several of these defects was achieved. A few triticale plants were found that had apparently outcrossed several generations before to an early-maturing, dwarf, Mexican wheat. These plants were early-maturing, semidwarf types that were completely fertile, although still possessing badly shrivelled grain. From that time onward, with funding by the Canadian Foreign Technical Assistance Program, the joint Mexican and University of Manitoba breeding program was greatly expanded. The large increase in the number of segregating plants handled each generation undoubtedly accelerated progress in the breeding effort.

Gradually improvements have been made in many characteristics. Among the newer varieties are types with suitable agronomic characteristics that are completely fertile. Many lines now have grain that is commercially acceptable. Last year in one of the International Wheat Yields Trials that was grown at 72 locations in different parts of the world under a wide range of soil and climatic conditions, a triticale variety outyielded all of the bread wheat varieties, including the best current, commercial, dwarf Mexican wheats.

Triticale is not intended at present to replace wheat as a commercial crop, except under very special conditions where it greatly outyields wheat. On very acid soils, such as the highly leached laterites of Brazil, Kenya, Ethiopia, Tanzania, the outer ranges of the Himalayas in northern India, Pakistan, and in Nepal, triticale will frequently yield twice as much as the best wheat. At present, in the best irrigated areas in Sonora, Mexico, -- the home of the Mexican dwarf wheats -- triticale yields as much as the best wheat varieties. In the same area ten years ago it yielded only 50 percent as much as the best wheat.

Triticale is currently grown on only a small commercial acreage in Canada, Mexico, Kenya, Spain, Hungary, the People's Republic of China, the U.S.S.R., and the U.S. The grain is currently used either for the production of specialty bakery products or as animal feed. There are considerable data that indicate that when compared at the same grain protein content, it is superior in nutritive value to wheat because of its higher level of the essential amino acid lysine.

Considering the tremendous progress that has been made in the improvement of triticale during the past decade, there is a good probability that it may become, within the next ten years, an important commercial crop in some areas of the world. If this happens, it will become the first major crop species produced by scientists -- women and men -- since all of our current major crop species were domesticated from wild ancestors by Neolithic women at the dawn of agriculture.

The progress made toward making triticale a new
important crop species already justifies, it would seem
to me, exploratory research into the possibility of
developing other useful crop species from wide crosses.

THE POPULATION MONSTER AND FOOD PRODUCTION PROBLEMS

The last section of the Appendix clearly indicates
the magnitude of the food production problem, interrelated problems of energy, land use, environment, and
competition of humans with other species that must be
coped with because of the anticipated relentless advance of the human population monster during the next
forty years. If human population growth continues at
its present rate, the 3.3 billion metric tons of food
that were produced by the world in 1975 must be increased at least to 6.6 billion tons by 2015. That
means that in the next forty years world food and fiber
production must be increased at least as much as it was
increased in the 12,000 year period from the discovery
of agriculture until 1975. We must expand our scientific knowledge and improve and apply better technology
if we are to make our finite land and water resources
more productive. This must be done promptly and in an
orderly way if we are to meet growing needs without, at
the same time, unnecessarily degrading the environment
and crowding many species into extinction. Producing
more food and fiber and protecting the environment and
a group of endangered species can, at best, be only a
holding operation while the population monster is being
tamed. Moreover, we must recognize that, in the transition period, unless we succeed in increasing the
production of basic necessities to meet growing human
needs, the world will become ever more chaotic. Civilization may collapse.

As we face the food production needs for the next
forty years, we must ask ourselves if the research
effort in agriculture is adequate to meet the challenge. Moreover, among the questions which must be
asked and answered: Are the current means of identifying the goals for research and technology relevant,
or is there an over-emphasis on basic research at the
expense of applied research? Are there in fact only
approximately a dozen granting agencies that are responsible for the funding and hence, in a large part,
for giving direction and emphasis to the majority of
the research effort in the United States? Has the
reorganization within the U.S. Department of Agriculture created more administrative bureaucracy rather
than making more research dollars available to the
scientists on the working level? Has the new competi-

tive grants program placed even more research dollars under the direction and influence of only a few grantors whose research-review panel members may often have similar and perhaps narrow views? Since grants to universities and the research done under such grants have a large impact on the numbers of people trained in the various areas of science, should there not be a greater concern focused on the appropriateness of how grant funds are allocated in light of future scientific manpower needs?

In the People's Republic of China (P.R.C.) the balance between applied and basic research and training of young scientists has shifted largely to solving the immediate problems of feeding, clothing, and housing 950 million people. In that country basic or fundamental research is regarded as a luxury which cannot be afforded at present. They will pay for the imbalance in the future when a base of knowledge to solve complex problems is not available. It could be devastating. In contrast to the P.R.C., has the pendulum in the U.S. at present swung too far in favor of basic research? Are we too often conducting research for the sake of research which often becomes chasing of academic butterflies without any regard for the possible benefits which might improve the quality of life? Grantsmanship and "publish or perish" are real and dominant factors in a researcher's thinking in U.S. institutions today. One learns quickly that the nature of one's research is directly related to where the grant funds might be obtained. Coupled with this directed research goes the "cloning" of future scientists. Is there a danger that we are now training more specialists to dig deeper into isolated academic wells, without worrying about educating generalists who can put the research findings together into a useful fabric? It is not uncommon to attend professional scientific meetings of related disciplines only to find that the participants cannot even communicate due to a lack of common terminology. However, all scientists today are very much aware of the importance of grantsmanship and the need to publish prolifically if they are to advance professionally.

Consequently, the responsibility of identifying researchable areas, allocating grant funds, and determining direction for all aspects of research must receive major attention. In making these judgments it must be remembered that scientific papers published in prestigious journals will not put food in the stomachs of the many millions of underprivileged, restive people.

APPENDIX

THE MAGNITUDE AND CHALLENGE OF PROVIDING
FOOD FOR A WORLD POPULATION OF EIGHT
BILLION BY THE YEAR 2015 OR 2035

THE MAGNITUDE AND DIFFICULTIES OF
PRODUCING FOOD FOR FOUR BILLION PEOPLE

The problem of producing adequate food for a world
population of four billion that is increasing at the
rate of about eighty million per year is more difficult
than placing a few men on the moon.

In the case of the moon landings, government gave
it top-development priority. All the financial support
that could be used effectively was provided. The
project also received outstanding psychological support
and encouragement, both from the government and the
general public. Moreover, the number of people directly involved probably did not exceed more than a few
thousand of the best scientists -- mathematicians,
physicists, engineers, metallurgists, chemists, and
bio-medics -- perhaps a few tens of thousands of highly
skilled technicians, machinists, mechanics, and finally, a few highly skilled astronauts. There were few
problems, if any, in communications among members of
the team. There was a clear-cut, sharply defined line
of responsibility and command for the entire project.
The scientists were in charge with a minimum of inter-
ference and confusion from the political leaders and
general public, who generally remained quiet because of
self-expressed ignorance of the problem at hand.

Now let us contrast this with the complexities of
producing food for four billion people. In this effort
one is confronted with dealing with the individual
efforts and lives of approximately two billion people,
mostly poor subsistence farmers. They constitute
approximately half the population of the world. The
situation is further complicated by the fact that the
agricultural population of the world is very hetero-

geneous in language, culture, education, and income. Worst of all, it has little political clout.

Although about half of the total world population is engaged in agriculture and animal husbandry, the percentage varies greatly from country to country. In the U.S., 3.5 percent to 4.0 percent of the population is engaged in agriculture. Yet they produce food for the entire population and, in addition, a vast amount for export. Contrast this to the situation in countries such as Bangladesh, India, and Pakistan where 70 to 75 percent of the population is in agriculture, or to that of many of the new African nations where 80 to 90 percent of their total population is engaged in subsistence agriculture. In these countries the rural farming and animal husbandry people, despite constituting the vast majority of the total population, have no political clout. They exist as the unorganized, dispersed, rural majority who are exploited by the political leaders for the benefit of the small nonrural sector.

Unfortunately, the mistaken belief is widespread that anyone has the talent to be a successful farmer or animal husbandman. But the truth is that in the developing as in the developed nations, most politicians, labor leaders, merchants, industrialists, lawyers, educators, and even scientists would starve if they were forced to try to make a living from farming.

In the developing nations the problem of trying to change agricultural production methods is further complicated by a general lack of education. Illiteracy makes it more difficult to reach the peasant farmer -- other things being equal -- than the generally well-educated farmers of the U.S. or other developed countries.

Neither farming nor agricultural science has any prestige in the developing nations, an idea which largely comes from the low order of importance given to agriculture in development programs by their governments. Consequently, anyone who has the opportunity and ability to study at the professional level wants to become a medical doctor, dentist, civil or hydraulic engineer, economist, chemist, or lawyer. The colleges of agriculture in the universities, if they exist at all, have little prestige and are inadequately staffed. Agricultural research institutions are very few and far between and both underfinanced and understaffed. With the extreme shortage of agricultural scientists and the poor salaries they receive in comparison with other professions, it is a slow process to build a well-trained, competent, dedicated scientific staff capable of helping farmers attack food production problems in food-deficit developing nations. These countries, in large part, must also depend on foreign universities to train their young scientists. With these handicaps is there any wonder why it is difficult to increase food production in the developing countries?

As I consider events in the U.S. after thirty-six years abroad wrestling to turn the tide of food production in many food-deficit developing nations, I become concerned by what I think I see happening to the hard-won prestige of American agricultural production, research, extension, and education.

The contributions of the American land grant colleges and universities through agricultural education, research, and extension programs assisted farmers and livestock producers to greatly increase production. Blessed with good soil, favorable climate, and creative, aggressive farmers and livestock men, the results of the research done at these institutions was applied to make the U.S. model of agricultural productivity by the middle of this century.

Unfortunately, as has been correctly pointed out by André and Jean Mayer, the very success of the colleges of agriculture, home economics, and forestry in the American land grant universities has led to their isolation as an "island in American academic life." This resulted largely from a rapid shift of the majority of the population from farms to large cities over the past four decades. As a consequence of this shift, the vast majority of the population lost contact with agriculture, with the land, and with the colleges of agriculture. Since food was always abundant and cheap, agriculture was soon taken for granted by the urban public. Colleges of agriculture, all too often maligned and referred to as "moo colleges" or "cow colleges," were considered of little significance to either nonrural or national well-being.

As a result of such an attitude during the past twenty-five years, all too many urbanites believed that food sprouted from the shelves in the supermarkets. In their urban eyes, farmers often were considered to be second-class citizens. But when food prices began to soar in 1972, following the drought in the U.S.S.R. and Asia, they were shocked and often indignant. During the previous twenty-five years, while wages and salaries in the industrial and business sectors increased greatly, farm prices remained stagnant or actually declined in real terms. The general public had forgotten that it takes investment in inputs to produce food.

American agriculture reached its apogee, from the viewpoint of public opinion, during World War II when it helped feed vast portions of the populations of Europe, Asia, and Africa when their agriculture was in disarray. Shortly after the end of the war, its prestige began to decline and rapidly fell into disfavor as agricultural surpluses accumulated. As a result, the political word was passed to officials of the colleges of agriculture -- and to U.S. Department of Agriculture research scientists as well -- to discontinue applied research and to emphasize extension, as much as possible, so as not to worsen the surplus problem. All were encouraged to shift the emphasis of agriculture. That shift shortly led to the twofold dilemma: (1) the growing shortage of funds for agriculture, and (2) the inability of administrators to evaluate the productivity of agricultural scientists under the new orientation. It soon begot the disastrous solution of "publish or perish," which in turn gave rise to an avalanche of publications of varying degrees of quality, certainly many of them of less than earth-shaking significance. These events gave rise to frustrations among many agricultural scientists, who found themselves on an isolated and sinking island.

Colleges of agriculture tried to defend their programs and budgets as best they could. But mostly it was a losing battle as agriculture was "turned off." Departments tried to obtain funds by including new subtitles that were in public favor as the mood of the time shifted. Programs and titles in isotope studies of soil and plant nutrition, mutation genetics, population genetics, and molecular biology gained favor. More recently, when the mad rush to correct the abuses of the environment got under way, colleges of agriculture and forestry were forced to include subtitles of environmental science, ecological genetics, ecological entomology, ecology of plant pathogens, conservation and the environment, etc. These shifts and drifts continued until the general public, including presidents of land grant colleges and universities, were shocked by the food crisis of 1972, and a year later by the energy crisis. It will take a long time, and progress will be slow, before the colleges of agriculture can reeducate the large urban population about the importance of agriculture and forestry to the American consumer, to the general economy, and to world stability, and how it fits into the total scheme of things in American life. Before this change can be achieved they will need to overcome the false image given to agriculture by the extreme environmentalists and neo-ecologists of being the great polluter.

What does the recent orientation of research and education programs in American colleges of agriculture have to do with food production in the developing, food-deficit nations? It asserts its influence indirectly in two ways: (1) the oversophistication of research programs generally tends to disorient the majority of inexperienced graduate students from developing nations who come to most American colleges of agriculture to study; and (2) since many of the younger professors in American colleges of agriculture, who are currently recruited by foreign American technical assistance programs to serve in developing nations, have had little or no practical experience in agricultural production, they often bring with them an oversophisticated approach toward trying to assist in solving production problems. The result is that they often become frustrated and unhappy and generally fail to be effective. This results in an unhappy experience for both countries.

THE IMPORTANCE OF AGRICULTURE AND FOOD PRODUCTION TO MANKIND AND WORLD ORDER

It is my personal belief that all who are born into the world have the moral right to the basic ingredients for a decent, humane life. How many should be born and how fast they should come on stage is another matter that I will explore with you next.

The basic ingredients for a decent life must include (1) adequate food; (2) adequate clothing and housing to protect one from the elements, which obviously will vary with different climates; (3) basic education to enable one to develop his talents; (4) a job to earn one's basic necessities; and (5) medical care when one is ill. In order to provide these basic necessities, it is obviously necessary for government, private enterprise, or joint ventures to develop the resource base and infrastructure required. This includes developing raw, natural resources and the systems of energy, transport, schools, hospitals, industrial plants, and other basic institutions required to deliver them. It involves the investment of huge sums of capital in these systems and in industries to provide opportunities for employment that are capable of producing the basic goods required by a modern society.

The Importance Of Food

We must consider the importance of food from three
different points of view: its biological importance,
its economic worth, and its political importance.

The significance of food for biological survival
should be self-evident. Without food a human being can
live at most for only a few weeks. Nevertheless, in
the affluent nations virtually everyone takes food
abundance for granted since most have never known
hunger. By contrast, in the food-deficit developing
nations, hunger and malnourishment are common and
widespread, and it reaches disastrous proportions in
years when crops are poor. Unfortunately, most political leaders in the developing nations have never
known hunger, or if they have, have long since forgotten about its degrading influence on the quality of
life. In my frustration at trying to encourage policy
decisions to stimulate food production during the past
several years of worldwide food shortages, I have
proposed that the world would almost certainly have
wiser agricultural and food policies and less hunger
and famine if political leaders -- including the heads
of states, their economic planners, the ministers of
finance, industry and agriculture and all their top
bureaucrats -- annually were locked in a conference
room and forced to go without food for two weeks, and
perhaps also forego all liquids for the last three days
while they debate and decide agricultural policy and
budgeting. On second thought, their attention might be
even more effectively focused on the urgency of the
problem if they also were subjected to the pleading
cries of hungry children as they debated these issues.

The economic worth of food varies widely from
location to location and from time to time. The price
people are willing to pay for food depends, in large
part, on how hungry they are and how long it is likely
to be before they will again be privileged to eat. For
food is the first and most important need for sustaining life. In a privileged, food-abundant country such
as the U.S., the average family spends a relatively
small percentage of its income, after taxes, for food.
In 1964 it was estimated to be about 25 percent of the
family budget. With food prices relatively constant
and income rising, the proportion of the budget spent
on food in 1971 had fallen to about 16 percent of
take-home income. Nevertheless, when prices for food
began to rise in 1972 and 1973 due to general inflation, which worsened and was further stimulated by
temporary worldwide shortages of certain kinds of food,
there were indignant outcries from the urban consumers

who had become accustomed to cheap food. Contrast this situation to the situation in the food-deficit nations of the developed world. In many of the nations in years when crops are good, the average family spends 70 to 75 percent of their income on food. In years when crops are poor and food becomes scarce, all of their earnings are spent on food and still they go hungry. For them the line between adequate food availability and hunger -- or even starvation -- is a delicate, narrow zone.

The political value of food cannot be evaluated in terms of monetary units under any system of government. Although political leaders in food-deficit nations all too often try to maintain prices for agricultural products and food at unrealistically low levels in order to pacify the minority group of "organized" urban consumers -- at the expense of the farm-sector -- there comes a point at which agricultural production stagnates or even collapses. In the long 1950 to 1971 period of large food surpluses in the food-exporting nations, political leaders in the developing nations, who used such shortsighted agricultural policies to the detriment of their own agriculture, often were saved from political ruin by obtaining large quantities of food on easy, long-term, concessional sales arrangements. However, when food reserves were sharply reduced in the exporting nations following the 1972 drought in Asia and Africa, and grain prices began to soar and concessional sales were greatly curtailed, the folly of such shortsighted policies became evident. Governments in the five Sahelian countries of Africa fell during the drought and food shortages of 1973 and 1974. Even the political lives of ministers of agriculture in the centrally planned, developed, socialistic nations are not spared when, through the neglect of agricultural development, they fail to produce food they need and are unable to obtain it except by vast expenditures in foreign exchange from food-exporting nations.

Unfortunately, agricultural production and development is not like a water tap that can be turned on and off in a second whenever someone wants a glass of water. The gestation period for agricultural development is long and complex before the payoff period of increased production is reached. Nevertheless, in times of food shortages, after many decades of neglect by political leaders, economic planners, sociologists, and demographers, agricultural scientists suddenly are asked to produce a miracle -- a Green Revolution -- that will produce an abundance of food and also permanently correct all of the social, economic, and political ills that have accumulated throughout the world since the time of Adam and Eve. Moreover, as if this were not difficult enough, agricultural scientists, while solving the food-production problem, are at the same time expected to produce a good standard of living for all small farmers who have been given one to three hectares of marginally productive land by bewilderingly idealistic, impractical, agrarian land-reform programs. Is there any wonder that we continue to have recurring chaos on the food-production front as long as such attitudes prevail?

The Food-Production Base

Most people who have not been involved in either agricultural production, research extension, or education -- and unfortunately also some who have had the benefits of such experiences -- have little comprehension of the limitations of the natural resource base on which we depend for food production for the present population of four billion, and for the future requirements of a rapidly growing world population. To most people, the earth is an enormous planet (and to many, unfortunately, still the center of the universe) with much unexploited land and water for expanding food production indefinitely as needed. The truth is that our earth is a medium-sized planet in our modest solar system which in turn is only a "speck" in the universe as most of us observe, but only vaguely comprehend, when we glance upward on a clear, moonless night and see the star-studded sky with many "solar systems."

As we begin to reflect on the natural resource base of our earth and its potential for food production, we see that the earth has a lot of poor real estate. Approximately 71 percent of the surface area is ocean and only 29 percent is land. Moreover, much of the land is of little value for agriculture and animal husbandry, as is indicated in Table 1. About 11 percent of the land area is classified as arable land suitable for agriculture or permanent nonforest tree crops, 22 percent is classified as permanent pastures or meadows, and 30 percent as forest and woodland. This leaves about 37 percent of the total land area in wasteland (subarctic and antarctic wasteland, rocky mountain slopes, tundra, or deserts) cities, industrial sites, highway, airports, etc.

This classification is, at best, only a tentative attempt to sort the world's land resources. In land-reform programs in several parts of the world, I have seen peasant farmers being allocated land that is incapable of producing food for a sizable population of

Table 1. World Food Production

From the land	1000 metric tons		Proportion % grand total 1977
	1975	1977	
Total cereals	1,362,153	1,459,012	42.0
Wheat	354,748	386,596	
Rice (Paddy)	359,653	366,505	
Maize	324,257	349,676	
Barley	150,003	173,094	
Other	173,492	183,141	
Total root crops	553,065	570,211	16.4
Potatoes	286,801	292,938	
Other	266,264	277,273	
Total pulses	44,451	47,959	1.4
Vegetables & melons	307,021	318,906	9.2
Total fruits	256,670	257,068	7.4
Grapes	60,440	57,005	
Citrus	49,546	50,328	
Bananas	34,209	36,868	
Apples	24,314	21,348	
Other	128,161	91,519	
Total nuts	3,614	3,595	0.1
Oil crops (oil equivalent)	42,695	45,225	1.3
Sugar	80,870	92,109	2.6
Coffee, cocoa, tea	7,602	7,506	0.2
Meat	119,853	126,086	3.6
Milk	430,108	450,713	12.9
Eggs	23,596	24,700	0.7
Total food from land sources	3,231,698	3,403,090	97.9
Total catch of fish, crustaceans and molluscans from ocean and inland waters	69,800	73,460	2.1
Grand total food from land and water	3,301,498	3,476,550	100

Sources: FAO, Production Yearbook, 1977, vol. 31 (Rome, 1978). FAO, Yearbook Fishery Statistics, 1977 (Rome, 1977).

grasshoppers, much less for a family of hungry people. Consequently, much essentially worthless land has been classified as arable for political reasons. Similarly, much land classified as forest and woodland is of little value, being in some cases vast extensions of sagebrush with an occasional lonely juniper or "pining pinion pine." Moreover, large areas of valuable agricultural land are being removed from agricultural use and converted to industrial and residential sites, highways, airports, etc., each year. It has been estimated that in the U.S. one million hectares are being lost to these uses annually. The truth is that the U.S. even now has no land-use policy and the situation is equally chaotic elsewhere in the world.

It is true that there are still opportunities for expanding the arable land area by irrigation of desert areas and by clearing forests in some areas, but these undertakings are both time-consuming and expensive. Moreover, we must weigh the advisability of clearing or not clearing forest lands for agricultural use or the world will soon find itself faced with a worsening shortage of forest products.

Where Does Our Food Come From?

Food is produced from three different sources: the ocean and inland water, the land, and, to a very limited extent, indirectly from microorganisms cultured under artificial conditions. The latter currently is of very limited importance and will not be discussed here.

Many people erroneously believe that the sea is a vast and largely untapped reservoir of food production. The truth is that, at the maximum level of fish and crustacean production in 1971, the harvest of the sea reached a level of approximately seventy million metric tons, or only about 2 percent of the tonnage of food harvested from the land. In recent years, the marine harvest has begun to decline despite improvements and expansion of the world's fishing fleets. A number of authorities, including Dr. Georg Borgstrom, have stated that already the ocean is being overharvested for many species and that it will be necessary to limit catches to a lower level to sustain yield. Consequently, it becomes apparent that we must not consider the ocean to be a vast, largely untapped food-production base.

Thus, it becomes obvious that, as in the past, the growing demands for food must largely be met by production on the land. This can be achieved by expanding the area cultivated, by increasing yields on the area now under cultivation, or by a combination of the two.

Since the first recorded history there have been many crises in food production leading to famines caused by droughts, plant diseases, or hordes of locusts. Each crisis was precipitated because the human population was approaching the carrying capacity of the land then under cultivation with the method then prevailing.

After each crisis more land was rapidly opened to cultivation -- for then land was plentiful -- to feed the growing populations. Population growth in those early times was slow because man had little control over the environment, his food supply, or his own diseases. In 1976 we celebrated two hundred years since the birth of this country; our lands were opened largely during that time. How much more land can we, and the rest of the world, open in the next ten years, or in the next two hundred years? It is true that in certain areas of the world the development of large irrigation schemes such as the Indus-Ganges-Brahmaputra drainage basin in South Asia, the Mekong in Southeast Asia, the Niger basin in Africa, and the Amazon and Parana River basins in South America could bring large areas of land under higher production. But this will require enormous capital investments that are beyond the capabilities of individual nations. Moreover, international agreements and international financing will be required to begin to develop the potential of these areas, and the gestation period between planning and implementation will be very long, extending to several decades. There are also vast tracts of land with good precipitation that gradually can be opened to cultivation in southern Sudan. Similarly, Brazil has vast tracts of lateritic, leached soils in areas with precipitation of 1,000 to 1,600 mm. Twenty-five years ago these areas, known as the campo cerrado and variously estimated to constitute an area of sixty to one hundred million hectares, were regarded as having little potential value for agriculture. However, with the introduction of the proper technology, this area in a period of ten years has become the second largest producer of soybeans in the world. Under the economic stimulus of the worldwide shortage of edible oil and meal, Brazilian soybean production increased from 350,000 metric tons in 1965 to ten million metric tons in 1975, truly a revolution in soybean production.

The grass savannahs of Central Africa, in many ways similar to the campo cerrado of Brazil and the savannah of Colombia and Venezuela, also offer opportunities for large increases in livestock production and cultivation, if the tsetse fly and trypanosomiasis can be controlled.

Even though opportunities for the future expansion
of cultivated areas in some parts of the world still
exist, they will be both slow and costly to develop.
Moreover, the problem of dealing with food shortages is
not only one of achieving adequate per capita production worldwide, but also one of distribution. The lack
of purchasing power both of nations and families is an
obstacle to proper distribution. Even in such basic
foods as cereals, only about 10 percent of the total
world cereal production moves in international trade.

In many countries, especially where yields are low
and population density great, there is little additional land that can be cultivated. Most of the increase in food production will, consequently, have to
be achieved by the development and application of
science and technology to increase food production per
hectare. We have done this successfully in the U.S.
largely through the development of research in land
grant colleges, the agricultural experiment stations,
the U.S. Department of Agriculture, and more recently
through research by the agri-industries. The new
information and technology have been adapted by the
American farmer and livestock producer to greatly
increase food production. Much of the rest of the
world has not had this same good fortune. The problem
then is to help them in the years ahead to rapidly
improve their yields. Various organizations and agencies have tried to do this over the last two decades,
with at best only modest success. All too often their
approach has been too short-term, too academic, and too
abstract. It has consisted mostly of counseling and
advising from the comforts of capital cities rather
than entering into the production fray in the mud,
dust, and sweat on farms in rural areas where the food
is produced. The approach used today with great success in the U.S. -- with its well-educated, sophisticated farmers -- is ineffective in the developing
nations handicapped both by shortage of trained scientists and by weak agricultural organizations. In the
future we must be more efficient in our research,
extension, and educational assistance programs in
developing nations so as to avoid causing even more
confusion while trying to assist them. Vast sums of
foreign assistance funds alone will not solve these
problems.

Food Requirements for Four Billion People

It takes a lot of food to feed four billion people. It is impossible in a brief review to discuss
individually the production of the many different foods
that enter into human diets. Cereals constitute the

largest and most important single group of foods, as indicated in Table 1. Of a total estimated production of 3.3 billion metric tons of all food produced in 1975, the cereals collectively -- wheat, rice, maize, barley, oats, sorghum, millet, and rye -- accounted for 1.362 billion metric tons, by far the largest tonnage of any group.

In attempting to provide a picture of the magnitude of the world food-production problem I will, therefore, use the cereal grains as a yardstick of total food production. This in no way means to denigrate the importance of potatoes, yams, cassava, sugar, beans, cowpeas, peas, lentils, chick-peas, soybeans, peanuts and other oilseeds, fruits, nuts, vegetables, meats, milk, eggs, and fish -- all of which enter into the total food supply.

Cereals are sown on about 50 percent of the total land area of the world. They directly contribute about 52 percent of the calories to the human diet worldwide, and about 62 percent of the calories of diets in the developing world. Moreover, they supply nearly 50 percent of the total protein intake. Indirectly, cereals also contribute greatly to both human protein and caloric intake, for about 40 percent of the world production of cereals is fed to livestock to produce meat, milk, eggs, etc.

The total world production of cereal grains in 1971 was about 1.2 billion metric tons -- equivalent to about forty-four billion bushels of wheat. Most of us cannot visualize so large a quantity. So let us imagine it as a highway of grain around the earth at the equator about seventeen meters wide and two meters deep. That is the amount of grain the world produced in 1971. But there are eighty million more people each year! If we are going to supply the same amount per capita, we must increase cereal-grain production by 2 percent and add another 0.5 percent for the more affluent eating habits which include more meat -- or an increase of thirty million tons annually. This means we must reconstruct the "World Pan-Equatorial Grain Highway" each year, for we eat it in its entirety, and begin to construct beside it a second highway of grain of the same dimension -- seventeen meters wide and two meters deep. We also must construct a new highway at the cumulative annual rate of 1,000 kilometers each year just to maintain per capita food production at the 1971 level. If we were to attempt to achieve this increase in production solely by increasing the land area under cereals, we would need to increase it by about 16.2 million hectares annually, based on the 1971 average world cereal yield of 1,850 kilos per hectare. This is a very sizable task and gives one a clear-cut picture of the magnitude of the problem. It also becomes evident that the best approach to increasing production for the near future is primarily through increasing yield per hectare on the land already under cultivation.

So if the world is to be fed during the next several decades, it will be necessary to transform much of the traditional low-yield, subsistence agriculture. This will require the development of appropriate technology, based on research, that is capable of increasing yields by 50 to 100 percent, with an acceptable level of risk. Moreover, once an improved, appropriate technology is available, sound economic policies must be developed by governments which will stimulate the adoption of the high-yield technology by farmers and thereby increase yields and production.

Improving Food Production Through Increasing Yields and Cropping Intensity

World food-production needs cannot be coped with even for the next four decades unless per hectare crop yields are greatly increased in the developing nations and especially in the densely populated Third World nations. Fortunately, in these nations where crop yields are still low, there exist excellent opportunities for both increasing per hectare yield, and in many areas for increasing cropping intensity through multiple cropping. It is almost certain that in the developed agricultural countries where current crop yields are already high, they cannot be further increased at the rate that occurred in these countries in the period from 1950-75. This being the case, far greater emphasis must be given to improving the agricultural output in the developing nations that have been, until recently, largely untouched by modern agricultural technology. Moreover, in countries such as India, Pakistan, Philippines, Turkey, Thailand, etc., that have developed and adopted high-yield technology on one or two crops during the past decade, the use of technology must be improved and extended to other geographic areas still untouched by these methods. Further improved technology must be developed for the other important crops that are still grown under traditional methods. I firmly believe that it is more advisable to aggressively develop new technology that has the potential to speed up the increase in yield and production of the "slow running" crops (e.g., grain legumes), rather than try to slow down the progress of the "fast-running" crops (e.g., wheat or rice), by adopting economic regulations that hinder overall progress.

Within the past decade, with the development of high-yielding crop technology, it has become possible to produce two or three crops in sequence on the same land in the course of a year in some tropical and semitropical areas which either have irrigation or rainfall distributed throughout the year. The People's Republic of China, the Republic of China, India, and the Republic of Korea have exploited multiple cropping effectively during the past decade. In other cases, the interplanting of crops in tandem has become popular. The increase in cropping intensity by either method has added greatly to the total food production per hectare per year. There is much opportunity for expanding food production through this approach during the next two decades. However, chemical fertilizer is essential before this approach is feasible.

Characteristics of Traditional Agriculture That Greatly Restrict World Food Production

The traditional agriculture of developing nations has certain characteristic defects in common. These must be corrected if yield and production are to be increased. Among these are:

Soil infertility. Without doubt the single most important factor limiting crop yield on a worldwide basis is soil infertility. The lack of one or more of the essential plant nutrients is the result of the joint effect of natural weathering followed by leaching, combined with extractive farming practices. In the former case phosphorous and potassium, and other water-soluble nutrients, have been gradually leached from the soil during geologic time to a level which, with subsequent continuous extractive farming, soon limits crop yield. Soil nitrogen deficiency is nearly universal. Virtually all traditional farming systems are highly extractive. They are essentially "mining" operations whereby each year crops are harvested and little or none of the crop residue or animal wastes, which would partially restore soil fertility, are returned to the soil. Soil fertility can only be restored by the application of the right kind and proper amounts of fertilizer. The proper kind and amount of fertilizer required to restore fertility will vary with soil type, climate, and crop. It can only be determined by field research. Until soil fertility is restored, improvement in crop cultural practices and the use of improved varieties will improve yield only marginally. Many political leaders and economic plan-

ners fail to understand or are unwilling to recognize
the necessity of making large investments in fertilizer
factories, which is the first step toward restoring
soil fertility and setting the stage for increased crop
yields and production. Often they are misled by poorly
informed groups who mistakenly believe the world's soil
infertility problem can be solved by the use of organic
manure alone or by bacterial nitrogen fixation. The
error of this over-simplified approach is indicated
later.

Crop varieties of low genetic yield potential.
Traditional crop varieties have generally evolved on
nutrient-depleted soils. They have low genetic grain
yield potential and have unfavorable grain-straw
ratios. When soil fertility is restored through the
proper use of fertilizer, such varieties respond poorly. They have structural, physiological and pathological weaknesses that become manifest under medium to
high soil fertility levels. They grow tall and have
weak straw, which lodges badly. They are vulnerable to
many epidemic diseases, e.g., rust fungi, when they are
grown on fertile soil that gives rise to dense stands,
which modify the microclimate within the grain field
and create an environment more favorable for the pathogens. Once soil fertility is restored, it is necessary
to develop and distribute a series of high-yielding,
disease-resistant crop varieties, with acceptable
agronomic and consumer-quality characteristics, which
are genetically capable of utilizing more effectively
and more completely the improved soil fertility conditions.

Poor agronomic practices. In traditional agriculture, where lack of plant nutrients limits yield,
agronomic practices generally receive little attention.
The traditional farmer has learned from experience that
he can do little to increase crop yields by manipulating agronomic practices. Consequently, seedbeds are
often poorly prepared, resulting in patchy stands with
poor spacing between plants. Little attention is given
to conserving moisture or to using irrigation water
efficiently. Inadequate attention is given to weed
control because weeds apparently are not highly competitive with the crop plant, since they too are suffering from plant nutrient deficiencies.

When soil fertility is restored, the use of an
improved set of agronomic practices becomes decisive
for exploiting opportunities for a big increase in
yields. These practices must be based on extensive
research conducted on farms throughout the area where
the crop is grown.

Once soil fertility has been restored, if the new opportunities for large increases in yield are to be seized, it becomes necessary to employ good seedbed preparation, proper seed rates and the correct dates of sowing for each of the improved varieties, proper conservation and management of soil moisture, and proper control of weeds. Under improved levels of soil fertility, weeds become aggressive and highly competitive; unless they are controlled either mechanically and/or chemically, one will harvest more weeds than wheat, and I have rarely seen, even in the most disorganized market, a situation where one can make more income selling weeds than wheat.

Inadequate or nonexistent control of disease and insects. In low-yield traditional agriculture, only in unusual years are ecological conditions sufficiently favorable for the pathogens and insects to produce serious destructive epidemics and infestations. When these conditions do occur, however, the losses are very severe, for there are no organized disease- or insect-control programs to advise and assist the farmer. In most years, however, the pathogens and insect-pest species, like the host plants, are all struggling for survival under difficult, unfavorable environmental conditions. The situation in high-yield, intensive agricultural systems changes dramatically, however, for when soils are fertilized and improved agronomic practices are used, this results in the development of thick, lush stands of crops. The ecology within these fields then becomes very favorable for the pathogen and insect pests. Disease- and insect-resistant varieties must be used to minimize the risks of crop losses; moreover, an integrated control program must be adopted insofar as possible -- which includes crop rotations, proper dates of planting, biological control (insofar as effective) and the regular monitoring of the pest population combined with the timely application of pesticides when necessary. These steps are necessary in order to reduce crop losses to acceptable economic levels.

The control of disease and insects becomes increasingly more essential with greater intensification of, and investment in, agriculture.

Nonavailability of production inputs. In many developing countries production inputs are either unavailable or available only in a few large cities far removed from the agriculture production areas where they are needed. It is necessary to establish an effective network to distribute seeds of improved crop varieties, fertilizer, herbicides, insecticides, and fungicides down to the village level if production is to be increased. The timeliness of distribution and appropriateness of the products being distributed are also of primary importance, but all too often they are hopelessly entangled in a web of bureaucratic inertia.

Government economic policies affecting agriculture. Whenever an attempt is made to provoke change in a primitive agriculture, an effective technological package of improved crop varieties and agronomic practices must be developed, based on research, to overcome the inherent defects and weaknesses in the traditional agricultural system. The improved technology must be checked for validity and demonstrated widely on many farms. The new technology must have the capability of increasing yields by 50 percent or more without involving unacceptable risks. Once an effective package of improved production practices has been developed and widely demonstrated with positive results, it must be "married" to sound economic government policy that will encourage its adoption, and thereby result in increased production.

The economic policy must assure the availability of proper kinds and amounts of inputs at reasonable prices at the village level. It must make credit available for their purchase, especially for the small farmers, so that they can participate in the use of the new technology. Finally, it must assure the farmer a reasonable price for his produce at time of harvest. Floor prices for grain of the important crops must be announced before time of planting. If prices in the free market at harvest fall below the announced floor level, the government must enter the market to assure the announced price level. The grain purchased by the government to stabilize prices at harvest can be fed into the market later to stabilize grain prices for the consumers.

Weak research and extensive extension programs. Agricultural research and extension programs in most developing nations are weak. They are handicapped by a shortage of trained people, inadequate budgets, and low prestige associated with agriculture. In reality, despite the fact that 70 to 90 percent of the total population in most developing nations is involved in agriculture, it occupies the lowest rung on the socioeconomic ladder. Consequently, many of the most talented young people with rural backgrounds want to forget about the hard work, drudgery, lack of prestige, and low incomes in agriculture. They seek careers in medicine, dentistry, law, chemistry, engineering, or business.

It is impossible to transform a traditional agriculture into a modern agriculture without the assistance of a large group of well-motivated and well-
trained scientists and technicians. Experience in a
number of countries where research and training programs were initiated when there were very few trained
people indicates that it takes twenty years to identify, train, and provide research experience for a
sufficiently large number of young scientists and
technicians so that a national research institute can
be organized and staffed effectively.

The recognition of the shortage of trained scientists in many developing nations, first by independent
foundations and subsequently by governments, has led to
the development of a network of nine international
agriculture research institutes. Seven of these are
devoted to the improvement of important crops and two
are devoted to research on animal production and diseases. The international agriculture research institutes are devoted to developing research materials,
methods, and data that may be useful in assisting
national programs. They are also deeply involved in
training large numbers of young scientists from the
developing nations.

There is no justification for complacency on the
food production front. I have already mentioned the
great progress that has been achieved by Mexico, India,
Pakistan, Tunisia, and Turkey in wheat production
during the past decade. The four latter countries have
not only become self-sufficient in wheat production,
but within the past two years they have also become
self-sufficient in all food grains. Reserve stocks of
wheat and/or rice in some of these countries have grown
to levels that now are causing concern in some government circles. Nevertheless, in each of these countries
there are large numbers of unemployed and low-income
people who lack purchasing power to buy the food they
need. It would appear to me that the sensible approach
is not to cut back production but rather to find ways
to increase the income and purchasing power of the
low-income people who need more food. Perhaps one
feasible approach is the establishment of a program of
public works which will be based in part on providing
"food for work" and thereby mitigate widespread unemployment and hunger, while building and improving the
agricultural infrastructure.

It appears to me that the agricultural scientists
in these countries have done their jobs exceedingly
well. This is indicated by the great increase in
production and the accumulation of so-called "surplus"

food-grain stocks. The record of the spectacular threefold increase in wheat production that has taken place in India over the 1966-79 period is shown in Table 2.

I now challenge the many well-trained economists and political leaders to find an imaginative, effective, and positive way to use these accumulating (so-called "surplus") stocks to banish hunger, and in the process improve the infrastructure and stimulate the economy within each of the countries involved.

Despite the progress that has been made in increasing food-grain production in a number of low-income, food-deficit nations during the past decade, there are warnings of danger ahead. These signals tell us there is no time to relax on the production effort. The continuing food-population problem is clearly brought out in Research Report No. 3, December 1977, of the International Food Policy Research Institute entitled, Food Needs of Developing Countries: Projection of Production and Consumption to 1990. This report indicates that the heart of the world food problem is now, and will continue to be, in the low-income, food-deficit countries, concentrated in Asia and Subsahara Africa. These countries have approximately two-thirds of the total population of the developing market economies.

The projections in this study are based on the assumption that food production will grow in the 1975-1990 period at the average annual rates of the 1960-1975 period. Consumption requirements are based on the United Nations medium-variant population projection and two levels of income. The data are presented in Figures 1 and 2 and Table 3.

This study indicates that the food deficit expressed in grain equivalents in the low-income, developing-market-economy countries is projected to increase from thirteen million metric tons in 1975 to seventy to eighty-five million tons by 1990. Just to maintain consumption at the 1975 per capita level would require thirty-five million metric tons more than projected production. The data in Table 3, for example, indicate India's deficit in grain will increase from the deficit of 1.4 in 1975 (it reached self-sufficiency in 1977) to between 17.6 and 21.9 million tons in 1990; Nigeria's deficit from 0.4 million tons in 1975 to an anticipated 17.1 to 20.5 million ton level in 1990 and, similarly, Bangladesh's deficit will increase from 1.0 million tons to 6.4 to 8.0 million tons in the same period.

Table 2. The Revolution in Indian Wheat Production

Harvest Year	Millions of Metric Tons	Harvest Year	Millions of Metric Tons
1966	10.4	1973	23.0
1967	11.1	1974	21.2
1968	16.4	1975	24.0
1969	18.2	1976	29.0
1970	21.3	1977	29.0
1971	23.3	1978	31.5
1972	26.5	1979	34.5 estimated

Table 3. Actual and Projected 1990 Cereal Grain Equivalent Deficit in Low-Income, Developing-Market-Economy Countries

Country	Actual 1975		Projected 1990	
	Millions of Metric Tons	% Consumption	Millions of Metric Tons	% Consumption
India	1.4	1	17.6 to 21.9	10 to 12
Nigeria	0.4	2	17.1 to 20.5	35 to 39
Bangladesh	1.0	7	6.4 to 8.0	30 to 35
Indonesia	2.1	8	6.0 to 7.7	14 to 17
Egypt	3.7	35	4.9	32
Sahel Group	0.4	9	3.2 to 3.5	44 to 46
Ethiopia	0.1	2	2.1 to 2.3	26 to 28
Burma	(0.4)+	(7)+	1.9 to 2.4	21 to 25
Philippines	0.3	4	1.4 to 1.7	11 to 13
Afghanistan	-	-	1.3 to 1.5	19 to 22
Bolivia and Haiti	0.3	24	0.7 to 0.8	35 to 38

Source: International Food Policy Research, Research Report 3 Food Need of Developing Countries (December 1977).

+ Surplus

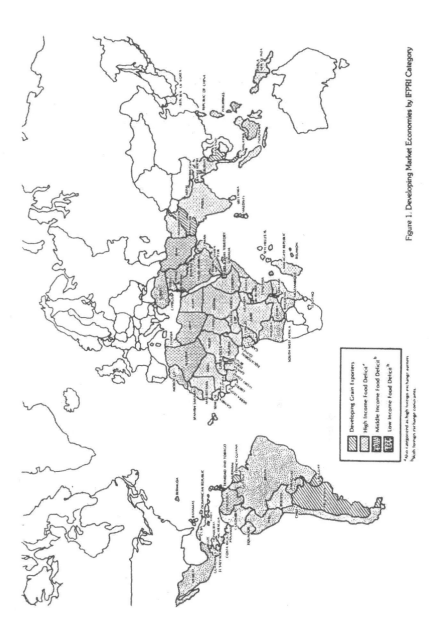

Figure 1. Developing Market Economies by IFPRI Category

Developing Grain Exporters

High Income Food Deficit[a]

Middle Income Food Deficit[b]

Low Income Food Deficit[b]

[a] Also categorized as high foreign exchange earners.

[b] With foreign exchange constraints.

142

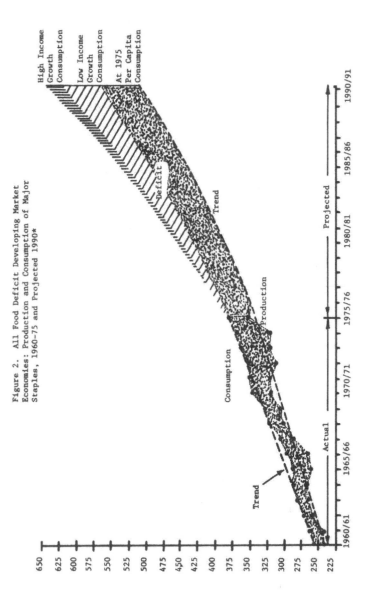

Figure 2. All Food Deficit Developing Market
Economies: Production and Consumption of Major
Staples, 1960-75 and Projected 1990*

International Food Policy
Research Institute
Report # 3 (Draft)
September, 1977

There may be some doubt about the validity of the interpretations of the projections for India, from my point of view, because of the use of the 1960-75 base period. The revolution in Indian wheat production did not significantly affect production until 1968. Its progress was interrupted in 1974 and 1975 by shortages of fertilizer and diesel fuel for the low-lift pumps in most areas. The dynamic upward trend in production has been reestablished during the past three years, during which time three successive record breaking crops have been harvested. Moreover, very significant increases in rice production were just beginning in 1975. Consequently, it may well be that the 1960-75 trend line does not accurately represent the current production trends which appear, to me, to be considerably higher. In the case of Bangladesh, recent increases in wheat production may, if continued, also reduce the projected deficit.

Nonetheless, the International Food Policy Institute's research report is a clear signal to government planners, political leaders, and to agricultural scientists that there is no time for complacency or relaxation on the food-production front. In the next two decades all food-deficit, developing nations with traditional agriculture must make every effort to revolutionize their agricultural production by developing suitable improved technology and economic policies that will implement its adoption. Moreover, the densely populated countries such as India, Pakistan, and the People's Republic of China, which have made recent break-throughs in production, must aggressively continue to expand food production or the improvements recently achieved will be overwhelmed by the relentless advances of the population effort. On the contrary, agricultural production must continue to be expanded aggressively, for a new opportunity is emerging which offers the possibility of significantly increasing the income, purchasing power, and standard of living of millions of small farmers who now, for the first time, are producing more food than is needed for their families and have in addition a surplus for sale in the market place. This increased purchasing power will in turn stimulate the general economy of the country.

The need for grain reserves. The shortage of food grains and soaring food prices which resulted during the 1972 to 1975 period should not soon be forgotten by political leaders and planners of food-deficit and marginally self-sufficient nations. Adverse weather over vast areas of the largest cereal-producing countries of the world resulted in a substantial reduction in grain harvests, especially during 1972 and 1973, and brought regional famine, widespread hunger, and worsening poverty to millions.

As world population increases, it becomes more and more important for each nation to carry a reasonable buffer stock of food grains as protection against poor harvests which result from the vagaries of weather and crop losses from insects and diseases. Moreover, world governments must jointly find some way to finance, build, and maintain an international reserve stock of grain that can be used as a second line of defense in emergencies in any part of the world whenever there is threat of famine. During the 1974 World Food Congress there was much rhetoric about the absolute necessity for establishing such a world reserve. However, as a result of three good harvests, worldwide interest has waned and no really functional second line world grain reserve has come into being.

ISSUES WHICH MAY INFLUENCE
AGRICULTURAL AND FOREST PRODUCTION

The production of food, fiber, and forest products, as well as the indirect noncommercial benefits such as watershed protection, aesthetics, wildlife, and recreation, involves the complex issues of land use in the broadest sense. Unfortunately, programs struggling to increase the production of food, fiber, and timber are today often handicapped in many countries either directly or indirectly by one or more lobby or political pressure group that insists that agriculture consumes too much energy, employs too much fertilizer, pollutes the environment unnecessarily, and endangers public health and many species of wildlife. These conflicts of interests are far more frequent and more often reach a level of confrontation in well-fed affluent societies than in developing nations.

Issues such as aesthetics, wildlife, and outdoor recreation become more important in affluent societies because many individuals have much leisure time. Certain individuals with much leisure time, money, and political influence feel the urge to take up a cause and launch a crusade to correct or alleviate some either real or imagined misuses of land, energy, water, or air -- and confrontations and lawsuits often result. Some of the objectives espoused by these movements are worthy causes in themselves, and they are being promoted by well-intentioned individuals and organizations. Unfortunately, much confusion results in many of the movements because of the failure of individuals and the general public to comprehend the complexities and the order of importance to the general public of many of the issues under debate. Most of these issues could be solved amicably and in the public interest if they were discussed and decided on the basis of benefit versus cost (or benefit versus risk), and how large a segment of the general public will benefit from a change in policy. All too often these conflicts degenerate into emotional diatribes. Unfortunately, unwise decisions made on certain of these basic issues in the affluent, developed nations may set the precedent for a similar action in food-deficit developing nations.

I have great faith in the good judgment of the general public in reaching the correct decision on complex issues in any country if they are given a balanced picture of the two sides of the issues. Unfortunately, in many affluent, developed countries in recent years, the general public is seldom provided this opportunity on many of the land-use and other environmental issues because of the extremely effective, biased, fear-provoking, lobbying programs of organizations representing, all too often, a rather small number of privileged individuals.

Among the important issues beclouded by such individuals, organizations, or movements, which either already have adversely affected or potentially could adversely affect agriculture and forest production, are the following:

Energy for Power

One now frequently reads that modern mechanized agriculture is too energy-intensive and requires vast quantities of scarce fossil fuel which cannot be justified in this time of energy shortage. These articles apparently imply that American farmers, and farmers from other developed nations, should go back to the "good old days" when food was produced with four-legged horse power or even further back to the drudgery of human muscle power, while scholars, academicians, lawyers, journalists, and bureaucrats would presumably continue to enjoy the pleasures of their two cars, air-conditioned offices and suburban homes. The facts are that the entire gamut of activities leading to putting food on the table are estimated to consume 12 to 15 percent of the total energy consumed in the U.S. However, depending on how the calculations are made, only about 3 percent of the total national energy budget is utilized to produce both the inputs and carry on the multitude of activities performed on the farm and place the food products at the farm gate. When one considers the energy input in food from the farm gate until it arrives at our urban tables, the energy inputs in transportation, processing, refrigeration, packaging, distribution and preparation, of course, mount greatly. These collectively represent 12 percent of the total national energy budget. But much of the latter expenditure of energy in the affluent nations results from special services demanded by the consumer and does not involve the farmer. Of course, the whole family transport system in a country such as the U.S. is the portion of the chain that really guzzles fossil fuel. It is convenient but hardly efficient. But how many families are willing to give up their second or third car, much less the first?

Therefore, what are the alternatives for a highly mechanized agriculture, such as that of the U.S., where in 1977 only 3.6 percent of the total population lived on farms compared to 42 percent in 1900? During the first decade of this century, twenty-two million head of horses and mules provided the source of power on American farms. The land area required to produce their feed was somewhere between twenty-nine and thirty-three million hectares. At present, if the U.S. were to attempt to go back to animal horse power, considering the intensiveness of present agriculture, it would probably need, as Dr. Earl Butz has calculated, about sixty-one million horses and mules. This would require about seventy-three million hectares of good crop land to produce their feed, taking this out of what is now available for other uses, including recreation, parks, forestry, etc. Moreover, it would require about thirty-one million workers on farms to handle the animal power. This is about four times the number of people who live on farms in the U.S. at the present time. Where would they come from? How many urbanites would voluntarily and willingly go back to living on farms and putting in the long hours of physical work and drudgery? The facts are that the drift of population, especially youth, from country to city has largely taken place and continues because of the drudgery and long hours of hard physical work at lower income levels on the farm compared to potential employment opportunities in the city, where there is hope for better opportunities for education, recreation, and medical care.

Recent General Accounting Office figures show there were about 22 percent fewer people living on U.S. farms in 1976 than in 1970. Moreover, despite eight revisions since 1933, government assistance programs designed to help the "family farmer" have definitely benefited the largest farms to a proportionately greater extent. Similarly, government tax policies have promoted the trend away from smaller, family-owned-and-operated farms while attempting to do the opposite. Federal income tax laws provide an excellent tax shelter for outside investors. The rush of outside "speculator/investors" to invest in land increases as inflation worsens. Moreover, recent estate tax laws will inhibit sale of farmland outside of the owning family and may thus create a "landed aristocracy." Meanwhile, with farmland prices inflated to levels far above their ability to produce a reasonable return on the investment, it makes it virtually impossible for young men to enter farming. Also, the farm population, in general, is getting both older and poorer. So the condition of agriculture in the U.S. is currently far from perfect from the standpoint of outlook for continued progress.

There also appears to be a generally mistaken belief among some academicians and especially among social scientists in American universities, judging from publications, that the so-called "Green Revolution Wheat Technology" introduced into Indian wheat production, is a transplantation of American mechanized production technology per se with a corresponding large increase in fossil fuel energy inputs. This is not correct. Most of the energy input in the so-called "Green Revolution Wheat Technology" in India continues to come from oxen power and manpower. The one very substantial increase in energy input from fossil fuels is that involved in production of nitrogenous fertilizers, and to a much lesser extent for diesel fuel for motors on shallow tube wells. Most of the irrigation, however, comes from gravity irrigation systems which also, in many cases, generate hydroelectric power. The land preparation, planting, and sowing is still probably 90 percent powered by oxen and man. The harvest operation is still responsible for much of the threshing by treading, although small diesel-powered, mechanical threshing machines are rapidly increasing in numbers.

Energy for Chemical Fertilizer Production

The belief seems to be common among intellectuals that the world can no longer afford to use scarce petroleum or natural gas for the production of nitrogenous chemical fertilizer. They seem to believe that if we simply resorted to the proper use of all of the natural organic fertilizer, there would be little need for large volumes of chemical nitrogenous fertilizer. I disagree with this point of view. In the first place, it must be emphasized that food production cannot be increased fast enough to keep pace with world population growth without the availability of adequate supplies of fertilizer, including nitrogenous ferti- lizers. Chemical fertilizers are especially essential to restore productivity to the "mined out, tired soils" which are typical of most of the developing nations.

What can the developing nations do to restore soil fertility? Agriculture is old in many of these na- tions. The land has been farmed for decades, hundreds or even thousands of years in some cases. Traditional farming has been a mining operation which has removed plant nutrients and returned little or none to the soil. In countries such as India, Pakistan, and many others, the grain is used for human consumption and the straw is frequently the only feed available for bul- locks, cattle and buffalo. The cattle and buffalo dung becomes the fuel to cook food because the forests have long since been cut in many of these densely over-popu- lated lands. Thus, few nutrients are returned to the soil.

There are some organic gardening enthusiasts who insist that the wise use of organic fertilizers could satisfy all of our fertilizer needs. It is true that organic manures are very effective for growing six beautiful, high-yielding tomato plants, six lovely rose bushes or a small vegetable garden in one's backyard. However, it does not necessarily follow that the same procedure can be effective for producing the food needed to feed four billion people in a land-hungry world. The amount of composted organic animal manure of forty-seven million metric nutrient tons of chemical nitrogen that was used in the world in 1977 would be about 3.2 billion tons -- quite a dung heap and quite an aroma -- were it available! If we assume a volume of three cubic meters per ton and about fifteen kilo- grams of nitrogen per ton, the dung requirement calcu- lates to be equivalent in volume to a highway of dung two meters deep and 117 meters wide around the earth at the equator. However, the total amount of animal manure produced each year throughout the world by all the large animals (excluding hogs, sheep, goats, and poultry) is approximately two billion metric tons, on a dry-weight basis.

To provide the theoretical equivalent amount of additional animal manure necessary to produce the forty-seven million tons of chemical nitrogen now used as fertilizer would require an increase of 60 percent in the cattle, buffalo, horse, and mule population, if all of it were available for growing crops. However, since about 75 percent of the animal manure is dropped on the range lands and pastures and recycled in this way, or burned as fuel, only the remainder that collects in barns and feedlots, where there are large concentrations of animals, would be available for agriculture. Therefore, it would be necessary to increase the present large animal population in the world about 2.8-fold to produce the manure that would be needed as agricultural fertilizer. But where would the feed come from for such a population of large animals? There are many areas of the world where overgrazing is now a severe problem. Imagine the widespread disaster from overgrazing that would result from a threefold increase in large animals were this attempted precipitously.

There is another complication for the use of organic fertilizer. Because of the low concentration of nitrogen, about 1.5 percent nitrogen on a dry-weight basis, it becomes prohibitively expensive to ship animal fertilizer any distance, except for specialty use. It is primarily because of this fact that dry cattle dung, excellent for organic gardening, sells in plastic bags in eastern U.S supermarkets at more than double the price of pasteurized milk on a pound for pound basis.

Certainly, up to the point where organic manures are available locally, they should be used to restore fertility to the soil and at the same time reduce pollution from feed lots. However, the Chinese, who are the most skillful people in the world at handling composted manures derived from human waste (from 900 million people), animal waste (from 250 million pigs and many large animals), and all types of plant wastes, are now making huge investments in chemical nitrogen and other types of fertilizer factories. As late as 1960 there was virtually no chemical nitrogenous fertilizer used in the People's Republic of China. By 1974 there were 1,200 small anhydrous ammonia fertilizer factories utilizing bituminous coal as a raw material. In addition, the country had also become the largest importer of chemical nitrogen in the world. When world petroleum prices soared in 1973 and exacerbated the nitrogenous fertilizer shortages with a consequent great increase in the price of chemical nitrogen, the People's Republic of China immediately took action to begin the installation of twelve 1,000-metric-ton-per-day anhydrous ammonia plants and the corresponding urea converter units of the most modern design. These plants probably cost about US$1.5 billion and have a rated annual production capacity of more than three million metric tons of nutrient nitrogen. When the cost of the infrastructure, i.e., opening new gas fields, pipelines, warehouses, and transport system, is added, it will probably represent a total investment of approximately US$7 billion dollars.

This action by the government of the People's Republic of China indicates they have learned that fertilizer can be effectively substituted for additional cropland as a means of increasing food production in densely populated, land-hungry countries. The late Dr. Raymond Ewell estimated in 1975 that the world would have to invest at the rate of US$7 billion dollars annually over each of the next five years to expand the nitrogenous fertilizer production facilities that will be needed by world agriculture. Although this at first glance appears to be an enormous sum to invest in fertilizer production, it is very modest compared to the more than $380 billion the world spends annually on armament in the collective name of national defense. Obviously, within the limits where it is practical and economical to do so, we must encourage the use of leguminous crops, which both produce food or forage and at the same time fix substantial quantities of nitrogen in crop rotations with cereals. Wherever feasible, green manure crops should also be encouraged in crop rotations. Unfortunately, this practice involves leaving valuable, scarce cropland out of food production for at least one crop cycle, which generally cannot be justified in food-deficit, densely populated, "land-hungry" nations.

One can hope that ingenious scientists will someday succeed in developing effective nitrogen-fixing symbionts, whether free-living, rhyzosphere, nitrogen-fixing organisms, or nodule-forming, rhizobium-like species on the roots of cereals, which will be capable of fixing a large proportion of the nitrogen required to produce high grain yields. Since the latter type of symbionts have long been known to exist on the roots of plants as diverse as all herbaceous plant and tree species of the family Leguminosae as well as on the roots of the genus Alnus, such a fantasy can, at least, be entertained.

In recent years, especially since the development of the acetylene reduction assay for measuring nitrogen fixation, it has become apparent that a large number of microorganisms associated with the root systems of nonleguminous plants do indeed fix nitrogen. Prior to the development of the acetylene reduction assay, their nitrogen fixation capabilities escaped recognition.

The organisms that now are thought to contribute substantially to nitrogen fixation, and that may have the potential for further increase in levels of fixation, include a considerable number of free-living (frequently in the rhizosphere) or intracellular or intercellular, root-inhabiting, bacterial species.

Dr. Johanna Dobereiner and coworker in Brazil have shown that Azospirillum spp. inhabit the roots and fix considerable but variable amounts of nitrogen in a number of tropical grasses, including Paspalum notatum, Digitaria decumbens. More recently they have found they are also sometimes present in the roots of maize, sorghum, and wheat. It has also been demonstrated that paddy rice has an important nitrogen-fixation system involving "free-living" blue-green algae and bacteria that are active in the thizosphere of the rice roots.

The rice farmers in southern China have effectively utilized for decades a symbiotic system based on the floating water fern Azolla spp., which harbors a nitrogen-fixing, blue-green alga, to fix substantial quantities of nitrogen. Azolla is reputed to add from twenty to forty kilograms of nitrogen per hectare to paddy soils in the short period of three to four weeks when it is grown on the flooded paddy fields prior to transplanting of the rice seedlings.

These new discoveries merit additional research to attempt to improve the efficiency of fixation and to increase the level of nitrogen production to a level where it can supply a part of the total nitrogen requirements for the cereal crop.

It is highly unlikely, however, that there will be any substantial quantities of the total world nitrogen requirements for agriculture that will be forthcoming from these approaches for at least the next five decades, if ever.

Today, we read articles written by well-intentioned but poorly informed people who ask whether good plant breeders cannot produce varieties of crop plants that are capable of producing high grain yields even when grown on nutrient impoverished soils and without the use of chemical or organic fertilizers. My only reply is that plant breeders will succeed in producing such varieties of crop plants about one year after utopian economists, sociologists, environmentalists, and lawyers produce a new race of man that needs no food to grow strong bodies, maintain health, work effectively, enjoy life, and speak and write eloquently. Often these utopians curse plant breeders for having been unable to produce a crop variety that will yield 100 metric tons from a ten hectare farm and at the same time with the same soil, climatic conditions, and technology produce only 100 metric tons on a 100 hectare farm. Were this made possible by the plant breeder, it would remove the need for the politicians and planners to face the unpleasant task of equitably distributing the food and income. Unfortunately, this, too, has been nonachievable up to now because plant varieties are both apolitical and impersonal.

In summary then, can the world continue to justify the use of scarce fossil fuels for the production of nitrogenous fertilizer? Let us look at the facts. When I was in Saudi Arabia in 1975, I learned that six billion cubic feet of gas were flared each day from wellheads and refineries in that country. The gas flared in that one country alone would supply the energy and raw materials for 167 anhydrous ammonia fertilizer plants and corresponding urea converters to produce 1000 tons a day each. That is enough to produce, at the rated plant capacity, forty-five million tons of nutrient nitrogen annually -- about as much as the 1975 world consumption of nitrogenous fertilizer. Vast quantities of gas are also being flared in twenty other countries. Also, nitrogenous fertilizer can be produced efficiently from coal when the need arises. Therefore, it is difficult for me to accept the judgment of those who insist the world cannot continue to use scarce natural gas, petroleum, or coal for the production of nitrogenous fertilizer -- which is essential to maintain the necessary expansion in food production -- while the world continues to ignore the tremendous wastage of fossil fuels in many other sectors of the economy where saving of energy and substitution of energy sources can be more easily accomplished.

Other Issues: Land Use, Agricultural and Industrial Production, Public Health, and the Quality of Life

During the past decade, the world has become increasingly more concerned about the impact of human life on the environment (and vice versa) in which we live. In several affluent, developed nations these complex issues have degenerated into heated debates and polemics, often followed by an avalanche of lawsuits over many aspects of land use or concerning the technology being employed by agriculture, forestry, and industry that affect the environment, health, recreation, wildlife, aesthetics, and sometimes the so-called quality of life in the broadest sense. Some of these confrontations have resulted in new legislation or new interpretation of old laws which have corrected unnecessary abuses of the past. Other legislation, such as banning the use of DDT in the United States, inadvertently has had a serious negative effect on the health, agriculture, and economy of many developing nations in the semitropical and tropical areas where malaria and many other vectorborne diseases are endemic.

In many cases where polemics have developed over these complex issues, emotions have prevailed over a reasonable interpretation of the scientific data and facts. Such irresponsible polemics have often confused the general public and misled political leaders.

In the remaining portion of this Appendix, I would like to explore with you some of the controversial issues that are likely to have a continuing impact on agricultural production, land use, and public health over the next several decades.

Since many of these issues either directly or indirectly are part of the broad environmental debate, I will begin here.

What is, after all, a good or bad environment? The concept of a good or bad environment is largely a matter of opinion. It is a subjective rather than an objective criterion. It is different for men, apes, sparrows, eagles, rabbits, rats, lions, elephants, cows, rattlesnakes, or mosquitoes. It depends upon the point of view or the eyes you look through. In an attempt to arrive at a tentative conclusion about the quality of the environment from the standpoint of Homo sapiens, one has to establish a realistic order of values and priorities. What are the factors in the environment that are of primary importance and what are those that are of secondary or minor importance? We must at the outset recognize that criteria concerning environmental quality are very different in the affluent and the developing nations, and they also follow an economic gradient within the affluent nations. For example, the criterion of a good or bad environment as viewed by a business executive, upper echelon government official, or university professor in an air-conditioned office of a large American city, who is privileged to live in a large house in a fashionable suburb and enjoy two late-model cars, is very different from the criterion used by a factory worker in Detroit, a small storekeeper in Minneapolis, a coal miner in Pennsylvania, a wheat farmer in Kansas, an unemployed laborer in a New Orleans slum, a minor government official in New Delhi, a landless peasant in a small village in Pakistan, a small hill farmer in Peru, or a hungry migrant laborer in Mexico. Therein lies one of the dilemmas of our time in trying to come to realistic grips with environmental issues on a worldwide basis.

During the past decades with higher incomes, more frequent and longer vacations, and better transportation facilities enjoyed by the majority of the urban population of the affluent nations, there has been a new and rapidly growing interest in the countryside and the beauties and fascination of nature. This renewed interest is a very positive development that has brought personal enjoyment into the lives of millions of urban and city dwellers. It has also had a positive effect in bringing to the attention of the general public the unnecessary past abuses of the environment committed by both urban and rural individuals, industries, organizations, and governments. Largely as a result of such interests and pressures, regulations have been adopted to reduce unnecessary pollution by industry, agriculture, forestry, and cities of rivers, lakes, oceans, and air. Similarly, action has been taken to reduce the littering by the general public of streets, roadsides, parks, and other public areas with all sorts of waste. Positive benefits (both direct and indirect) have accrued from the environmentalist movement.

However, there have been some adverse, indirect side effects which have come from it. Among these has been a failure by many of the extremist environmentalists and "neoecologists" to understand the complexities of ecosystems and the many difficulties encountered by industry, agriculture, and forestry in taking action to improve the environment while still maintaining industrial, food, and fiber production. Unfortunately, in their zeal to correct past abuses, some of the more fanatical in the movement have added more confusion than enlightenment to some of the basic issues that confront management of our farms, ranches, forests, and industries. In their zeal for the new environmental cause -- that has sometimes taken on the fervor of a new religion -- they have given the popular press and indirectly through it the general public the impression that ecology is a simple and exact science and that in a given ecosystem all the factors affecting the environment can be accurately and rapidly identified and their impact quantified. Anyone who has worked with ecosystems, whether on a farm, ranch, or forest, knows this is not so. The truth is that ecology is one of the most complex but inexact of the sciences.

At any given time, even in a relatively simple ecosystem consisting of a few thousand hectares, it is

impossible to determine the number of complex reactions
that are going on in the soil between the innumerable
species and biotypes of microorganisms (e.g., fungi,
bacteria, insects, viruses, microfauna), between roots
of herbaceous and woody plants and trees, and between
rodents and other burrowing animals. Above-ground
relationships are equally complex. There are often
large numbers of species of flowering plants, shrubs,
trees, and other lower plant forms competing for space,
light, and plant nutrients, as well as innumerable
microfauna, microorganisms, insects, and viruses, both
parasitic and saprophytic, living on or attacking
different parts of the different plants. Each of the
higher plants is being utilized or influenced in one
way or another by a large but varying number of higher
animals, including man. And all of these organisms are
interacting with each other and with the environment in
a bewildering number of complex reactions. Because of
these complex associations, we are unable to determine
whether there are thousands, tens of thousands, or
millions of reactions going on in the system at any one
time. Moreover, the system is dynamic, not static. <u>If
we are unable to even determine the number of reactions, how then can we with any reasonable degree of
accuracy quantify the potency and significance of all
of these reactions even with the best of modern computer programs</u>? Nevertheless, today in the U.S., for
example, our professional foresters in charge of
managing public lands -- as well as many scientists in
other fields -- are forced to spend much of their
working time in preparing detailed, unproductive,
largely meaningless Environmental Impact Statements to
justify the present or planned changes in management of
their operations so as to satisfy the Council for
Environmental Quality and the Environmental Protection
Agency regulations. I am certain everyone agrees with
me that general studies and some form of environmental
impact statement are needed periodically concerning
management of public or private lands and industries,
and especially when there are apparent abuses. Currently, however, the avalanche of Environmental Impact
Statements relating to agriculture, forestry, and
industrial management, it appears to me, all too often
have been carried to unreasonable extremes and now
serve primarily to justify the existence and expansion
of one of the fastest growing and already-bloated
bureaucracies in the nation's capital. This well-
intentioned legislation has given rise to an avalanche
of lawsuits that have mainly served to increase lawyers' incomes. Moreover, as a result, in recent years
much of the decision-making concerning sound agricultural and forest-management practices has been, or is
being, removed from the hands of farmers, ranchers, and

agricultural scientists, foresters, and wildlife biologists who are working in the fields and forests and are best qualified to make the management decisions. Instead, these decision making powers are being placed in the hands of bureaucrats located in the capital hundreds of thousands of miles from the scene of action. This trend, if continued without moderation, is certain to have a negative effect on agricultural, forest, and industrial production in the future.

I am, for the present at least, not deeply concerned about some of the drastic shortsighted actions that are being imposed under the lobbying pressures of the utopian environmentalists within the affluent food-surplus nations such as the U.S. and Canada. They can afford the luxury of overkill for the short term. However, I would caution political leaders, planners, and scientists of the developing nations not to precipitously adopt similar environmental legislation or decrees without carefully weighing the benefits versus the costs (social, health, and economic). We have seen the sad results of developing nations blindly following the lead of the U.S. and Sweden in one environmental issue.

In 1971, before the use of DDT was banned in the U.S., I pointed out in the McDougall Lecture the consequences that banning of DDT would almost certainly have on public health and food production in semitropical and tropical countries where malaria and other vector-borne diseases are endemic, but where they had largely been brought under control during the 1960s by the use of this insecticide. The polemics that resulted prior to and following the banning of DDT in the U.S., together with the pressures indirectly brought to bear on scientists and political leaders in the developing countries by extremist lobby groups in the U.S., provoked fear that mankind and wildlife were on the verge of being poisoned out of existence. As a consequence, most governments in the developing world discontinued their mosquito-control programs with subsequent disastrous consequences.

During the past five years, malarial epidemics have returned with a disastrous vengeance wherever mosquito control programs had been discontinued. A year ago the World Health Organization announced that in 1977 1.5 million people died from this disease. Many tens of millions more are infected. This is a costly price in human suffering and death that many developing countries are paying for following the decisions and suggestions of some utopian environmentalists and politicians in affluent, temperate zone countries.

Protection of Agricultural Crops and Forests from Diseases, Insects, and Other Pests

Crop protection is one of the areas of agriculture and forest production that is generally poorly understood by the general public. In recent years this subject has received much attention in the popular press and television, but unfortunately, the information often has been over-simplified, inaccurate, and sometimes biased. Very few people, and this includes many scientists, have much grasp of the great difficulties involved in protecting our crops from diseases and other pests. It appears to be a simple undertaking when viewed from the sidewalks and offices of our cities and towns by those who have little comprehension of the taxonomic diversity and genetic variability and instability in the disease pathogens and insects and other pest species that continue to threaten crop and forest production.

My personal experience, as well as that of many other scientists, indicates clearly that the best opportunity for keeping crop losses from diseases and pests at acceptable economic levels is through the utilization of a package of integrated control practices. This includes: (1) the choice of crops that are ecologically well-adapted to the environment under consideration, and consequently generally less vulnerable to severe economic losses from its major disease and insect enemies; (2) the use of disease resistant (and where possible insect resistant) varieties; (3) the use of improved cultural practices including crop rotations, proper timing of plow-down of the previous crop residue, and proper dates of planting so as to reduce the inoculum and break the biologic cycle of the disease organisms and pests; (4) the management of crops in such a way as to maximize biologic control of insects by parasites and predators; (5) the use of male sterile techniques for control of insects where feasible; (6) the experimentation to explore the feasibility of the use of pheromones or sex attractants in combination with trapping, or to guide spray programs; and (7) the proper use of insecticides, fungicides and herbicides.

No single individual practice alone can be relied upon to control the diseases and pests that are of economic importance on a given crop. The best combination will also vary from crop to crop, and place to place, depending upon the different spectrum of diseases and insects that must be coped with, and upon the environments in which one is working. For example, autogamous crops such as wheat, oats, barley, and rice

are self-pollinated and notoriously vulnerable to
airborne epidemic diseases, but allogamous crops such
as maize and virtually all of the forest tree species
are cross-pollinated and less vulnerable to epidemics
of indigenous diseases as long as they are maintained
as genetically variable populations.

The Debate Over the Use of So-called Crop Monocultures

In recent years, one frequently reads articles
that are critical of the use of the so-called "monoculture" in modern agricultural and forest production.
Generally, these points of view are being propounded by
theoreticians, generally laymen, but sometimes by
scientists, who have little comprehension of the complexities of agricultural production and even less
comprehension of the present and worsening human population pressures for food in different areas of the
world.

In their crusades against the use of so-called
"monoculture" in modern agriculture or forest, they
declare that this is not nature's way of growing plants
and, therefore, such methods are unacceptable. They
imply that the cultivation of one species or crop
(e.g., wheat, rice, soybean, maize or Douglas fir) over
large areas is always undesirable because it makes our
food production or forest production more vulnerable to
adverse weather or to epidemics of diseases, insects,
and other pests. They ignore the fact that pure even-
aged stands of Douglas fir and jack pine and vast
expanses of a single species of grasses, covering many
millions of hectares, had been established by nature
long before agricultural man arrived on the scene.

In the case of our food crops, it implies any crop
can be sown successfully and grown economically in
mixed populations in competition with different crop
species. This is not true. Virtually all of our
food-crop species are sun-loving or shade-intolerant
species that develop poorly under shading and in competition with other species. Their ecological requirements have apparently been like this since they were
first domesticated thousands of years ago. The lack of
competitive ability with other species undoubtedly
explains in a large part why the wild progenitors of
one of our most important crop plants, corn (maize),
has disappeared from nature, or why in other cases such
as wheat, rice, and soybeans, the present wild forms
are restricted to scattered isolated patches in special
econiches. To genetically modify the physiologic and
ecologic characteristics of our major crop plants so
that they would tolerate shading and competition with

other species, while retaining high yield potential, would require, at best, perhaps millions of years of genetic manipulation and selection, and in all probability, could never be accomplished.

In recent years, however, in labor-intensive, high-yield, small-plot agriculture in developing countries, a number of successful intercropping systems have been worked out. This involves the careful choice of crop species that are to be intercropped, the proper spacing between rows, and proper dates of planting of the different species. This is very different from the traditional mixed cropping, which is difficult if not impossible to use in mechanical, large scale agriculture. In any case, mixed cropping (as opposed to scientific intercropping) is characteristic of low-yield traditional agriculture which we are trying to replace.

The fact that some of our agricultural crops such as corn, wheat, soybeans, or rice are grown on vast contiguous expanses in certain geographic regions does not imply that this is being done without due regard for ecological requirements of the crop species involved or without due respect for its major pathogens and insect pests. The truth is that the major agricultural crops that are currently widely grown in any geographic and ecological zones of the advanced agricultural countries are located there because through experimentation and extensive cultivation over many decades, or even centuries, they have been found to be the crops best suited to that given environment, and year after year they have produced the highest product yield and highest farm income with a minimum of risk of losses caused by the vagaries of weather, diseases, insects and other pests.

This is well illustrated by corn in the U.S. One hundred years ago wheat was "king" in what is today the Corn Belt. Wheat, at that time, was grown on a larger area and outyielded corn in today's principal corn-producing states. It was the experience of repeated losses from diseases, especially stem rust, leaf rust and scab in the period from 1880 to 1920, that gradually shifted wheat production from the moist Corn Belt to the drier areas further west where these diseases are less of a problem. Concurrent improvements in corn production technology, developed through experimentation combined with the practical experience of farmers, gradually increased corn yields and income to a level where it outperformed wheat and, consequently, facilitated the shift to this crop. In reality these were changes made on the basis of ecological, pathological,

agronomic, and economic considerations. They were
based primarily on farmers' experience with important
assistance from agricultural scientists. As a result
corn, and within the past four decades, soybeans, have
become the major crops in the present day Corn Belt.
This does not mean that one or two crops are grown
exclusively in such a region, but rather that two or
three ecologically well-adapted major crops, as well as
several other crops of minor importance, are grown in
sequential rotations on most farms. Moreover, many
genetically different hybrids and varieties of each of
the crops are available to the farmers in these so-
called "monocultures," which are referred to by some
scientists as being genetically vulnerable.

It should be self-evident that, in a highly mechanized agriculture in a country such as the U.S. where
3.6 percent of the total population live on the land,
it is not feasible now to provide protection against
crop losses by growing small plots of many different
crops, as was done with animal traction and hand labor
eighty years ago, or as it is done effectively today in
some of the densely populated developing countries. To
attempt to do so would necessitate forced recruitment
of millions of urban youth to carry out the farm work
in such an undertaking. Although this procedure is
used successfully today under different systems of
governments in some countries, one can imagine the
public outcry and the social and political chaos that
would result if such a program were attempted in present-day, well-fed, affluent America.

The Use of High-Yielding Disease- and Insect-Resistant
Crop Varieties to Reduce Crop Losses

Over the past fifty years, agricultural scientists, employing the sciences of genetics and plant
breeding, have produced varieties and hybrids of all
the important crops. These improved varieties have
increased greatly the yields of the usable products --
grain, fiber, forage -- per unit of cultivated area,
especially when such varieties are used in combination
with improved cultural practices such as the proper use
of fertilizer, moisture, and control of weeds, which
then permits them to express their high genetic yield
potential.

The development and use of improved high-yielding
disease- and insect-resistant varieties have not only
assisted in increasing yield and production, but also
have helped to stabilize yield by reducing -- but not
eliminating -- crop losses from diseases and insects.
The successful use of disease-resistant varieties,

unfortunately, has given the general nonrural public and many scientists the false impression that their use is all that is needed to protect our crops against diseases, insects, and other pests. They have little insight into the complexities of incorporating genetic resistance to the major disease pathogens and insect pests into our crop varieties. And they have no comprehension whatsoever about the ephemeral nature of the resistance and how precarious the protection of our crops is despite our best scientific efforts.

The development and use of disease- and insect-resistant varieties of self-pollinated species. The complications and limitations of developing and effectively utilizing disease- and insect-resistant varieties of self-pollinated crop species will be illustrated with wheat.

There are more than thirty different diseases, and about half that number of insects, capable of causing economic losses in wheat. The problems are equally complicated for other important self-pollinated crop species. Among the major wheat diseases are those caused by a variety of fungi, bacteria, mycoplasmas, and viruses. The most serious of the diseases are caused by fungal pathogens. Among these are three different species of Puccinia, the causal organisms of the rust diseases. Each species reaches the optimum development under slightly different ecological conditions. Each can explode into devastating regional epidemics when conditions are favorable. Each of the three species of Puccinia, moreover, is made up of hundreds of pathogenic races, morphologically identical, but differing from one another in their ability to attack different varieties of wheat. All three of these pathogens are highly unstable. They mutate and/or hybridize frequently to give rise to many new pathogenic races, having new combinations of virulent genes. In addition, these organisms have high biotic reproductive potentials. One spore, upon infecting a susceptible plant, can produce a pustule containing 50,000 to 150,000 spores within a week to ten days. Since these spores can be carried by the wind in a viable form for long distances, and since wheat is grown from central Mexico to central Canada, over a range of 2,400 miles, an opportunity exists for the production of from ten to fifteen generations of rust, involving untold trillions of spores in a single continental-wide crop season, whenever varieties are susceptible and ecological conditions are favorable for the pathogen.

This high biotic reproductive potential, coupled with modest mutation rates as well as opportunities for hybridization on the alternate host in two of the species, is constantly giving rise to new physiologic or pathogenic races of the rust pathogen. Some of these new races will possess a combination of genes for pathogenicity that at some point in time will enable them to attack commercial varieties of wheat that were bred, multiplied, and distributed to farmers as being resistant to rust. This illustrates why the problem of breeding to develop and maintain rust resistance in commercial varieties is a never-ending, baffling, frustrating job in self-pollinating (autogamous) species like wheat.

When a new strain of rust pathogen or insect evolves which is capable of attacking a wheat variety formerly resistant to this same rust or insect, the pathogen or insect in effect has become resistant or tolerant to the unidentified biochemical substance that formerly imparted resistance to the variety. This is a phenomenon equivalent to that of an insect building up tolerance or becoming resistant to an insecticide which formerly provided protection to a crop on which it was applied. The change in virulence of the pathogen or in resistance of the insect is both the result of mutations or new hybrid forms that have occurred in the organisms, even though most antichemical activists either fail to recognize this fact or will not admit it.

The rust diseases can be kept under control only through effective research programs that monitor the changes in the pathogenic races of the rust pathogens taking place in the commercial wheat fields throughout a large geographic area, and opportunely multiplying and distributing new varieties which provide protection against both the old races and the potentially destructive new races. The need for change of varieties must be anticipated and timed to avoid rust epidemics and serious crop losses, before disaster strikes. In northern latitudes where severe winter breaks the rust cycle, varieties may retain their rust resistance for ten to fifteen years. In the semitropics where the rust pathogen can remain active on wild hosts or volunteer wheat plants during the noncommercial crop season, a variety will retain its resistance for a shorter period -- generally only three to eight years. The aforementioned difficulties described for developing and maintaining varietal rust resistance against any one species of _Puccinia_ is equally difficult for the other two species. The development of rust-resistant cultivars is further complicated by the need to simul-

taneously combine in the same variety resistance to two
Septoria species, both of which are extremely variable
to scab (Fusarium), to mildew (Erysiphe), and to a
number of important insects and viruses, while maintaining the resistance to the three shifty species of
rust fungi. In summary, the breeding program becomes a
very difficult, frustrating, never-ending task.

Nevertheless, the widespread use of disease-resistant varieties of wheat, even while recognizing its
limitations, has been of enormous benefit in reducing
losses and stabilizing production over the last fifty
years. It has been the standard method used in protecting self-pollinated cereal crops from diseases.

In recent years a number of scientists, most of
them cloistered in American universities, have raised a
cry of alarm about genetic vulnerability in our crop
plants. They have shocked the public and convinced
many laymen that the breeding methods being used to
improve our crop varieties have a very narrow genetic
base and therefore are becoming genetically more and
more vulnerable to epidemics of diseases and insects.
In my opinion this is a gross exaggeration and distor-
tion of the facts. If it were true, the criticism
theoretically might be most justifiably levelled at the
breeding programs in such self-pollinated crops as
wheat, rice, barley, and oats.

The breeding methods employed in these crops over
the past sixty years were by necessity geared to meet
the demand by farmers and industrialists for varieties
with uniformity for a number of different characters.
This results in all individual plants in such a variety
having essentially the same genetic makeup.

Nonetheless, I disagree with those who say that
the current methods of varietal breeding in self-pollinated crop plants are decreasing genetic variability
and increasing genetic vulnerability. I contend that
the present methods being employed on wheat, rice, and
barley are actually increasing genetic variability more
than has ever been done in the history of plant-breed-
ing programs. This has certainly become the case
during the past decade since the International Maize
and Wheat Improvement Center (CIMMYT) and the International Rice Research Institute (IRRI) were established.

In the case of wheat, the CIMMYT program in Mexico
alone makes a minimum of 60,000 different new crosses
each year. These new crosses are based on combining
the desirable characteristics and high yield potential
of the best current commercial varieties from all of

the important wheat-producing countries of the world
with a widely diversified group of progenitors chosen
on the basis of their performance against disease and
insect pests in cooperative international nurseries, at
more than 100 locations in more than sixty different
countries of the world. In addition, large numbers of
crosses, numbering in the tens of thousands, are also
made each year within the national wheat-breeding
programs of India, Pakistan, Iran, Iraq, Lebanon,
Egypt, Turkey, Tunisia, Algeria, Argentina, Chile,
Brazil, Guatemala, Mexico, and U.S. with which CIMMYT
collaborates. The segregating lines from this enormous
number of crosses are freely interchanged and fed into
national breeding programs in all parts of the world
for selection. This type of cooperative breeding
effort was pioneered by the predecessor of CIMMYT --
The Cooperative Mexican Agricultural Program which was
a joint dependency of the Mexican Ministry of Agriculture and the Rockefeller Foundation -- more than
twenty-five years ago.

Subsequently, the most promising lines determined
on the basis of performance in national yield nurseries
in the home country are identified and submitted to
CIMMYT where they are multiplied for inclusion in the
International Spring Wheat Yield Nursery, which is a
replicated plot test sent for evaluation to more than
one hundred locations in the world each year. The
results are summarized and published in a report which
is sent to wheat researchers throughout the world.
These tests conducted over a very broad geographic area
identify those lines which have a broad level of disease and insect resistance combined with broad adaptation, desirable agronomic characteristics, and high
grain yield.

The net effect of this entire cooperative effort
is such that in any one year the amount of genetic
variation now being incorporated into wheat and the
number of lines combining outstanding combinations of
disease resistance is greater than would have taken
place in several thousand years in the past.

However, the danger of losses from epidemic diseases in self-pollinated crop varieties persists. When
an excellent high-yielding variety with good rust
resistance, for example, is developed and released to
farmers, and it outyields all other varieties, it will
soon spread and cover large areas to the exclusion of
most other varieties. This happens because of the
grower's own choice. When a new race of rust appears,
either through mutation or hybridization, as is certain
to happen sooner or later, the entire area sown to this

variety becomes susceptible, and if ecological conditions conducive to development of an epidemic occur
before the variety can be replaced, disaster can result
since all plants in the variety have essentially the
same genetic makeup. Currently, the best protection
against epidemics is provided through constant monitoring of the rust population in commercial fields.
This will identify potentially dangerous new races as
early as possible and must be combined with a dynamic
seed multiplication and distribution program of new
varieties from the breeding program whenever danger is
imminent. It is precisely because of this inherent
vulnerability of pure-line breeding that CIMMYT has
pioneered the research and development of multiline
varieties aimed at diversifying the resistance within a
variety.

During the past few years, CIMMYT has been assisting the Indian wheat-breeding program in developing a
multiline variety in a Kalyansona (8156) background.
Kalyansona and its closely related sister varieties,
known under many different names, were formerly cultivated on an aggregate twenty-five million hectares in
many parts of the world. It is still grown on large
areas in many countries because of its high grain
yield, broad adaptation and farmers' preference. The
object in developing multiline varieties is to provide
additional protection against such airborne epidemic
disease as the rusts.

Basically, the approach is to develop a large
number of lines through back-crosses and double-crosses
that are phenotypically similar (look alike) for plant
type, head type, grain, maturity, height, and industrial quality, but which differ for resistance to the
races of the rust pathogens.

The multiline variety that a farmer will grow at
any time, employing such an approach, will be a mechanical mixture of a number of lines (ten to twenty) that
carry different genes for resistance to the races of
the rust prevalent in a large geographic area where the
variety is to be released.

A large number of phenotypically similar lines,
each with different genes for rust resistance, can be
developed and maintained in a viable condition in a
seed bank. These lines can be used to reconstitute the
multiline variety as needed through removing susceptible lines from the mixture and replacing them with
ones which are resistant. The rust population in
commercial fields throughout the region must be moni-
tored each year. Whenever new races appear with the

genetic capability of attacking part of the population of the variety, the multiline variety can be reconstituted. However, in all probability whenever a change in races occurs, only one or a few lines in the multiline variety will become susceptible. The majority of the plants in the population will still be resistant and, consequently, there will be little danger of a destructive epidemic developing. The rust increases very slowly in multiline populations where the majority of the population is resistant; this provides protection for two or three years while the newly reconstituted multiline variety is being multiplied and distributed. This approach to varietal improvement can provide the maximum protection possible in self-pollinated crop plants against airborne diseases. In effect, it is an attempt to mimic the conditions naturally occurring in the cross-pollinated crops.

Recently I have been pleased to learn that three multiline wheat varieties have been released for commercial use. The successful introduction and use of such multiline varieties will usher in a new era in breeding of self-pollinated crop plants.

Breeding disease- and insect-resistant cross-pollinated species. Cross-pollinated species of plants such as corn and most forest tree species are less vulnerable to epidemics of indigenous diseases than are varieties of self-pollinated species, other variables being equal. Moreover, the opportunity exists for progressively increasing the level of resistance of a superior level or different kind of resistance. Subsequently, superior families are identified and crossed to other superior families and later sibbed and reselected for several generations, which gives rise to a reconstituted improved population. A superior population can be improved for resistance to a number of different diseases or insects and/or for an agronomic characteristic or grain type by identifying a "donor" population with useful genes and intercrossing the two and employing the aforementioned procedure. There is little probability of suffering a reduction in resistance to other diseases or insects or retrogression in agronomic type in the original population so long as the genetic base is kept broad (many families used) and so long as the families and improved populations are tested over an adequate range of sites and environments.

Utilization of Cultural Practices to Minimize Losses from Diseases, Insects, and Weeds

Crop rotations are often used effectively in annual field and vegetable crops to reduce losses from root rot organisms, soil insects, and nematodes.

Control of date of planting and prompt postharvest plow-down of crop refuse is of great value in many field crops. This practice breaks the biologic cycle of many insects and reduces the amount of carry-over inoculum of many pathogenic fungi.

The control of dates of cotton planting, imposed in many areas, has been decisive in reducing insect damage to an economically acceptable level when combined with other elements of an integrated control program. To be effective, the control of planting date must be applied over a large agricultural area to reduce the effect of migration of insects or inoculum. In the Valle del Yaqui in Sonora, Mexico, thirty years ago, twelve to fifteen applications of insecticides were required to produce a poor crop of cotton. During the past twelve years with strict control over dates of planting and destruction of stalks, with good monitoring of the population dynamics of predator and parasite species and of the buildups of destructive insects, and with the timely and wise use of insecticides, the number of applications of insecticides required to produce a good crop of cotton has been reduced to an average of 2.5. By contrast, in the Tampico area of the Gulf of Mexico where no control of date of planting and plow-down of wither or corn was practiced, it became economically impossible to grow cotton because of the buildup and damage caused by the bollworms Heliothis virescens and Heliothis zea. The factor that made the production of cotton impossible in that area was not the buildup of resistance of the insects to chlorinated hydrocarbon insecticides, as was erroneously reported in the press by the antichemical activists, but rather the insistence of small farmers on planting small patches of corn, which is an excellent host of Heliothis, throughout every month of the year. That, coupled with the failure to establish rigid appropriate dates of planting and plow-down for cotton, spelled the doom for this important crop.

In some irrigated areas of the tropics, with the introduction of the use of chemical fertilizer, insecticides, and high-yielding rice varieties, there is an increasing tendency to plant two or three crops of rice on the same land in a year. In many of these areas there are rice paddies in all stages of development throughout the year. This will almost certainly lead to increasing crop losses from leafhoppers, aphids, and viruses in the future. There have already been several cases of breakdown of new resistant varieties of rice to both viruses and to leafhoppers. The problem is likely to worsen unless control on date of planting and destruction of crop residue is imposed.

Biological Control of Pests

There are today a considerable number of people, including some scientists, who look forward to the day when chemical pesticides will not be needed and harmful insects and pests will be controlled biologically. I, too, dream such dreams, but in my real world know this will never happen. Nevertheless, it is obvious that predator, parasites, and pathogens in the form of certain insects, bacteria, fungi, viruses, nematodes, reptiles, amphibians, birds, and animals sometimes play an inconspicuous but, nonetheless, important biological role in assisting to maintain many harmful plant and animal pests at an acceptable level of control. This, for example, is especially true of indigenous host-insect-predator and/or parasite biological systems, wherein all biological entities of the system have been associated with one another for long periods of time and have reached a certain uneasy evolutionary equilibrium. The wheat-aphid-predator/parasite (e.g., lady beetle, lace-winged fly, syrphid flies) system is a good example of such a complex association. In most years the aphids are held in check by the predators and parasites so that they fail to reach a population level where they provoke economic damage to the wheat crop. In certain years, however, slight shifts in climatic factors that are poorly understood adversely affect the predators and parasites. When this occurs, aphid populations explode to high levels early in the crop cycle, and this results in severe reductions in wheat yields, unless the population buildup is controlled by insecticides. Attempts have been made many times to rear large numbers of lady beetles and other predators and distribute them in areas where aphids and predators are out of balance with the hopes of bringing infestations under control before economic losses occur. These attempts to improve biologic control by rearing and releasing predators have mostly been unsuccessful, and in most cases chemical control has been necessary to avoid serious losses.

In the wheat-aphid-predator/parasite biologic systems, it is essential to monitor on a weekly basis the balance between the aphid-predator/parasite population, so as to determine if and when the biologic

balance has been upset to a level that justifies the
timely and proper application of insecticide. We must
recognize and respect the tremendous reproductive
potential of aphids. It has been estimated that a
female cabbage aphid has an average of forty-one young
per generation and sixteen generations in one crop-
growing season. Therefore, if all individuals survived, the progeny of a single female aphid would be
$1/56 \times 10^{24}$ by the end of the season.

Sometimes exotic species of animals, birds, weeds,
insects, or fungi have been introduced into another
continent (or distant ecosystem) where they encounter a
congenial host and no natural predators or parasites.
When this happens, there is usually an explosive increase in the population of the introduced species to a
level where it becomes a serious economic pest. When
such an imbalance occurs, the introduction of one or
more natural parasites or predators from the center of
origin of the imported pest species is often of at
least some temporary value in bringing the pest under
control.

The case of the introduction of the European
rabbit (Oryctolagus cuniculus L) into Australia illustrates the biologic imbalance and economic losses
resulting from an introduction species which became a
major pest. It also clearly indicates the biologic
complexities involved in attempting to bring an introduced pest under control.

The introduced rabbit developed into the most
serious agricultural pest in Australia. Enormous
populations of rabbits competed for grass with sheep
and greatly reduced mutton and wool production. Indirectly they also contributed to overgrazing and deterioration of the ecosystem. A continuous battle was
carried on for decades by government biologists and
ranchers to reduce the population of rabbits to an
acceptable economic level through the use of trapping,
hunting, poisoning, and destruction of burrows, but
this was never very effective. Serious economic losses
continued. In 1950 the mysoma virus was transferred
from the American rabbit to Australia in an attempt to
control the European rabbit. The results during the
first two or three years were spectacular. The virus
spread more than one hundred miles from the point of
release within the first year. It decimated the rabbit
population over large areas and many people, including
some scientists, felt a safe permanent control was at
hand. In some local areas the virus caused a 99.8
percent mortality and an 80 percent reduction throughout vast areas. However, this spectacular biological

control was ephemeral. Within two to four years the
rabbit mortality began to decline markedly. Less
virulent mutant strains of the virus appeared which
resulted in:

1. Lower mortality in the rabbit population
   which resulted from stimulating antibody
   production and the recovery of a part of the
   rabbit population.

2. The less virulent mutant strains survived
   better in the wild rabbit population than the
   virulent strains and quickly replaced them.

3. Surviving rabbits, after several generations
   of interbreeding, developed a considerably
   higher level of genetic resistance to the
   viruses than was present in the original wild
   rabbit population. Currently, after more
   than twenty-five years of skillful dedicated
   research by an excellent team of scientists
   and extension workers, Australia has again
   reverted to an integrated program involving
   fencing, trapping, hunting, poisoning, and
   burrow destruction combined with the use of
   mysoma virus in a frustrating, never-ending
   struggle to maintain the damage from rabbit
   populations at acceptable economic levels.
   The mysomatosis experience clearly indicates
   the limitation of relying exclusively on
   biologic-control methods.

Insect Control Methods Using Sexual Trickery and Deceit

The use of male steriles. Within the past fifteen
years, a fascinating new approach to insect control was
pioneered and developed by Drs. E. F. Knipling, R. C.
Bushland, and colleagues who checked and later eradicated the screwworm, Cochliomya hominivorax Coquerel,
from vast areas of the southern United States and
northern Mexico.

The screwworm is an economically important parasitic fly that develops its larval stage in living
tissues around wounds or unhealed navels of newborn
animals.

The insect has a wide geographic distribution
throughout virtually all semitropical North and South
America. It cannot survive severe winters but migrates
northward in the spring from overwintering areas. It
is an insect that has an extremely wide host range
attacking all domestic animals including cattle,

horses, mules, burros, hogs, sheep, goats, and dogs, as
well as man. It takes a heavy toll among many wildlife
species, especially deer. As late as the early 1960s
before the eradication campaign was initiated, in
certain years it was reputed to kill up to 60 percent
of the fawn crop in Texas.

In 1954 an eradication campaign employing the
male-sterile technique was launched in the Caribbean
island of Curacao to establish the feasibility of this
approach. Within the first year the screwworm was
eradicated from the island. In 1957 a similar campaign
was launched in Florida and neighboring southeastern
states, and by November 1958 the insect had been eradicated throughout that vast region. Similar campaigns
were launched in Texas and neighboring states in 1963,
and by early 1964 these areas were free of screwworms.
By 1966 screwworms had been eradicated from the U.S.
and from a 300-mile strip attempting to eradicate the
screwworm from the entire area of Mexico north of the
Isthmus of Tehuantepec.

The successful eradication campaign was based upon
developing satisfactory techniques for: (1) rearing
massive numbers of screwworm insects on artificial
media originally made from ground fresh meat, whole
blood, and water; (2) subjecting the pupae to a proper
dosage of gamma rays from Cobalt 60 to induce damage to
the cells in the developing reproductive organs of the
pupae; and (3) dropping irradiated pupae from airplanes
in sufficient numbers to "saturate" an area for several
generations with male sterile insects, and thus, to
assure that eventually all the "wild" females would
mate with one of the gamma-ray-induced, mutant, sterile
males. The result is that, although the irradiated
male mates satisfactorily with the "wild" female and
the sperm fertilizes the egg satisfactorily, because of
chromosome and gene damage transmitted through the
sperm from the irradiated male, the malformed embryo
dies within the egg, thereby "breeding" the insect into
extinction throughout the area.

The spectacular success that has been achieved in
eradicating the screwworm from large areas has resulted
from rearing, irradiating, and releasing approximately
sixty billion flies since the program was initiated.
The eradication of screwworm has greatly reduced animal
losses and increased the income from livestock over
vast areas. It has also resulted in a large increase
in the deer population of Texas.

This fascinating new approach to insect-control
methods ushers in a new era. It has an advantage over

the use of chemicals in that it has no impact on other species or on the environment. This method of insect control, however, apparently <u>can</u> <u>be</u> <u>used</u> <u>successfully</u> <u>only</u> <u>on</u> <u>those</u> <u>species</u> <u>which</u>: (1) have low populations in nature; (2) can be reared cheaply and in big numbers in laboratories; and (3) the female mates only once.

Currently, the use of the male-sterility technique is being explored for use on a number of other species that possess the aforementioned characteristics. These include: fruit fly, melon fly, pink bollworm of cotton, boll weevil of cotton, oriental fruit fly, corn ear-worm, tobacco hornworm, tobacco budworm, fall armyworm, and tsetse fly.

<u>Insect hormones and pheromones</u>. Research in re-cent years has begun to explore the fascinating new field of insect hormones and pheromones. A rapidly increasing group of hormones that control the meta-bolism growth, metamorphosis, and reproduction of insects is being identified and studied with the aim of finding ways to distort hormonal action and, thereby, break the life cycle of insects. Among the hormones under study are: those that cause sterility by pre-venting ovary development; juvenile hormones that prevent normal metamorphosis; neurohormones that con-trol hunger and digestion; and hormones that disrupt diapause and consequently throw the insect's biologic development cycle out of harmony with the climate.

Although none of this very diverse group of na-turally occuring chemicals at present has been used commercially in insect control programs, several show considerable promise. Among the most promising are the female attractants, or "female perfumes," which in some species are so powerful that, when emitted by the female at the time of mating, can lure the male from great distances. Research also is being done on food attractants -- the pheromones which serve as physiolo-gic radar-like systems which guide insects to their preferred food sources. Hormone attractants offer promise for use in future insect-control programs by provoking insects to mate with artificially sterilized insects, to eat harmful substances, or to concentrate them in large numbers where they can be trapped, killed by insecticides, or destroyed by hormones. Although this diverse group of interesting biologic compounds offer promise for use in insect-control programs in the future, it would appear to me that there is no assur-ance that many of them would not fall prey to being classified as carcinogenic under the unrealistic Delaney Clause of the Food and Drug Act of 1958. It also must be remembered that some of these hormonal compounds are among the most potent chemicals known, considering the physiologic reactions provoked by infinitesimally tiny quantities (e.g., one billionth of a gram). Is there any assurance that just because they are natural biologic compounds they will not be found to be highly toxic to some other species, including men, when they are used in much larger quantities in insect control programs? Let us not be led into the same trap into which the organic food faddists have fallen, namely that natural or biologic is always superior and safer to synthetic.

Chemical Control of Pests

Today there are some in the antichemical and environmentalist movement who apparently believe that modern science and technology is the real cause of the insect, disease, and weed problems that threaten our crops and forests. They seem to feel that before the advent of agricultural chemicals -- and especially pesticides -- nature maintained a biologic balance that prevented serious crop losses. Of course, we know this is not necessarily so because the earliest records, beginning with the Old Testament, tell of famines attributed to disastrous crop losses caused by locust hordes and to epidemics of mildew, blasting, and rusts. We know there were widespread famines in Europe during the mid-1880s caused by the loss of the potato crop from Phytopthora late blight. When the pioneers moved into the virgin lands of the Great Plains of the U.S., about 125 years ago, their crops were sometimes seriously damaged or destroyed by devastating grasshopper plagues and their wheat and oats were often hard hit by rust epidemics. One devastating grasshopper plague in Nebraska in the 1870s left the fields as barren as if they had been burned over. The invading cloud of grasshoppers averaged a half mile in depth (height) and covered an area one hundred miles wide and 300 miles long. When the Mormons settled in Utah, their first crops were seriously damaged by a cricket which is now known as the Mormon cricket.

We can see that crop losses from pests and diseases were an old story long before the invention of modern technology, including the use of pesticides. On numerous occasions in the distant past, natural or biologic control failed to hold pests in check, just as it cannot be relied upon alone today, to the exclusion of the use of improved disease-resistant varieties, crop rotations, cultural practices, and chemicals. Mother Nature has had a way of getting out of balance quite often down through the ages. Many species disappeared because they did not possess the genetic variability which would permit them to adapt rapidly enough to changes in the environment or to new diseases and pests.

It seems to me that the environment and utopian neoecologists who are advocating going back to the balance-of-nature movement as a remedy for correcting the world's ecological and biological imbalances are inconsistent in the application of this ideology. When they become ill, most of them apparently rush to a medical doctor, just as I do. But I often wonder what happens to their ideology when they are informed that they have a bacterial infection or an internal parasite. This certainly must present them with both a physchological and moral-ethical dilemma. Is it morally right for them to intervene with medicine and kill the poor little, happy, flourishing bacterial colony that is peacefully inhabiting their intestinal tract while producing that not-so-enjoyable phenomenon popularly known as the "tourist trots?" Or, is it morally and ecologically right to take medicine that kills or expells that long tapeworm or the innocent looking roundworm that has invaded their body in search of a cozy, pleasant habitat?

If our utopian neocologists really believe in a world order based on the balance of nature, as they proclaim, then it would seem to me they should carry this philosophy to its logical conclusion on all fronts. That being the case, when they are sick they should not go to the doctor. Neither should they vaccinate their children or grandchildren against any of the contagious diseases that have been brought under control by medicine during the last forty to fifty years. It logically follows that when their children or grandchildren are sick, they should not be taken to the doctor. And if they are really convinced of this philosophy, it seems they should convince all friends, neighbors, colleagues, and people throughout the world to take similar action.

If this were done and they were successful with their mission, we would see that there would be a new shift in the balance of nature. It would indeed help solve the world population problem. Such widespread world action, in all probability, would reduce the world's human population to one-third or one-fourth of what it is today within four or five decades, perhaps faster. This would solve both our food production and energy problems for at least a couple of hundred years. It almost certainly would reduce environmental pollution as well. It would greatly reduce human population pressure and permit the survival of thousands of species of birds, animals, fishes, amphibians, snakes, and plants that are currently being placed on the endangered-species list in the U.S. It even might permit the survival of the endangered small-pox virus, which I futilely have tried to place on the prestigious list. And imagine how the buzzards and other scavengers would flourish once again. But what chaos! It is very certain that under this stress fanatical environmentalists, to their consternation, would soon discover that not only is the world's ecosystem fragile, as they constantly argue, but also that the world's social and political systems are even more fragile and would erupt in flames. Is this the kind of world we want to live in? Not me. Nor do I believe there are many who want to revert to placing human destiny in the hands of the balance of nature.

I want it to be perfectly clear that I am opposed to the reckless and careless use of chemicals either industrial or agricultural. However, we must also recognize that we need agricultural chemicals to produce and protect our food, fiber, and timber. Using chemicals is like using medicine. If you are ill, you go to a doctor so that he can diagnose what is wrong with you. If it is a bacterial infection, he will probably prescribe the proper antibiotic or sulfa drug. If you use the medicine as directed, it may cure you. If you use it improperly, it may kill you.

We oversimplify. We want utopian environments with zero risks. And it should be obvious to everyone this will never be achieved.

There are those who believe we are on the verge of being poisoned out of existence. They visualize poisons and carcinogens in the air, water, in every bit of food we eat, and in everything we touch. Despite such criticisms, we live a longer and more pleasant life than our parents or grandparents. The life expectancy of women in the U.S. is now eighty-one years and for men, seventy-three. Life expectancy has increased spectacularly in nearly all developing countries since the end of World War II.

Since most infectious diseases that formerly took the lives of many in childhood, youth, and middle age have been brought under control within the last forty years, many people die at a later age from other diseases, such as cancer and heart troubles. Many who a generation ago would have died at an early age, because of susceptibility to one or another infectious disease, now survive as the result of improved vaccines, antibiotics, or other medicines only to die at a later

date, often from the still bewildering and poorly
understood group of cancer diseases. This gives me the
layman the feeling, in large part caused by sensa-
tionalism in the press designed primarily to stimulate
the sale of newspapers, that there is a startling
increase in the incidence of cancer. The public,
consequently, over-reacts in near panic. Yet, in the
case of the most clearcut proven carcinogen of all --
that of tobacco and its capacity to substantially
increase the incidence of lung cancer -- many millions
of Americans nonchalantly continue to smoke and enjoy
their two packs per day. The Surgeon General has
reported that cigarette consumption in the U.S. has
increased from 511 billion per year in 1964, to 600
billion in 1976. On another front, privileged youth
search for even bigger "kicks," and conceivably better
carcinogens for the future, via the use of marijuana,
heroin, hallucinogenic plants and fungi, and a host of
pep pills. The American Medical Association has reported that there are now 192 different herbs commercially available and used for smoking in the U.S.

The present hysteria about cancer in the environment and food stems largely from the fact we seem to
forget the self-evident truth that all that is born
must sooner or later die. It also raises the question
whether it is necessarily wise to perpetuate life
beyond the time when life remains enjoyable. Each
generation, however, continues to look for the "foun-
tain of youth," just as Ponce de Leon did and many
before him. We will never find it, but imagine the
disaster that would result on the human population
front, to the environment, and to wildlife, were
eternal youth attainable.

Reflecting on the present condition of men and
women in the affluent societies of the world today, one
can better appreciate Will Durant's keen insight into
the curious behavior of man which he expressed many
years ago: "Man's capacity for fretting is endless,
and no matter how many difficulties we surmount, how
many ideals we realize, there is a stealthy pleasure in
rejecting mankind or the universe as unworthy for our
approval."

It occurs to me that this is one of the worsening
dilemmas of the affluent societies where all of the
basic needs and conveniencies are available and where
there is much leisure time which, when improperly
utilized, often becomes boredom. In previous generations this was a problem largely restricted to sons and
daughters of wealthy families. Today, however, because
of the more general affluence of society in many of the

developed nations, this malaise of boredom is widespread. Under these conditions certain restive, privileged individuals must find a cause and develop a
crusade to support the cause in order to be happy. If
the cause chosen is a soundly-based, just one for the
majority of the population, it may advance and improve
the society. All too often, however, the causes cham-
pioned are ill-chosen, irrelevant trivia not based on
trustworthy data and facts and primarily of concern to
only a small sector of the society. When this happens,
these crusades are at best mischievous, but more often
add confusion rather than enlightenment to basic
issues. The American people and western Europeans seem
to have been subjected to an unusually large number of
poorly chosen and ill-founded crusades during the past
decades. Is this the price that societies pay when
life becomes too easy?

A FUNDAMENTAL RESEARCH PROPOSAL DESIGNED
TO PRODUCE A RUMINANT STRAIN OF HOMO SAPIENS

At the Third International Wheat Genetics Symposium a decade ago, in summarizing the complexities of
the food-population problem, with tongue in cheek, I
made the following proposal:

In closing, I would like to leave this final
proposal for someone who wishes to delve into
fundamental research: I suggest that you direct
your efforts to save our civilization by establishing an irradiation genetics and breeding
program designed to develop a new race of Homo
sapiens. This new improved race of man should
have the enzyme cellulase in his gut which will
permit him to eat, digest, and grow fat on the
mountains of paper and red tape which are being
produced in ever-increasing quantities by the
world's planners, bureaucrats and the press.
Moreover, let this mutant gene for cellulase
production be tightly linked to the following
three additional genes that are essential for the
improvement of the human species: (1) a gene for
compassion for his fellowman, which seems seldom
to have existed in the wild (normal) type; (2) an
additional gene which will provide massive doses
of common sense, since the original gene for this
character has been badly eroded and has lost its
effectiveness in the wild type; and (3) a gene
that will assure a low level of human reproduction. This seems to be a gene that has never
existed in the wild type, for history shows that
man has repeatedly bred himself into misery and
famine.

Should it be impossible to identify a potent gene for a low-biotic, reproductive potential, then we cereal breeders had better discontinue our efforts to incorporate improved nutritional quality, i.e., better amino acid balance and high-protein levels, into cereals and turn our attention instead to locating and incorporating into wheat, rice, maize, sorghums, and millets a potent gene for estrogen production which would produce the same effect as the 'pill' in a more acceptable form.

Today the general world situation appears to be considerably worse than it was ten years ago. There are about 700 million more people today than a decade ago. The absolute number of poverty-stricken, hungry, miserable people has undoubtedly increased on a worldwide basis, as has the number of unemployed and illiterate. The situation has deteriorated from an energy viewpoint. On the environmental front we repeatedly hear we are about to be poisoned out of existence by an ever-increasing number of pollutants.

The only sectors of society that seem to be prospering and growing vigorously are the national and international brotherhoods of bureaucrats. This group continues to multiply itself both through fission and accretion. It spews forth in ever greater numbers an avalanche of edicts, decrees, regulations, and orders to guide the masses to the promised land. But, as always, they only succeed in further entangling the unimportant minority who, despite the worsening social parasitism, are still idealistically embued with trying to produce the basic necessities, including food, for more and more people.

During the last decade we have seen wonderful advancement made by the world's surgeons in successfully transplanting many vital organs into Homo sapiens. I now seriously urge that a serious effort be made to transplant the rumen from cows into both man and woman. If this can be done, mankind may still reach the promised land. Once the transplants are made, we geneticists must then get busy and try to resurrect and put into effective operation the pre-Mendelian ideas of Darwin and Lamarch and the post-Mendelian theory of Lysenkoi on the transmission (inheritance) of acquired characters. This, if successful, might even at this late date save mankind from both starvation and suffocation by bureaucratic mountains of paper and red tape. Hopefully, while the new race of Homo sapiens chews contentedly on its cud of paper, those of us working on the food production

front -- geneticists, agronomists, plant pathologists,
entomologists and soil scientists -- might be permitted
an occasional day off so we could go fishing. But
should this noble experiment be successful, imagine the
plight of the foresters who would then become responsible for producing more pulp wood to amply supply the
new food chain: from pine tree to pulpwood to paper to
bureaucrat to printed decrees, regulations, and orders -- the food for the new ruminant strain of Homo
Sapiens.

THE HUMAN POPULATION MONSTER

Although I have been primarily concerned with
agriculture and food production, I have, by necessity,
developed interest in the broad fields of land use --
or misuse -- and in demography.

If one is involved in food production, it naturally follows that one must be concerned about the
land base upon which we depend for food production and
the number of people that land base must feed. Ninety-
eight percent of the tonnage of food production, worldwide in 1975, was produced on the land. Anyone engaged
in attempting to increase world food production soon
comes to realize that human misery resulting from world
food shortages and world population growth are part of
the same problem. In effect, they are two different
sides of the same coin. Unless these two interrelated
problems and the energy problem are brought into better
balance within the next several decades, the world will
become increasingly more chaotic. The social, economic, and political pressures and strife are building
at different rates in different countries of the world,
depending upon the human population density and growth
rate, and upon the natural resource base that sustains
the different economies. Poverty in many of the developing nations will become unbearable. There is also
every likelihood that standards of living in many of
the affluent nations will stagnate, or even in some
cases, retrogress. The terrifying human population
pressures will adversely affect the quality of life, if
not the actual survival of the bald eagle, stork,
crocodile, wildebeest, wolf, moose, caribou, lion,
tiger, elephant, whale, monkey, ape, and many other
species. In fact, world civilization will be in
jeopardy.

Unfortunately, even in privileged, affluent,
well-educated nations such as the U.S., we have recently concerned ourselves unduly with symptoms of the
complex malaise that threatens civilization rather than

with the basic underlying causes. In recent years we seem to have become determined to attack all of these ugly symptoms by passing new legislation or filing lawsuits against companies, individuals, or various government agencies for polluting the environment. Most of these lawsuits just fatten the incomes of lawyers without solving the basic problems. Most of us are either afraid or unwilling to recognize, confront, and effectively fight the underlying cause of most of this malaise -- The Human Population Monster. The longer we wait before attacking the primary cause of this problem -- worldwide -- with a serious, intelligent, unemotional, effective, and humane approach, the fewer of our present species of fauna and flora will survive.

There is still heated debate about the date man or "near man" appeared on the planet Earth. Evidence indicates he has been roaming the earth for at least three million years.

About 12,000 years ago, humans discovered agriculture and learned how to domesticate animals. World population at that time is estimated to have been approximately fifteen million. With a stable food supply, the population growth rate accelerated. It doubled four times to arrive at a total of 250 million by the time of Christ. Since the time of Christ, the first doubling -- to 500 million -- occurred in 1,650 years. The second doubling required only 200 years to arrive at a population of one billion in 1850. That was about the time of the discovery of the nature and cause of infectious diseases and the dawn of modern medicine which soon began to reduce the death rate. The third doubling of human population since the time of Christ, to two billion, occurred by 1930 -- only eighty years after the second doubling. Then sulfa drugs, antibiotics, and improved vaccines were discovered. They further reduced deaths spectacularly.

World population doubled again in 1975 to four billion people. That took only forty-five years and represents an increase of 256-fold -- or eight doublings since the discovery of agriculture.

It is obvious that the population/food land ratio as well as the competition between species is worsening dramatically as the numbers of humans increase so frighteningly. And the interval between doublings of human population continues to shorten. At the current world rate of population growth, population will double again in the next forty years, reaching eight billion souls by 2015.

This means that in the next forty years world food and fiber production must be increased more than it was increased in the 12,000 year period from the discovery of agriculture up to 1975. It also clearly indicates the urgency of dealing effectively and humanely with The Human Population Monster.

This is a tremendous undertaking and of vital importance to the future of civilization. Failure will plunge the world into economic, social, and political chaos. Can the production of food and fiber be doubled in the next forty years? I believe it can, providing world governments give high enough priority and continuing support to agriculture, aquaculture (fish, crustaceans and mollusks), and forestry. It cannot be achieved with the miserly and discontinuous support that has been given to agriculture and forestry during the past fifty years.

We must expand our scientific knowledge and improve and apply better technology if we are to make our finite land and water resources more productive. This must be done promptly and in an orderly way if we are to meet growing needs without, at the same time, unnecessarily degrading the environment and crowding many species into extinction. Producing more food and fiber and protecting the environment can, at best, be only a holding operation while the population monster is being tamed. Moreover, we must recognize that in the transition period, unless we succeed in increasing the production of basic necessities to meet growing human needs, the world will become more and more chaotic. Civilization may collapse.

"Human rights" is a utopian issue and a noble goal. But it can never be achieved as long as hundreds of millions of poverty-stricken people in the world lack the necessities of life. The "right to dissent" doesn't mean much to a person with an empty stomach, a shirtless back, a roofless dwelling, the frustrations and fear of unemployment and poverty, the lack of education and opportunity, and the pain, misery, and loneliness of sickness without medical care. My work has brought me into close contact with such people, and I have come to believe that all who are born into the world have the moral right to the basic ingredients for a decent humane life. How many should be born and how fast they should come on stage is another matter. This latter question requires the best thinking and efforts of all of us, in my opinion, if we are to survive as a world in which our children and their children will want to live and, more important, be able to live in.

Those of us who work on the food production front,
I believe, have the moral obligation to warn the political, religious, and educational leaders of the world
of the magnitude and seriousness of the interrelated
arable land/food/population problem which looms ahead.
If we fail to do so in a forthright unemotional manner,
we will be negligent in our duty and, inadvertently
through our irresponsibility, contribute to the pending
chaos.

5

THE CONTRIBUTION OF LIVESTOCK TO THE WORLD PROTEIN SUPPLY

John A. Pino
Andres Martinez

SUMMARY

The analysis presented in this paper on "The Contribution of Livestock to the World Protein Supply" unfolded a set of conclusions relative to the world-protein situation and the contribution of animals to the protein supply. These conclusions are summarized below:

-- Sufficient protein is now produced in the world to satisfy the requirements not only of the present population but also those of the projected population for the year 2000. In spite of this surplus, protein deficiency does occur among some population groups -- particularly among the poor in developing regions. Redistribution of food supplies from rich countries to poor countries is generally not an acceptable long-term course of action since it will only perpetuate food-import dependency and drain scarce foreign exchange in the developing countries. Rather, the developed countries should provide long-term assistance in the development of the food-production process to those areas where food deficits exist.

-- Food from animal products is of extreme importance to the protein supply; the meat, milk, and eggs produced in 1977 contained sufficient protein and essential amino acids to meet 80 percent and 96 percent, respectively, of the world's population requirements for these nutrients. Significant improvements in overall animal production and productivity can yet be realized, particularly in developing areas. Inadequate feeding, sanitation, and breeding and selection programs, and lack of established animal management and marketing systems are constraints that need the attention and necessary corrective action from leaders in developing countries.

-- Animals do not necessarily compete for food with man. Only through ruminants (cattle, buffaloes, sheep, goats, and others) can the vast grassland areas, agroindustrial by-products and crop residues be converted from nonfood materials to high-quality, protein-rich food. In the developing world, pigs and fowl kept largely by small farmers depend on waste food and feed for most of their nutrient needs. The extent to which grain, or for that matter any other feed, is fed to livestock depends on prevailing economic conditions. In the developed regions, such as North America, this practice is feasible because of a large supply of feed grains (corn, barley, and sorghum) and a strong demand for animal products. In developing countries very little grain is fed due to a short supply and widespread use for human consumption, which makes grain relatively expensive as an animal-feed source.

-- Animal products will continue to be desirable food items in the foreseeable future. In virtually every analysis, the proportional increase in demand for livestock products, as income increases, is much greater than for crop products. The compound annual growth rates in consumption for 1970 to 1985 in the developing countries, for example, are 3.9 percent for meat and 3.2 percent for grains.

-- The per unit cost of animal protein is surprisingly low. It costs the American consumer less to obtain his/her daily protein requirement from poultry meat, eggs, cheese, or milk than from bread, rice, or potatoes. However, properly balanced diets that include mixtures of both plant and animal products are likely to be more economical and nutritious in the long run.

-- The United States is confronted with a series of important issues related to food supply which will affect not only our own welfare but that of other nations as well. Among these are: (a) the need to establish a forward-looking policy on the use of land and other resources for food production; (b) the need to sustain our agricultural research; (c) the need to evaluate the disposition of farm products; (d) the necessity of adequately addressing international food and agriculture issues; and (e) the urgency to crystallize United States food and agriculture policies.

INTRODUCTION

The amount of literature addressing the world food problem is monumental, but even more impressive is the diversity of expressed solutions to the problem. This paper does not intend to deal with the myriad of opinions offered on this subject nor take sides on any of the issues; rather, it focuses on one aspect of the food chain -- production of protein from animals, excluding marine resources. It is hoped that an objective assessment of the contribution of animals to the welfare of humanity will assist those involved in decision-making on agricultural issues.

In the interest of establishing common grounds relative to the world protein situation, the first sections of this paper identify some basic reasons underlying regional protein deficiencies by comparing requirements with available supplies. Subsequent sections deal explicitly with various animal production and productivity aspects as they relate to the food supply.

PROTEIN IN HUMAN NUTRITION

THE ROLE OF PROTEIN AND OTHER NUTRIENTS

Recent advances in the fields of physiology, genetics, biochemistry, and nutrition have contributed substantially to our knowledge and understanding of biological systems. Yet a host of recognized (or suspected) biological, environmental, and socioeconomic interactions still cloud some of the issues that relate directly or indirectly to the adequate nutrition of the world's population. In discussing the contribution of a given nutrient (in this case protein) to the overall nutritional status of individuals, acknowledgement must be given to the role of other essential nutrients which must be supplied in amounts necessary to maintain good health. To this end, a brief summary of the basic functions and requirements of each of the five groups of nutrients recognized as essential for maintaining adequate body functions follows.

Protein

Proteins are complex chemical compounds composed of amino acids. There are some twenty amino acids available in nature; half of these -- the essential amino acids -- must be directly ingested by mammals in their diets. The remaining ten can be manufactured in the body and therefore need not be present in the food. Different proteins contain varying quantities of total and individual amino acids. The same amino acid may be present several times in the same protein chain. The quality of protein can be estimated by comparing the amounts of essential amino acids in the protein food with the amounts required by the body. Proteins of animal origin, for example, are considered of higher quality because they contain a higher proportion of the essential amino acids.

Amino acids are required by the body to synthesize tissue proteins. Within the daily body processes, proteins are broken down and resynthesized. Amino acids present in excessive amounts are not stored; some are reutilized to build proteins, others are rapidly broken down. The nitrogenous components resulting from the degradation of amino acids are excreted, while the organic acids are used as a source of energy or converted to carbohydrate or fat. In energy-deficiency situations, considerable amounts of protein may be used as energy sources. This can result in a net protein deficiency although the intake may be adequate.

Human protein requirements are affected primarily by the level of energy in the diet, age, and sex of the individual, physiological activities such as pregnancy and lactation, and the availability of other nutrients in the diet. Requirements are usually expressed in grams of protein per day. The United States Allowance (requirement plus a margin of safety) for a seventy-kilogram man is fifty-six grams of mixed protein per day and forty-six grams for a fifty-eight-kilogram woman. Severe deficiency of protein causes Kwashiorkor, a disease that affects millions of young children in the tropics, particularly in Africa.(1) Pregnant and/or lactating mothers and children living in developing regions are considered vulnerable groups; that is, they are more susceptible to protein deficiency.

Energy

Energy is required by the body for biochemical and physiological activities. The sources of energy are primarily carbohydrates and fats although proteins can also be used for this purpose. Carbohydrates include the starches commonly found in grains, roots and tubers, and the various sugars found in plant and animal products. Fats are derived from both plant and animal sources and have a higher concentration of energy per unit of weight than carbohydrates. In the United States, carbohydrates and fats contribute about equally to the energy supply of the average national diet.

Food energy values are often expressed in kilo-calories. The amount of energy required by the body depends on the size of the individual, the age, the sex, the amount of activity, and the climate. Women require additional energy during pregnancy and lactation. Daily energy expenditures of mature women and men engaged in light occupations in the United States are 2030 and 2740 kilocalories, respectively.

Vitamins

Vitamins are complex organic compounds necessary for normal human metabolic processes. Based on their solubility, vitamins are classified into fat-soluble vitamins (A, D, E, and K) and water-soluble vitamins (B-complex and C). Some vitamins, especially those of the B complex group, are synthesized by the intestinal flora and, under normal conditions, contribute a por-tion of the daily requirements. The daily requirement for each of the vitamins differs and is influenced by factors such as age, sex, disease, lactation, preg-nancy, and others. Vitamin A deficiency is widespread in the Western Pacific, the Far East, semiarid zones in Africa, and parts of Latin America. Some 50,000 to 100,000 children become blind every year due to Vitamin A deficiency.(2) Beri-beri and pellagra are other examples of vitamin deficiencies.

Minerals

Minerals are inorganic elements essential to numerous body functions. For example, minerals are structural components of the skeleton (e.g., calcium and phosphorus), they function in acid-base balance (e.g., sodium, potassium and chlorine), are part of enzyme systems (e.g., copper and zinc), or some may have more than one function. Minerals required in excess of one hundred milligrams per day are commonly referred to as macroelements (calcium, phosphorus, sodium, potassium, chlorine and magnesium); those required in much lesser amounts are called trace ele-ments (e.g., copper, iron, zinc, cobalt, manganese, iodine, fluorine, chromium, molybdenum and a few others). Deficiency of any of the essential minerals results in some type of metabolic disorder. Among the most common are anemia and goiter caused by insuffi-cient iron and iodine, respectively. In developing countries approximately 20 to 25 percent of the chil-dren, 20 to 40 percent of the adult females, and 10 percent of the adult males suffer from nutritional

anemias of which iron-deficiency anemia is the predominant type. Goiter affects some 200 million people
in the world.(3)

Water

Water is of extreme importance to biological
processes although it is not commonly considered a
nutrient. It accounts for one-half to three-fourths of
body weight and its functions are numerous. Needless
to say, severe restriction of water intake results in
cessation of body functions more rapidly than restriction of other nutrients. The recommended dietary
allowance (RDA) for water is one milliliter per kilo-
calorie for adults.

— — —

Not one specific naturally occurring food item has
all the essential nutrients in the allowances recommended for support of good health in the adult. That
is, some food items may be high in one nutrient and low
in another (or others); sugar, on one extreme, is high
in energy (99 percent carbohydrates) but has no protein
or vitamin A, while milk has adequate protein and
vitamin A but has practically no iron. Combinations of
various food items are needed to balance the daily
nutrient intake since over or under consumption of one
or more nutrients may be detrimental to health (e.g.,
excess energy may lead to obesity and lack of calcium
can cause rickets).

When discussing a specific nutrient (or sources of
that nutrient) it should be kept in mind that biological systems are complex. Poorly understood interactions among nutrients, genetic variations of individuals or groups of individuals, and differences in
environmental and socioeconomic factors are but a few
obstacles facing nutritionists and policymakers in
their efforts to improve the nutrition of the human
race.

PROTEIN SOURCES FOR HUMANS

A variety of plant and animal products contributes
to the supply of protein available to man. The Food
and Agriculture Organization (FAO) groups plant food
sources into six categories (i.e., cereals and grain
products, starchy roots and tubers, dry legumes and
legume products, nuts and seeds, vegetables, and
fruits) and animal food sources into four (i.e., meat
and poultry, eggs, milk and milk products, and fish and

fish products).(4) The quantity and quality of protein in each food item need to be examined to determine the nutritive value of the source in the human diet.

Animal Products

The protein content of representative foods from each group and subgroup in terms of normal moisture content and on a dry matter basis (i.e., the amount of protein in the food after the water has been removed), is presented in Table 1. The data show that animal food products are generally higher in protein than most plant food products. Approximately one-half of the dry matter in poultry and beef meat and in hen eggs is protein. The lower protein content of pork is due to its relatively higher proportion of fat in the edible portion; lean pork meat, however, approaches the protein content of beef. Among the animal food products, milk dry matter has the least protein (about 27 percent) while fish dry matter is close to seventy-five percent protein. The average dry matter protein content of animal products is about 45 percent.

Plant Products

An arbitrary classification based on the amount of protein in the dry matter of plant foods (Table 1) is presented for discussion purposes fully recognizing the variability that exists between the subgroups and within each subgroup. High protein plant sources are those that contain more than 20 percent protein in the dry matter. Included in this class are dry legumes and legume products and some leafy vegetables. Intermediate sources contain between 10 and 20 percent protein in the dry matter. Cereals and grain products, nuts and seeds, and some vegetables fall in this class. Low protein sources contain less than 10 percent protein and include fruits and starchy roots.

Quantity and Quality of Protein

The quantity (or concentration) of protein in a food item becomes increasingly important when food sources with a low total-protein content are evaluated. An extreme, but illustrative, example will help to clarify this point. Assume that the daily energy and protein requirements of an active, adult male living in the tropics are in the neighborhood of 3000 kilocalories and 50 grams, respectively. If this individual had only cassava available, it would require 3.12 kilograms (6.88 pounds) to provide the total protein required. This amount, however, supplies 10,546 kilocalories, which is 3.5 times the energy requirement.

Table 1. Protein Content of Selected Food Items

Food Item	Protein %	Water %	Protein in Dry Matter %
Plant Sources			
Cereal & grain products			
Barley, pearl	8.2	11.1	12.1
Corn, flour	7.8	12.0	13.0
Wheat, flour	13.3	12.0	13.8
Starchy roots, tubers			
Potatoes, raw	2.1	79.8	10.4
Cassava, meat	1.6	13.1	1.8
Yam, raw	2.4	72.4	8.7
Dry legumes & legume products			
Beans	22.1	11.0	24.8
Cowpea	23.4	11.0	26.3
Groundnut	25.6	5.2	27.0
Soybean seed	38.0	8.0	41.3
Soybean milk	3.2	92.0	40.0
Nuts and seeds			
Almonds	16.8	4.8	17.7
Cashew	17.4	7.6	18.8
Sunflower, seed	12.6	5.5	13.3
Vegetables			
Broccoli	4.3	85.7	30.1
Carrot, root	1.1	88.6	9.7
Lettuce, leaves	1.3	94.8	25.0
Tomato, fruit	0.9	93.8	14.5
Fruits			
Apple	0.4	84.0	2.5
Banana	1.2	71.0	4.1
Date, dry	2.7	17.2	3.3
Peach	0.8	86.6	6.0
Animal Sources			
Meat			
Poultry, edible flesh	20.0	66.0	58.8
Beef, edible flesh	17.7	61.0	45.4
Pork, edible flesh	11.9	42.0	20.5
Eggs			
Hen	12.4	74.0	47.7
Milk			
Cow	3.5	87.3	27.6
Fish			
Most types, fresh	18.8	74.1	72.6

Sources: FAO, Amino Acid Content of Foods and Biological Data on Proteins, (Rome, 1970). B.K. Watt and A.L. Merrill, Composition of Foods. Agriculture Handbook No. 8, (Washington, D.C.: USDA, 1963).

Since the amount of food consumed is determined largely by the requirement for energy (not protein), it is likely that insufficient amounts of the low-protein food will be eaten to meet the protein requirement.(5) Thus, to meet his energy requirements, the man in question must eat some 887 grams of cassava; this results in an intake of only 14.2 grams of protein or less than one-third of his requirement.

The quality of a protein depends largely on its essential amino acid content. Other factors, such as the concentration of energy, vitamins, and minerals in the diet, the degree of availability of the amino acids, the physiological status of the recipient individual, and the environmental conditions surrounding him are also important. Incomplete digestion or absorption of amino acids reduces availability. Generally, over 90 percent of the amino acids in animal proteins are absorbed, while the value may be 80 percent or less for the amino acids from some plant proteins.(6) The availability of the amino acids can also be reduced by excessive heating during processing or home cooking.

Comparison of the essential amino acid content of a protein with a reference pattern of amino acids (amounts of essential amino acids needed to meet the requirements) provides a measure of quality known as amino acid score or chemical score. The joint committee of the Food and Agriculture Organization (FAO) and the World Health Organization (WHO) developed the provisional amino acid pattern shown in Table 2.

Table 2. Provisional Amino Acid Scoring Pattern

Amino acid	Suggested level	
	mg per g of protein	mg per g of nitrogen
Isoleucine	40	250
Leucine	70	440
Lysine	55	340
Methionine + cystine	35	220
Phenylalanine + tyrosine	60	380
Threonine	40	250
Tryptophan	10	60
Valine	50	310
Total	360	2,250

The formula for calculating the amino acid score of a protein or a mixture of proteins is:

$$\text{score} = \frac{\text{mg of amino acid in 1 g of test protein}}{\text{mg of animo acid in reference pattern}} \times 100$$

The amino acid score identifies the most limiting amino acid in the test protein. It follows that the amino acid in shortest supply determines the nutritive value of a protein since all amino acids must be present at the site of protein synthesis in adequate amounts for protein synthesis to proceed; an equal percentage deficit of any essential amino acid would limit protein synthesis to a comparable degree.(7) Figures 1 and 2 graphically compare the essential amino acids of a number of plant and animal products with the FAO/WHO reference amino acid pattern.

Hen eggs, beef, lamb, and fish contain all the essential amino acids in excess of the reference pattern; consequently, their amino acid score is one hundred. A slight deficiency of sulfur containing amino acids (methionine and cystine) in cow's milk results in an amino acid score of ninety-five, while threonine limits the value of poultry meat to ninety-nine.

Amino acid scores of plant proteins are usually much lower than for animal proteins. Rarely do plant proteins score higher than seventy-five, the bulk scoring between fifty and sixty-five (Figures 1 and 2). Lysine is the limiting amino acid in most cereal grains, some nuts and seeds, and some vegetables. Dry legumes and legume products such as soybeans, beans, lentils, cowpeas, and peas are generally low in sulfur amino acids. Lysine, leucine, isoleucine, threonine, methionine, and cystine are among the limiting amino acids in fruits and tubers. Some of the higher scoring plant food items are soybeans (74), banana (67), rice (66), groundnuts (65), and barley (64); among the lowest are tomatoes (34), turnips (36), sorghum (37), and oranges (38).

In addition to the amino acid score, Figures 1 and 2 show a value for an adjusted protein content. This value, used by Schmitt (8) in estimating more adequately the contribution of individual food items to the total protein supply, represents the corrected amount of protein in a food based on the amino acid score. It is obtained by multiplying the protein content (%) times the amino acid score and dividing by one hundred. For example, if wheat contains 12.2 percent protein and the amino acid score is fifty-three, the adjusted

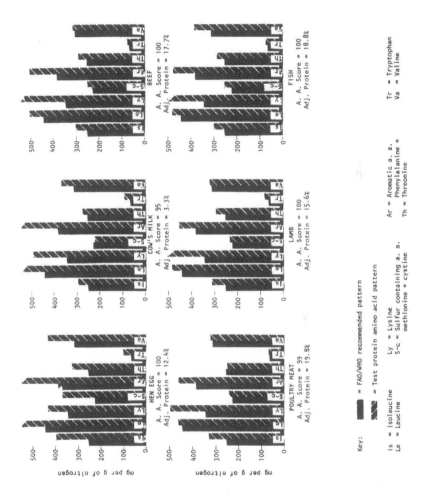

Figure 1. Essential amino acid content of animal products compared to the FAO/WHO recommended pattern.

193

mg per g of nitrogen

BEEF
A. A. Score = 100
Adj. Protein = 17.7%

FISH
A. A. Score = 100
Adj. Protein = 18.8%

COW'S MILK
A. A. Score = 95
Adj. Protein = 3.3%

LAMB
A. A. Score = 100
Adj. Protein = 15.6%

HEN EGG
A. A. Score = 100
Adj. Protein = 12.4%

POULTRY MEAT
A. A. Score = 99
Adj. Protein = 19.8%

Key:

▮ = FAO/WHO recommended pattern

▨ = Test protein amino acid pattern

Is = Isoleucine
Le = Leucine

Ly = Lysine
S-c = Sulfur containing a. a.
      methionine + cystine

Ar = Aromatic a. a.
     Phenylalanine +
Th = Threonine

Tr = Tryptophan
Va = Valine

194

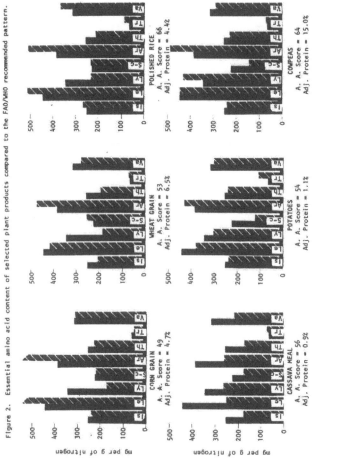

Figure 2. Essential amino acid content of selected plant products compared to the FAO/WHO recommended pattern.

Key:

▓▓ = FAO/WHO recommended pattern

▨▨ = Test protein amino acid pattern

Is = Isoleucine
Le = Leucine

Ly = Lysine
S-c = Sulfur containing a. a.
methionine + cystine

Ar = Aromatic a. a.
Phenylalanine +
Th = Threonine

Tr = Tryptophan
Va = Valine

protein content is 6.5 percent (12.2% x 53 ÷ by 100).
The high quality of egg and milk proteins has led to
their use as reference proteins; consequently, amino
acid scores of foodstuffs or food mixtures may vary
depending on the amino acid pattern used as reference.

As pointed out earlier, this paper emphasizes the
need for diets that contain a mixture of food items.
Assuming other nutrients are supplied in adequate
amounts, the complementarity of proteins from diverse
sources is illustrated by the following example. The
amino acid scores of whole wheat and fluid milk are
fifty-three and ninety-five, respectively. A mixture
(on a weight basis) of eight parts of milk to one part
of wheat yields a protein mixture of higher quality
(amino acid score of 100) than either component. In
this case shortages of amino acids in one protein are
supplemented by excesses in the other and vice versa.
Wheat contributes sulfur amino acids (limiting in
milk), while milk contributes lysine, leucine, isoleucine, threonine, and valine which are in short
supply in wheat protein. Similar results are obtained
by mixing approximately three parts of milk to one part
of corn. Thus, the common practice in the United
States of adding milk to breakfast cereals is nutritionally sound.

PROTEIN REQUIREMENTS AND SUPPLY

Traditionally, protein allowances have been expressed in terms of total grams of protein needed daily
by average individuals in a given population and environment. The rule of thumb that at least one third
of the total protein should be derived from animal
sources has also been useful in suggesting diets.
However, as knowledge of protein metabolism increases
and protein analysis techniques improve, the trend is
to recommend allowances for each essential amino
acid -- or amino acid patterns -- rather than total
protein. This approach seems logical since proteins
are hydrolized (broken down) to their component amino
acids in the gastrointestinal tract before absorption
into the bloodstream takes place. This section estimates the current and future (year 2000) protein and
essential amino acid needs of the world's population
and attempts to determine the contribution of animals
to the protein supply.

Protein and Amino Acid Requirements

The Joint FAO/WHO Committee on Energy and Protein
Requirements (9) has proposed a requirement of twenty-

nine grams of egg or milk protein per capita per day for a population of proportionally mixed ages. Diets containing protein of lower quality relative to that of egg (e.g., from 60 to 90 percent depending on the amino acid score) must be adjusted as shown in Table 3. If the average diet contains a mixture of proteins with about 60 percent of the value of egg protein (10), some 106 million tons of protein will be needed in the year 2000, compared to 72 million in 1977, to satisfy the requirements of the world population.

Table 3. Protein Requirements of World Population 1977-2000

Protein Value[1]	Requirement/ Caput/day	Total Protein Required in 1000's MT	
		1977[2]	2000[1]
100%	29.0[3]	43,430	63,510
90%	32.2	48,256	70,567
80%	36.3	54,288	79,388
70%	41.4	62,043	90,729
60%	48.3	72,383	105,850
50%	58.0	86,860	127,020

[1] Refers to the quality of protein compared to that of high quality protein such as egg or milk protein.

[2] Based on a world population of 4,103 million in 1977 and 6,000 million in the year 2000.

[3] Requirement suggested by FAO/WHO Committee for a population of mixed age groups.

The estimated essential amino acid requirements shown in Table 4 follow the pattern suggested by the Joint FAO/WHO Committee.(11) Currently, over fifteen-million tons are required annually to meet the needs of the people. This amount will increase proportionally as population increases reach close to twenty-three million by the year 2000. It is imperative that all essential amino acids be present at the site of protein synthesis in sufficient quantities for synthesis to occur. The values presented here assume this is the case; hence, the problem of distribution of amino acids has been precluded.

Table 4. Essential Amino Acid Requirements of World
          Population for Years 1977 and 2000.

Amino Acid	Amount in Protein, % 1/	Requirements in 1000's MT 1977 2/	2000 2/
Isoleucine	4.0	1,737	2,540
Leucine	7.0	3,040	4,446
Lysine	5.5	2,389	3,493
Methionine + cystine	3.5	1,520	2,223
Phenylalanine + tyrosine	6.0	2,606	3,811
Threonine	4.0	1,737	2,540
Tryptophan	1.0	434	625
Valine	5.0	2,172	3,176
Total	36.0	15,635	22,864

1/ Based on FAO/WHO, Energy and Protein Requirements.
FAO Nutrition Meetings Report Series No. 52 (Rome,
1973).

2/ Based on population estimates from Table 3.

Protein and Essential Amino Acids Supply

        Base data for 1972-74 (Table 5) show that the
protein supply was 68.3 grams per capita per day of
which 44.6 grams were supplied by plant products and
the remaining 23.7 grams by animal products. Assuming
a protein quality equivalent of 60 percent for plant
sources and 98 percent for animal products, the supply
of high-quality protein available would be in the order
of 26.8 and 23.2 grams, respectively. This indicates
that plants alone contributed 92 percent of the high-
quality protein required, while animal products alone
contributed 80 percent. A mixture of the available
supply of both plant and animal products would exceed
the population protein requirements by more than 1.8
times. Interestingly, the 1974 supply is sufficient to
meet the requirements of the projected population in
the year 2000 if surplus protein consumed by some
groups were redistributed to protein-deficit populations.

        A look at the essential amino acid requirements of
the population and the available supply from animal
food sources (Tables 4 and 6) suggests that animals
alone can meet even a larger part (96 percent) of the
human protein needs. All amino acids are supplied in

excess of the requirements except valine and sulfur-containing amino acids which account for the reduction to 96 percent. If no increases in production or productivity of livestock were to occur by the end of the century, animal products could supply an impressive 67 percent of the essential amino acids required by the estimated year 2000 population of 6,000 million people.

Table 5. Food Supply: Grams of Protein Per Capita Per Day, 1972-1974 Average

Region	Total Protein	From Vegetable Sources	From Animal Sources
Developed	95.1	39.5	55.6
North America	104.1	32.9	71.1
W. Europe	93.0	41.6	51.4
Oceania	100.7	33.7	67.0
Other	84.5	45.9	38.6
Developing	53.9	42.7	11.2
Africa	52.4	43.0	9.4
Latin America	64.8	39.8	25.0
Near East	67.9	54.2	13.7
Far East	48.8	41.5	7.3
Other	50.5	31.1	19.4
Centrally Planned	73.5	50.6	22.9
Asia CPE	61.6	49.7	11.9
E. Europe & USSR	102.3	52.8	49.5
World	68.3	44.6	23.7

Source: FAO, The Fourth World Food Survey. Food and Nutrition Series, No. 10 (Rome, 1977).

The Protein Gap

Although the data presented above indicate that sufficient protein is currently being produced to satisfy the requirements of the population, FAO estimated the number of people with insufficient protein/ energy supply in 1970 at about 460 million (excluding Communist Asia), or close to 13 percent of the population. The major issue, as pointed out by the Fourth World Food Survey (12), seems to be one of distribution of protein among the population -- particularly among the vulnerable groups such as the poor. However, even the issue of distribution is only a reflection of the more fundamental problem of control and participation in the production process itself.(13)

The nature of the distribution problem is exceptionally complex; a few comments on the supply data and the protein sources may shed some light on the subject. The supply figures calculated by FAO (Table 5) represent only the average supply for the population as a whole and do not indicate what is actually consumed by individuals, nor do they give any indication of differences that may exist in the diets of different population groups (e.g., different socioeconomic groups or ecological and geographical areas within a country) or seasonal variations in the food supply. A case in point is Brazil, which has, by all standards, an adequate per capita protein supply (63.3 grams per day). However, the Dirigente Rural (14) reports that the population of large areas in the northeast (which is plagued by frequent droughts) has a 47 percent protein deficit according to a 1971 survey.

Table 6. Percentage of Essential Amino Acids Available from Meat, Milk and Eggs in 1977

Region	Iso	Leu	Ly	S-E	Arom	Threo	Trypt.	Va-line	Total
Developed	288	291	336	250	339	768	293	252	291
North America	391	346	430	320	415	325	358	313	366
W. Europe	289	289	337	247	351	328	305	258	296
Oceania	777	763	955	671	924	726	792	684	796
Other	99	180	218	89	110	86	97	85	96
Developing	49	43	106	38	52	40	45	38	44
Africa	29	28	35	17	33	27	29	25	29
Latin America	123	120	148	126	144	112	123	107	125
Near East	59	59	70	51	71	54	62	53	61
Far East	22	22	25	18	27	29	23	20	20
Other	23	21	26	19	25	22	24	20	23
Centrally Planned	111	107	128	96	131	101	115	97	112
Asia CPE	43	32	47	38	44	39	41	35	26
E. Europe & USSR	283	287	232	243	349	261	302	255	293
World	110	108	129	96	130	100	113	97	112

Furthermore, the supply data indicate only what is available to the population at the retail level. Losses during home storage, cooking, or as left-overs

are not taken into consideration. These can be substantial, particularly in developing areas where refrigeration for perishable food is nonexistent and where
abundant numbers of rodents, insects, and other pests
compete with man for food.

It is interesting to note that the large difference in total protein supply between the developed and
developing regions (ninety-five versus fifty-four grams
per capita per day) is largely due to a much higher
supply of animal products in developed countries. The
plant-protein supply does not vary widely among the
regions although it is somewhat higher in developing
areas where protein deficiencies of large population
groups have been reported (Table 5). The calculated
availability of essential amino acids derived from
animal food products further suggests this is the case
in developed countries, East Europe, and the Soviet
Union which produce 300 percent or more of the requirements for their respective populations (Table 6).
Developing areas such as Africa, the Far East, and the
Asian centrally planned economies, on the other hand,
produce only one-third or less of their required,
essential amino acid needs from animal products. Plant
proteins must be relied upon in these areas to meet the
remaining two-thirds of the essential amino acids.
Since consumption of animal products, particularly
meat, is characteristic of higher income groups within
a population, the supply in developing areas is acquired by a small percentage of the population in
relatively larger proportions (Table 7). This obviously results in decreased availability of high-
protein animal food to the bulk of the population and
increases their dependence on plant food products.

It is probable that those vulnerable groups depending primarily on a few staples of low-protein
quality and quantity (e.g., cassava) cannot consume
sufficient amounts to meet the daily protein requirements and that the small amount of animal protein
available to them may be consumed on such an irregular
basis that its potential contribution is considerably
diminished. If these vulnerable groups are to better
their nutritional status, improvement of the local
long-term-food-production capability appears to be a
more plausible solution to the protein-deficiency problem rather than attempting to redistribute worldwide
available supplies. Referring to the question of world
food sufficiency and meat consumption, Svedberg (15)
concludes that a reduction of food consumption in the
rich countries can only in the very short run, and
under special circumstances, be warranted as a means of
relieving food deficits in the poor countries.

201

Table 7. Beef Expenditures of Various Income Strata
          in Cali, Colombia

Item	I	II	III	IV	V
Income, US$/year	353	677	1074	1937	4588
Beef expenditure/family/year,US$	37	68	129	185	473
Avg. family size	5.9	6.3	6.8	6.3	6.6
Avg. beef expenditure/person,US$	6.3	10.8	19.0	29.4	71.7
Beef/person/year, Kg	4.2	7.2	12.7	19.6	47.8
Beef/person/day, g	11.5	19.7	34.8	53.7	130.9
Protein from beef/person/day,g	1.98	3.49	6.16	9.50	23.18

(Income Strata span columns I–V)

Source: Adapted from Pinstrup-Anderson P., "Decision-making on Food and Agricultural Research Policy: the Distribution of Benefits From New Agricultural Technology Among Consumer Income Strata," _Agricultural Administration_ 4 (October 1977):15.

ANIMALS AND FOOD PRODUCTION

CONTRIBUTION OF ANIMALS TO MANKIND

Animals provide man with a diversity of products and services. Bowman (16) and McDowell (17) have classified these into various groups summarized below:

Food: meat, milk, eggs, gelatin.

Fiber and Skins: wool, hair, hides and pelts.

Traction: cartage and traction power for crop production, irrigation pumping, and threshing.

Waste Products: fuel, feed, fertilizer, construction material, methane gas production, and livestock feed.

Storage: storage of food supply or capital and seasonal excess of feeds.

| Pest Control: | biological control of brush, plants, and weeds along road-sides and waterways (irriga-tion canals), predator con-trol. |

Research: biological research in phys-
                    iology, nutrition, genetics,
                    biochemistry, and product-
                    safety control.

Religion: sacrificial offerings.

Sports/recreation: racing, exhibitions, hunting,
                    fighting, and companion ani-
                    mals.

Miscellaneous: tallows, hooves, horns, medi-
                    cines, bone meal, tankage.

By far the greatest quantifiable contribution of animals to the well-being of man is in the form of food -- meat, milk and eggs (Table 8). The estimated total value of these three products in the 1977 world market was about US $311,060 million. This represents an annual average per capita production figure of US$76 compared to roughly US$43 for the aggregate of all cereal-grains production. More important, however, the total production of meat, milk, and eggs presumably supplied 19, 80, and 96 percent of the 1977 world's population needs for energy, protein, and essential amino acids, respectively. Other products and services contributed by animals are detailed in Appendix I.

Table 8. World Production of Meat, Milk and Eggs in
          1977.

Region	Meat	Milk	Eggs
	1000 metric tons		
Developed	60,991	216,667	11,622
Developing	23,813	91,193	4,160
Centrally Planned	41,283	142,483	9,307
World	126,086	450,343	25,089

Source: FAO, _Production Yearbook_, _1977_, vol. 31 (Rome, 1978).

ANIMAL RESOURCES FOR FOOD PRODUCTION

Thus far this paper has referred to animals and animal production in general. The foregoing section focuses on two specific groups of animals -- ruminants and nonruminants -- kept for the production of food, fiber and/or draft. The classification is based on the anatomical structure of the digestive tract which, in turn, determines the type of feed the animal can utilize.

Ruminant Animals

Description. These even-toed, hooved mammals belong to the order Artiodactyla and the suborder ruminantia. The word "ruminant" means to chew over again; the more common reference is to "chewing the cud." An important characteristic of ruminants is their multicompartment stomachs which enable them to utilize fibrous feeds. The first two compartments -- the reticulum and the rumen -- are inhabited by innumerable bacteria and protozoa. These microorganisms produce enzymes that attack plant materials (cellulose, hemicellulose and xylan) degrading them into sugars. These sugars are subsequently attacked by other microbial enzymes yielding volatile fatty acids, which are the ruminant's major source of energy. The functions of the third compartment, the omasum, have not been clearly defined, but the abomasum (fourth compartment) appears to have functions similar to that of simple stomach or monogastric (nonruminant) animals.

Rumen microorganisms are also able to utilize nonprotein nitrogen in the synthesis of microbial protein. This protein subsequently is broken down into its constituent amino acids in the lower gut and, after absorption into the bloodstream, the amino acids are utilized by the various tissues in the synthesis of animal protein. Various preparations containing urea, a nonprotein nitrogen source, are extensively fed to ruminant livestock in developed countries.

The ability of ruminants to convert cellulose and nonprotein nitrogen, both of which are nonhuman foods, into high-quality food places them in a unique position in the food chain. The role of ruminants in harvesting the vast grasslands of the world and effectively utilizing agricultural and industrial by-products, and in many occasions human food wastes, is indeed an important one.

Ruminant animal resources. Existing ruminant animals are divided into five families. Most commercially important ruminants -- cattle, buffalo, sheep and goats -- belong to the Bovoidae family which consists of forty-seven genera. Camels, llamas and alpacas, of importance in some regions, are members of the Camelidae family. Other ruminants such as giraffes, deer, antelopes, and many wild species form part of the other three families. This paper deals only with domesticated ruminants, fully recognizing the importance of wild animals in certain societies, particularly in Africa.

World stocks and distribution of ruminant livestock are presented in Table 9. Cattle and sheep are by far the most numerous, each numbering well over 1,000 million in 1977. Stocks of goats, buffaloes, and camels were approximately 410, 131 and 14 million, respectively. During the past decade the population of camels has remained stable, but sheep numbers have declined slightly (less than 1 percent). Cattle, buffalo, and goat stocks, on the other hand, increased 10.6, 6 and 4.5 percent, respectively, during the same period.(18)

Table 9.   World Stocks of Selected Ruminants by Region, 1977.

Region	Cattle	Buffalo	Sheep	Goats	Camels
			1000 Head		
Developed	295,326	146	330,405	16,820	12
Developing	702,770	96,874	433,257	319,351	11,850
Centrally Planned	214,766	33,843	264,197	74,171	1,909
World	1,212,861	130,863	1,027,859	410,343	13,771

Source: FAO, Production Yearbook, 1977, vol. 31 (Rome, 1978).

The largest number of domesticated ruminants is found in the developing regions (Table 9). Approximately 58 percent of the cattle, 74 percent of the buffaloes, 42 percent of the sheep, 78 percent of the goats and 86 percent of the camels are distributed throughout the Near and Middle East, Africa, and Latin America. The developed countries have about 25 percent of the cattle, 32 percent of the sheep, 4 percent of the goats and insignificant numbers of buffaloes and camels. The balance, 18 percent of the cattle and goats, 26 percent of the buffaloes and sheep, and 14 percent of the camels are found in the centrally planned economies region.

205

Contribution to the food supply. Although the developing regions claim the largest population of ruminants, their contribution to per capita food supply is comparatively low because of the very large human population in the area and the low level of animal productivity.

The ratio of ruminants to humans is important in that it reflects, to a great extent, resource alloca- tion and per capita food supply potential. Oceania, for example, has more than ten times the number of cattle per person than the Near or Far East and thirty times more than the Asian centrally planned economies (Table 10). The substantial proportion of grasslands in Oceania (60 percent of the land base compared to 4 percent for the Far East) is undoubtedly a factor responsible for the very high per capita stocks.

Table 10.   Ruminant Animals to Human Population Ratio in Developed and Developing Regions, 1977.

Region	Cattle	Buffalo	Sheep	Goats	Camels
	Number of head per 100 persons				
Developed	38.5	–	43.1	2.2	–
Developing	34.6	4.8	21.4	15.7	0.6
Centrally Planned	16.4	2.6	20.2	5.7	0.1
World	29.6	3.2	25.0	10.0	0.3

Source: FAO, Production Yearbook, 1977, vol. 31 (Rome, 1978).

Per capita ruminant meat and/or milk production represents more closely the contribution of these animals to the food supply. The ruminant meat supply in developed countries is close to thirty-five kilo- grams per capita per year as opposed to only 7.5 kilo- grams for developed countries (Table 11). The high- quality protein supplied by these amounts would meet about one-half of the the daily requirements of persons in developed countries and close to one-eighth of the requirements in developing regions, respectively. Per capita milk production follows a pattern similar to meat production -- highest in developed regions, lowest in developing regions, and intermediate in centrally planned economies (Table 12). The amount of milk produced in developed countries is sufficient to satis- fy nearly 90 percent of each individual's daily re- quirements for protein. In the developing countries, however, milk production can only meet about 14 percent of the requirement.

206

Table 11. Per Capita Ruminant Meat Production in 1977:
Kg/Capita/Year

Region	Beef and Veal	Buffalo Meat	Mutton & Lamb	Goat Meat	Total
Developed	31.5	-	2.9	0.1	34.5
Developing	5.7	0.3	0.9	0.6	7.5
Centrally Planned	8.0	0.5	1.2	0.3	10.0
World	11.3	0.3	1.4	0.4	13.4

Source: FAO, Production Yearbook, 1977, vol. 31 (Rome, 1978).

Table 12. Per Capita Milk Production in 1977.

Region	Cow	Buffalo	Sheep	Goat	Total
		Kg/Capita/Year			
Developed	277.3	-	3.2	1.8	282.3
Developing	28.5	12.9	1.6	2.1	45.1
Centrally Planned	106.2	0.9	1.3	0.6	109.0
World	99.7	6.7	1.8	1.6	109.8

Source: FAO, Production Yearbook, 1977, vol. 31 (Rome, 1978).

Animal productivity -- the amount of meat and/or milk produced by an animal in a given period of time -- is also reflected in the food supply. A measure of productivity, often referred to as offtake, represents the percent of animals slaughtered for meat from the total stocks. Table 13 shows that the offtake from cattle, sheep, and goats is generally higher in developed regions and lowest in developing regions. The offtake for cattle, for example, is over three and one-half times higher in developed than in developing regions. The causes for these large differences are numerous but generally are related to poor nutrition, disease, uncontrolled breeding, lack of economic incentives, and marketing factors present in developing regions. However, the extremely low offtake for cattle in the Far East (3 percent) has to do more with Hindu restrictions on cattle slaughter and the high proportion of animals kept for draft purposes. Excluding India, cattle offtake for the Far East is about one-half that of the developed regions.

Table 13. Offtake from Ruminant Animals in Developed
and Developing Regions in 1977.

Region	Cattle	Buffalo	Sheep	Goats
	% slaughtered from total stocks			
Developed	36.6	0.7	41.3	55.8
Developing	10.2	5.2	29.7	36.2
Centrally Planned	28.3	11.5	37.8	32.5
World	19.8	6.9	35.5	36.4

Source: FAO, Production Yearbook, 1977, vol. 31 (Rome, 1978).

Nonruminant Animals

Description. This group of animals lacks the characteristic pregastric digestion found in ruminants. Anatomically, their digestive tract resembles that of man; however, certain modifications in the lower gut allow them to utilize, to varying degrees, materials indigestible to man. For purposes of this discussion, nonruminant animals are subdivided into herbivorous and omnivorous. Herbivores, or forage consuming animals (e.g., horses, mules, asses, and zebras), have a well-developed caecum where fiber digestion takes place through the action of microorganisms. Omnivores, such as pigs and fowl, depend on a variety of feeds because they lack well-developed, gastrointestinal structures suitable for the digestion of relatively large propor-tions of fibrous feeds.

Nonruminant animal resources. A wide variety of nonruminant animals are used as sources of food and other products throughout the world. Among the most common are pigs, fowl (chickens, turkeys, ducks, geese, and other birds), horses, rodents (rabbits and guinea pigs) and some wild species such as the zebras. Empha-sis is placed on pig and chicken production because the contribution of the other classes to the world's food supply is largely unknown.

World stocks and distribution of nonruminant animals are shown in Table 14. Chickens are most numerous, with well over 6,300 million head in exis-tence, followed by some 666 million pigs, 155 million ducks, 115 million horses, asses, and mules, and 90 million turkeys. Unlike ruminants, the largest pro-portion of nonruminants is found in the centrally

planned economies. Approximately 57 percent of the
pigs, 42 percent of the chickens and turkeys, and 50
percent of the ducks are distributed throughout the
Asian centrally planned economies, East Europe, and the
Soviet Union. The developed countries share only 26
percent each of the pig and chicken stocks. The
balance is found in the developing regions of which
Latin America has the largest proportion.

Table 14. World Stocks of Selected Nonruminant
Animals.

Region	Pigs	Chickens	Ducks	Turkeys	Horses, Asses and Mules
		1000's Head			
Developed	173,887	1,673,046	13,902	29,923	15,167
Developing	114,755	2,022,521	63,137	22,026	67,867
Centrally Planned	377,632	2,639,517	78,128	37,883	32,148
World	666,274	6,335,085	155,167	89,832	115,182

Source: FAO, Production Yearbook, 1977, vol. 31 (Rome,
1978).

Worldwide and regional trends in the changes in
numbers of nonruminant animal stocks during the past
decade are shown in Table 15. Generally, the overall
population of herbivores -- horses, mules and asses --
has declined slightly. Significant decreases in the
number of mules and asses in developed regions can be
explained by the declining use of these animals as
beasts of burden. The unexpectedly small increases in
the population of these animals in the developing
world, which relies heavily on them for transport of
people and goods, is perhaps a reflection of expanding
mechanized transportation and/or the result of the
heavy migration to urban centers where animals cannot
be kept. The rather limited and regionalized (most in
Europe) consumption of horse meat is unlikely to have
significant effect on the horse population; however,
the increasing popularity of horses for recreational
purposes is likely to result in stable if not increased
numbers in some developed countries.

Table 15. Changes in Nonruminant Animal Stocks During
the Past Decade

	Region			
			Centr.	World
	Developed	Developing	Planning	Aver.
Animal		% Change		
Horses	+ 4.0	+ 5.1	-14.7	- 1.8
Mules	-38.4	+ 4.4	- 4.0	- 0.5
Asses	-24.0	+ 1.6	+ 1.6	- 0.5
Pigs	+ 3.7	+17.1	+12.4	+10.7
Chickens	+ 4.3	+25.8	+21.6	+17.7
Ducks	+28.9	+20.6	+24.4	+23.2
Turkeys	+14.8	+52.4	+44.1	+34.4

In contrast, during the past decade omnivorous
animals -- pigs and fowl -- have increased considerably
throughout the world, particularly in the developing
regions. The centrally planned economies and developing regions shared the largest increases in pigs and
chickens while only minor increases occurred in the
developed world where commercialized, intensive-pro-
duction systems for these animals are well established.
Overall stocks of turkeys and ducks have increased at
an impressive annual rate of about 4 and 3 percent,
respectively, during the past eight years. The general
upward trend in the pig and fowl populations within the
developing regions and the centrally planned economies
is indicative of the growing dependency on these animals as sources of desirable food -- meat and eggs.

Contribution to the food supply. The proportionately large human population together with the char-
acteristically low animal productivity in the developing world results in low per capita food production
from nonruminants (Table 16). Per capita poultry meat
production in the developed world, for example, is ten
times greater than in developing areas and over 3.5
times that of centrally planned economies. Low animal
productivity seriously limits the potential food supply
in underdeveloped areas; the calculated offtake rate
for swine, as an example, is 150 percent in developed
areas compared to only 57 percent for developing countries.(19) Similarly, per bird production of eggs is
close to 3.5 times greater in developed countries.
This disparity in productivity levels suggests that
there is ample room for improvement in the field of
animal production in developing nations once local constraints are identified and appropriate technological

and managerial practices are developed for each situation.

Table 16. Per Capita Meat and Eggs Production From Nonruminant Animals in 1977.

Region	Pig Meat	Horse Meat	Poultry Meat	Total Meat	Eggs
	Kg/Capita/Year				
Developed	25.1	0.3	18.2	43.6	15.1
Developing	1.8	0.1	1.9	3.8	2.1
Centrally Planned	16.0	0.1	5.1	21.2	7.1
World	10.7	0.1	5.9	16.7	6.1

Source: FAO, Production Yearbook, 1977, vol. 31 (Rome, 1978).

LIVESTOCK PRODUCTION SYSTEMS

There are two basic livestock production systems -- intensive and extensive. The extent to which each one is applied in a given situation depends largely on the resources available and the environmental conditions. A brief description of each follows:

Intensive Production Systems

Under intensive production systems animals are housed in relative confinement and are closely supervised and managed. High animal-health standards and controlled breeding are the rule rather than the exception. Feed mixtures, balanced to meet the specific requirements of each productive stage, are customarily brought to the animals and fed free choice. In developed countries and in some developing areas, intensive systems are commonly used for the production of fed beef and lamb, milk, eggs, poultry meat, and pork. These systems generally require high capital, technology, and management inputs in exchange for high-quality and quantity outputs.

Intensive production systems, particularly those that utilize significant proportions of grain in the feed mixtures, are currently being criticized by some authorities primarily because: (a) animals are competing directly with man for food; and (b) the efficiency of converting plant (grain) protein to animal protein is relatively low. Suffice it to say at this point that feeding grain, or for that matter any other feed fed to livestock under intensive systems, depends almost entirely on prevailing economic conditions. Large surpluses of grains in developed countries tend to encourage their use as animal feed as long as the price ratio for feed grains to animal products remains favorable. Typically, this ratio is much more favorable for feeding grain to cattle and sheep in developed than in developing countries but the reverse holds for poultry, eggs, and milk.(20)

Extensive Production Systems

Turner (21) defines extensive production systems as those in which animals gain most of their nutritional requirements by grazing or browsing in the open. Extensive systems require lower per animal capital, technology, and management inputs than intensive systems, but productivity and total outputs are also lower.

In developed countries extensive systems are, for the major part, limited to beef cow-and-calf and sheep operations that utilize range and/or improved pastures as the primary source of feed. Conversely, in developing countries the extensive system of production predominates; however, there is considerable variability in the degree and type of management from one area to another and among the different species involved. The majority of ruminant and nonruminant animals found in developing areas are kept by small farmers. These animals receive little or no health care, breeding is for the most part uncontrolled and the feed supply is often seasonal. The vast majority of small producers regard their pigs, sheep, goats, and chickens as a store of capital which can be easily transformed into much needed cash to purchase a few commodities. Alternative production systems adapted to local economic and environmental conditions together with sound livestock marketing systems need to be developed and implemented in many developing situations in order to increase animal productivity and income levels of small farmers.

FEED RESOURCES FOR LIVESTOCK PRODUCTION

Both ruminant and nonruminant animals can meet their nutritional requirements for maintenance and productive functions from a wide variety of feedstuffs or combinations of feedstuffs. The extent to which nutritional needs can be met, however, depends largely on the quantity and quality of the feeds available to the animal. To simplify this discussion, feeds are

grouped into three broad categories based primarily on
their source of origin and their end-use; the categories are: (a) forages, (b) agroindustrial by-products
and (c) grains. The overall availability of these
groups of feeds is discussed in terms of current and
potential utilization by ruminants and nonruminants.

Forages

This group of feeds is characterized by its high
fiber content and its unsuitability as human food.
Plants such as grasses, legumes, bushes, and a few
others are the sources of most forage. Because ruminants have the capability of converting these fibrous
materials into high-quality food, the forage-ruminant
interrelationship is of enormous economic and nutritional importance in the food production chain. Nonruminant herbivores also depend heavily on forages as
sources of nutrients, but their major contribution to
the welfare of man is in the form of transportation
and/or packing. Omnivorous animals (pigs and chickens)
in developed countries are fed only small quantities of
high-quality forages such as alfalfa; however, in
developing regions these animals, particularly pigs,
consume substantial amounts of fibrous feeds and waste
materials.

Pastures and meadows are the largest single source
of forages. Approximately 23 percent (over 3,000
million hectares) of the world's land area is classified as permanent meadows and pastures where herbaceous
forage crops are cultivated or grow wild (Table 17). A
recent study by Fitzhugh et al. (22) indicates that the
forage produced in these vast areas has the potential
of meeting over 80 percent of the energy needs of the
world's cattle, buffalo, sheep, goat, and camel stocks.
They also indicate that the potential of this huge
reservoir for ruminant production remains largely
untapped inasmuch as advanced technology has been
applied to less than 8 percent of the total permanent
pasture and meadow area. An additional source of
forage, yet to be assessed, is the more than 4,100
million hectares of forests and woodlands.

The ruminant livestock production potential from
the arid and semiarid rangeland and from the humid and
subhumid tropical grazing lands has been evaluated by
Bula et al.(23) This group concluded that under specified institutional, technological, and managerial conditions the potential for doubling production in the
arid and semiarid rangeland is very good. The potential production of humid and subhumid grazing lands,
particularly in Latin America, can increase at an

annual rate of 4.2 percent for meat and 4.6 percent for
milk through 1985 provided a drastic reallocation of
resources to ruminant production is made.

Table 17.   Permanent Pasture, Forest, and Woodlands
            Available for Ruminant Livestock in 1976

Region	Permanent Pasture	Forest & Woodland	Total Area
	— — — — — — -1000 Ha- — — — — — —		
Developed	889,302	914,592	1,803,894
N. America	264,961	616,129	881,090
W. Europe	72,290	124,430	196,720
Oceania	469,127	144,450	613,577
Other	82,924	29,583	112,507
Developing	1,435,963	2,076,826	3,512,789
Africa	688,057	542,966	1,231,023
Latin America	523,603	1,026,433	1,550,036
Near East	190,417	138,932	329,349
Far East	33,247	326,983	360,185
Other	639	41,512	42,515
Centrally Planned	732,818	1,153,797	1,886,615
Asia CPE	344,102	204,783	548,840
E. Europe & USSR	388,716	949,014	1,337,730
World	3,058,083	4,145,215	7,203,298

Source: FAO, Production Yearbook, 1977, vol. 31 (Rome,
1978).

Although productivity of permanent pasturelands
can be increased significantly through the use of
appropriate management techniques (e.g., fertilization,
overseeding with legumes, and proper grazing), the constraints that limit production of cultivated crops in
these areas can be overcome only in certain situations
and at great expense.(24) The lack of plausible land-
use alternatives proposed for permanent pastureland
strongly implies that without ruminants these areas
would be of little value for food production.

Forages, other than those derived from rangelands
and permanent pastures, include alfalfa, corn and
sorghum. These are grown in significant quantities
almost exclusively in developed countries and often
conserved in the form of hays and/or silages to maintain a constant supply of forage feed throughout the

year. It is true that these crops compete with other
crops for land use since they require fertile soils.
However, the alleged competitiveness becomes less of an
issue when the total amount of land allocated to these
three crops and the alternative land uses are examined.
In the United States, where these crops are used extensively for feeding livestock, an average of about 10
percent of the total cropland is routinely in these
crops. Alfalfa is grown on about 7 percent of the
cropland not only as part of sound crop rotation systems to preserve soil fertility but also as a high cash
return crop. The 3 percent of the cropland planted to
forage corn and sorghum is less than the amount of
cropland left idle each year (e.g., 13 percent in 1969
and 5.5 percent in 1974). This fact implies that the
latter forage crops are not impinging on the United
States food supply.

Agroindustrial By-products

This category encompasses a wide variety of by-
products resulting from harvesting or processing of
crops, animals, and some chemicals. This analysis
groups by-products into four broad classes: crop residues, oilseed meals, waste materials and chemicals.

Crop residues. This group includes: (a) residues
left after crops are harvested such as straws, corn
stover, sugar beet tops, peanut and pea vines, and
others; and (b) residues from processing of crops like
sugar beets, citrus and pineapple pulps, various fruit
and vegetable cannery wastes and culls, milling by-products from cereal grains, sugar cane bagasse and molas-
ses, coffee pulp, waste bananas, and several more. The
extent to which each crop residue is used depends
largely on its nutritional quality, the class of animal
to which it is fed, the animal production system, the
cost, and the availability.

The potential contribution of crop residues to
livestock production has not been fully explored,
particularly in the developing world. Fitzhugh et al.
(25) estimates indicate that crop residues could potentially meet close to 42 percent of the energy requirements of all domesticated ruminants. In view of this
tremendous potential, intensification of research on
the utilization of residues (including research with
nonruminant animals) is warranted since these by-
products are not suitable for human consumption.

Oilseed meals. These cakes or meals are the
residual material after the oil has been extracted.
Cottonseed and soybean meal are the most important in

terms of quantity; sunflower, peanut, linseed and copra meals are of lesser importance worldwide. Oilseed meals are rich in protein and energy and can be used in ruminant and nonruminant rations as a protein supplement. Fitzhugh et al. (26) concluded that only in North America and Europe do oilseed meals make up a significant portion of ruminant diet. This is probably the case also with nonruminant animals.

Oilseeds, soybeans in particular, are the subject of intensive research as possible protein supplements for humans because of their high-protein content and amino acid scores. Meat analogs and milk substitutes manufactured from soybeans have gained some acceptability. The degree of competition between humans and animals for oilseeds as a food will depend largely on: (a) acceptability of the products by the target populations; and (b) the cost of the product to the consumer. The latter factor is important if the target populations are in the humid tropics where the present high-yielding varieties of soybeans do very poorly. This means that the raw material must be obtained in the international market at a competitive price or the finished product must be imported. Either solution perpetuates a food-import dependency situation accompanied by a corresponding outflow of foreign exchange, rather than the attainment of self-sufficiency and full participation in the production process.

Waste materials. Feeds in this class include waste materials that when adequately processed can be fed to livestock. Bone meal, tankage, feather meal, garbage, poultry litter, animal manures, and others have been successfully fed to livestock. The nutritive value of these feeds varies considerably with the type of animal being fed, the raw materials, and the processing techniques. Bone meal, tankage, and cooked garbage are commonly fed to swine in developed countries, while poultry litter and animal manures are limited to ruminants.

In developing countries utilization of by-products is often overlooked. Barat (27), for example, points out that in developing countries there is a general lack of awareness of the potential worth of slaughter house by-products, particularly as these high-protein feeds would contribute substantially to the poultry and pig development programs in many countries.

The potential uses of waste materials have not been fully explored. In developed countries disposal of manure and some processing by-products such as whey are a challenge to researchers. Production of biogas and single-cell-protein concentrates (for animal feed) from these wastes are promising alternatives.

Chemicals. Grouped in this class of feeds are nonprotein nitrogen, mineral supplements, and feed additives. As mentioned earlier, ruminants can utilize nonprotein nitrogen (urea) as a source of protein. The practical recommended level is limited to about one-third of the total protein; however, under careful management this level can be increased some, but when improperly fed, urea can be toxic to ruminants.

In developed countries urea and other nonprotein nitrogen-containing compounds are important components of ruminant diets. The amount of nonprotein nitrogen fed in 1973 in the United States had approximately the same protein equivalent of the oilseed meals fed to ruminants in 1977. The quantity of nonprotein nitrogen fed in the developing world is unknown. Active research is being conducted in various countries (e.g., Venezuela and Brazil) on the use of nonprotein nitrogen as a supplement in diets containing large amounts of cassava or sugar cane by-products. The development of feeding systems utilizing these high-carbohydrate feeds, usually in large supply in the tropics, will undoubtedly improve animal productivity.

The important role of minerals in nutrition is recognized in the developed countries where the mineral supplement industry is well-established; however, in many developing countries this is not the case. Too often animals are not even provided with salt on a regular basis because of unavailability or ignorance.

Feed additives are not considered nutrients, but they promote growth (gain) by stimulating appetite, reducing health problems, or increasing the efficiency of digestion. Controversy has plagued the use of some feed additives in the United States (e.g., diethylstil-bestrol) and strict regulations control their feeding. The practicality of using feed additives under the prevailing extensive systems in developing countries has yet to be determined.

Grains

The term feed grains in this paper refers to grains that are commonly fed to livestock in the United States -- corn, sorghum and barley -- although any one of these may be a staple for humans in another country.

Feed grains are desirable because they contain over one and one-half times the energy of most forages.

Ruminant animals under intensive production systems have a high genetic potential for productive functions [meat (growth) or milk] that require energy inputs which cannot be satisfied entirely by forages because of their bulk (low energy density). Grains are therefore combined with forages in such proportions that the resulting mixtures supply all the daily nutrient requirements in amounts that can be physically consumed by the animal. High-producing swine and poultry are more dependent than ruminants on high-energy feeds because of their limited ability to utilize fibrous materials.

An issue of current debate centers on the desirability of feeding grain to livestock in the developed as well as developing countries. The argument is that the grains themselves could more efficiently be fed directly to humans. The earlier discussions in this paper throw light on several aspects of this debate. First, the vast majority of feed consumed by ruminants is not in direct competition with man for food. Second, the protein-quality characteristics of animal products favor a mixed diet with some animal products. The animals in effect convert low-quality plant protein into high-quality animal protein. A third factor relates to animal nutrition, particularly for ruminants, where the inclusion of grains into the diet may improve digestibility by only 5 to 7 percent but pushes the energy intake to 1.5 to 2.0 times maintenance. Thus, a small input in grain leads to a large jump in meat or milk production. Finally, the overall governing factor is economics. In the United States, grain prices are low relative to animal-product prices because of high levels of purchasing power for animal products. Conversely, the developing countries produce, on a per capita basis, only about fifteen percent as much feedgrains as the developed countries (Table 18), and these grains are more expensive because of their widespread use as human foods. Many animal products, on the other hand, tend to be cheaper in the developing countries because of much lower purchasing power. Therefore, very little grain is fed to animals in developing countries, the main energy sources for intensive feeding being by-products.

Cassava is included in this section because of its potential use as high-energy feed in tropical countries or as an export commodity. Cassava (manioc) is a starchy root that grows in the tropics year-around and requires simple agricultural techniques. World production of cassava in 1977 was approximately 110 million metric tons. The single largest producers were Brazil (24 percent), Zaire (11 percent), Indonesia (11 percent), Nigeria (10 percent), and Thailand (10 percent).(28) Yields vary from about three to thirty tons per hectare, but normally average eight to ten tons.

Table 18. Per Capita Feed Grain Production by Region, 1977

Region	Corn	Barley	Sorghum	Total
		Kg/Capita/Per Year		
Developed	268	98	29	394
North America	690	86	84	860
W. Europe	81	140	2	222
Oceania	24	170	57	251
Other	68	3	3	73
Developing	38	8	16	62
Africa	40	9	22	70
Latin America	125	5	36	166
Near East	27	41	13	81
Far East	14	3	9	27
Other	1	-	1	2
Centrally Planned	512	64	-	114
Asia CPE	38	17	-	55
W. Europe & USSR	84	181	1	265
World	85	42	14	141

The European Economic Community is the largest importer of cassava pellets for animal feed. In 1978, it imported six million metric tons, and FAO projects that by 1985 world net import requirements will be around 13.2 million tons of which 11.5 million will go to the European Economic Community.(29) The overall production potential for this crop in the tropics has not been adequately assessed; however, extensive work has resulted in development of improved varieties that can yield up to forty tons per hectare.

CONSTRAINTS TO LIVESTOCK PRODUCTION

The constraints affecting livestock production are diverse, but the large majority fall within the areas of nutrition, health, breeding and selection, production economics, and marketing. Special attention is given in this section to some of the problems limiting production and productivity in the developing regions.

Animal Nutrition

Inadequate nutrition is among the most important factors limiting animal productivity throughout the developing world. Extended periods of little or no rain in the vast semiarid and arid regions severely limit forage growth. Under these conditions the small quantities of feed available (usually of poor quality) to each animal cannot meet even the maintenance requirements. The resulting prolonged nutritional stress adversely affects growth, reproduction, and survival of the offspring. The problem is compounded in many areas by gross overstocking and overgrazing. Most rangelands in Iran, for example, have over twice the number of animal units they can support.(30) Unless effective grazing control measures are taken, the ecozone will continue to deteriorate to the point of total uselessness. Rehabilitation of degraded rangeland is an expensive long-term commitment; policies directed to the conservation of this resource through appropriate range and animal-management techniques are without doubt a more fruitful investment.

A somewhat different situation prevails in the humid and subhumid tropical grazing lands where, with some significant exceptions, these areas are not overgrazed.(31) However, the overall low nutritive value of native grasses, particularly during the dry seasons in the vast savannahs of Africa and Latin America, tends to slow growth and reproductive performance. Complete mineral supplementation of cattle grazing native savannah pasture in Colombia, for example, resulted in increases in cow liveweight and calf production.(32) Judicious burning, sowing of adapted legumes or improved grasses, fertilization where economically possible, and strategic energy and/or protein and mineral supplementation can undoubtedly increase animal production in these areas.

Millions of cattle, sheep, goats, pigs, and chickens kept by small farmers and landless families in developing countries are often referred to as scavengers. The term is quite suitable since these beasts must fend for themselves to a large extent in the procurement of whatever feed or waste materials they can find along roadsides, harvested fields, and backyards. The nutritional status of these animals is unknown, but it can be theorized that unless the feed supply is constant and relatively abundant (both unlikely occurrences), nutritional deficiences are apt to occur. These animals are best suited for a more intensive integration within the cropping systems so that full advantage can be derived from crop residues and other by-products. There is a need to develop sound feeding systems based on local feeds that provide the required nutrients for maintenance and productive functions throughout the year.

Animal Health

Disease is a factor responsible for low offtake in developing areas. The moist, warm conditions of the humid tropics are conducive to the rapid proliferation of pathogens. The situation is aggravated by the lack of national or regional programs designed to provide animal health services to small producers at low cost. Frequently, the producer is not aware of the symptoms of a disease or does not have the financial means to purchase services and medicines. Animal diseases lower overall production and productivity through retardation of growth, cessation of reproduction, or, most obviously, death.

Conspicuous epidemics involving large numbers of animals and costing millions of dollars have occurred from time to time. An outbreak of foot-and-mouth disease in the United States from 1914 to 1916 covered twenty-two states and involved over 172,000 animals.(33) In 1978, African swine fever all but decimated the pig population of the Dominican Republic. The catastrophic result of this outbreak is the potential reduction of the meat supply by 20 to 30 percent to the Dominican population which is already suffering from protein/calorie deficiency.(34)

Annual losses in products and revenues caused by diseases not manifested in large-scale epidemics or by parasites are largely unknown, although some estimates suggest they may be up to 20 percent in countries with active disease-prevention programs and as high as 40 percent in others.(35) Preliminary results of on-going research by Allonby (36) in East Africa suggest that the detrimental effects of internal parasites (helminths) on productivity have been underestimated. His research on the use of parasite-free pastures resulted in a sevenfold return in terms of extra meat production from each ewe. The practicality of the system is now under investigation. Control of widespread diseases such as trypanosomiasis would significantly expand overall livestock production and increase productivity in large areas of Africa.

The root of the disease and nutrition problems confronting the small producer in developing regions does not seem to be the lack of technology to deal with these problems, but the high cost of technological

services and drugs. Most small holders operate on
marginal returns and cannot justify the added expense.
Governments must therefore unfold comprehensive programs through their extension services or in other ways
to make these services available at a minimum cost to
the producers.

Animal Breeding and Selection

Low productivity appears to be a common characteristic of most indigenous breeds throughout the developing world. A diversity of disease and nutritional-
stress conditions and varying environmental and managerial situations further accentuate low productivity.
Generally, native breeds are well-adapted to their
respective environment but natural selection has resulted, for the most part, in the establishment of
undesirable production traits for meat, milk, or egg
production.

The wide gap in animal productivity between the
developed and developing countries discussed earlier
indicates there is ample opportunity for improvement
through well-planned breeding and selection programs.
According to Stonaker (37), productivity in the tropics
can be increased 1 to 2 percent per year through better
directed selection and perhaps another 15 to 20 percent
through planned crossbreeding. Production traits to be
selected for in meat and/or milk animals include rapid
growth, high birth weights, offspring survival, overall
fertility, high milk yields, conformation and carcass
quality, resistance to parasites and disease, and
others. Crossbreeding plans should encourage crossbreeding between improved native breeds and/or improved
and exotic breeds.

Successful development of breeding and selection
programs for most parts of the developing world has
been hampered by the consistent unavailability of
reliable base data on performance of local breeds of
livestock. This information is essential to the decision-making process of a program planning phase.
Governments must be strongly encouraged to devote sufficient resources and mechanisms through which dependable performance data can be gathered, analyzed, and
properly inventoried.

Credit and Marketing Systems

Insufficient credit and limited marketing options
seriously constrain livestock production by those who
own the majority of livestock in developing countries -- the small producers.

Where small producer associations or cooperatives are weak or nonexistent, banking or government-lending institutions consider the individual small producer a bad risk. This situation encourages the proliferation of "opportunistic" moneylenders and their concomitant high interest rates. The resulting indebtedness precludes any opportunity for expansion or change in production systems. A handful of nonprofit organizations in Central and South America has been successful in extending credit to small farmers on a limited scale with excellent results. The scope of the credit needs, however, requires involvement at the national level. The allocations of sufficient resources to meet the needs and the establishment of credit outlets to reach the majority of small producers are essential components of effective credit programs.

Livestock marketing systems are poorly developed in many developing areas. Too often the producer has little choice but to sell his animals to middlemen who in turn sell the animals in urban centers at large profits. The lack of meat-quality standards or incentives for higher quality animals preclude selection pressure on the animal population. Planners should encourage the establishment of marketing cooperatives that provide small producers with an opportunity to obtain fair prices for their livestock and promote quality stock.

SELECTED ECONOMIC ASPECTS OF ANIMAL PRODUCTION

EFFICIENCY OF LIVESTOCK PRODUCTION

The comparative efficiency of animals in producing human food is subject to a number of factors, the major two being (a) whether efficiency is related to an individual productive unit or to the entire group of animals making up the producing unit or (b) whether protein or energy is being used as the major input into the system. These comparisons are further complicated by basic differences in the gastrointestinal structure between species. Thus we differentiate between simple-gutted animals (man, rat, pig, dog, chicken), nonruminant herbivores (rabbit, horse, donkey), and ruminants (sheep, goat, cow, water buffalo, and deer). Some of the large differences in efficiency reported in Tables 19 and 20 are a direct result of these differences in gastrointestinal structure and thus on the ability to utilize high-quality feed needed for high production levels.

In general, digestible energy serves as the driving force for animal production and thus serves as the denominator for most efficiency measures. First, however, two commonly used measures of animal efficiency are discussed. The first section of Table 19 presents some measures of female reproductive efficiency and the commonly accepted biological ceilings for these measures. Interestingly, sheep appear to have the greatest potential for improvement with current levels about 15 percent of the biological maximum while cows and pigs are each operating at 41 percent of the ceiling level. Another common measure is feed conversion efficiency (second section, Table 19) which considers product produced per unit of a well-balanced ration during the growth cycle of an individual animal. The figure for milk is less than one because milk is primarily water as opposed to the other products. The ruminant meat products (beef, lamb) are much less efficient because of the basic differences in digestive processes; however, these animals can compete with simple-gutted animals because they efficiently digest low-quality roughages as opposed to the latter species which require high-energy, low-fiber diets to achieve the stated efficiencies.

Next, the efficiency with which plant protein is converted into animal protein is summarized along with currently accepted biological ceiling levels. Again, milk is the most efficient animal protein source at both current levels and biological ceiling levels. In terms of energy produced per unit of energy consumed (bottom Table 19), pigs currently perform slightly better than milk cows with poultry, eggs, and meat next.

Probably the most appropriate measures of efficiency, for purposes of this paper, are those taken from Reid and White (38) who look at protein produced per unit of energy input on the basis of a producing unit (herd, flock) where energy required for reproduction, maintenance, and others is accounted for (Table 20). These results also point out the efficiency of milk production relative to other sources of animal protein although the high efficiency levels of egg and broiler production are about equal to current conversion efficiencies of milk production under United States conditions.

Table 19. Some Efficiency Measures Summarized for Farm Animals

Measure	Species/Product

Female reproductive efficiency[a]

	Cow	Ewe	Sow	Hen
Existing standard				
No. litter	1.0	1.5	9	n/a
No./year	0.9	1.5	18	220
Biological ceiling				
No./litter	2.0	5.0	20	n/a
No./year	2.2	10.0	44	365

Trends in feed conversion efficiency[b]

	Pork	Beef	Milk	Sheep	Poultry	Eggs
1940	3.9	10.0	1.2	-	4.5	3.5
1970	3.3	7.5	0.7	8.0	2.2	3.0
1985	2.5	5.0	0.4	4.0	1.8	2.5

Efficiency of protein conversion[c]

	Milk	Beef	Sheep	Poultry	Eggs	Pig	Rabbit
Presenting level	.28	.06	.04	.20	.20	.12	.11
Ceiling level	.46	.10	.12	.33	.31	.18	.20

Energy produced/energy consumed[d]

20	7	3	13	15	23	-

[a] Source: University of California Food Task Force, A Hungry World: The Challenge to Agriculture (University of California, 1974), p. 141.

[b] Ibid., p. 139.

[c] J. C. Bowman, Animals for Man (London: E. Arnold Publishers Ltd., 1977), p. 63.

[d] Edible energy (Kcal) per 100 Kcal metabolizable feed energy consumed. Source: University of California Food Task Force, A Hungry World: The Challenge to Agriculture (University of California, 1974), p. 139.

Table 20.   Efficiency With Which Farm Animals Produce
Food Protein[a]

Food Product	Level of Output and/or Degree of Intensivity	Protein Production Efficiency[b] (g/Mcal of DE)
Eggs	200/eggs/year	10.1
	250 eggs/year	13.7
Broiler	1.59 kg/12 wk; 3 kg feed/1 kg gain	11.9
	1.59 kg/10 wk; 2.5 kg feed/1 kg gain	13.7
	1.59 kg/8 wk; 2.1 kg feed/1 kg gain	15.9
Pork	91 kg/8.3 mo; 6 kg feed/1 kg gain	5.0
	91 kg/6.0 mo; 4 kg feed/1 kg gain	6.1
	91 kg/4.4 mo; 2.5 kg feed/1 kg gain	8.7
	91 kg/3.7 mo; 2.0 kg feed/1 kg gain; No losses; biological limit (?)	12.1
Milk	3,600 kg/yr; No concentrates	10.5
	5,400 kg/yr; 25% of energy as concentrates	12.8
	9,072 kg/yr; 50% of energy as concentrates	16.3
	13,608 kg/yr; 65% of energy as concentrates	20.5
Beef	500 kg/15 mo; 8 kg feed/1 kg gain	2.3
	500 kg/12 mo; 5 kg feed/1 kg gain	3.2
	Highly intensive system; no losses; biological limit (?)	4.1

[a] These data represent the overall efficiency with which dietary energy is converted to food, as they include the energy cost of reproduction, rearing of breeding stock, and mortality as well as that of production itself.

[b] Efficiency is expressed as grams of protein produced per megacalorie of DE ingested.

Source: J. T. Reid and O. T. White, "Energy Cost of Food Production by Animals." A. M. Altschul and M. L. Wilke, eds. New Food Protein Sources, vol. 3 (New York: Academic Press, 1978), p. 132.

DEMAND FOR LIVESTOCK PRODUCTS

A variety of estimates have been made for demand of livestock products by country, by region, and at world levels. These are generally based on growth in demand caused by population increase and per capita income growth as it affects livestock product demand. In virtually every case, the proportional increase in demand for livestock products, as income increases, is much greater than for crop products.(39) Therefore, the uniform characteristic of the available demand projections is the relatively rapid increase in demand projected for livestock products relative to foodgrains for the developing countries. Compound annual-growth rates in consumption given in the USDA's Grain-Oil-seeds-Livestock (GOL) model (40) under the Alternative 1 (Baseline, most likely scenario) for 1969-71 to 1985 are 3.9 percent for meat and 3.2 percent for grains in the developing countries. Some estimates for 1985 on a per capita basis are given in Tables 21 and 22 as well as projected total estimated consumption for the year 2000.

Table 21. Estimated Per Capita Demand (kg/yr), 1970 and 1985

Region, Year	Meats	Beef & Veal	Mutton & Lamb	Pig	Poultry	Eggs	Milk
World							
1970	29.1	10.7	1.9	9.5	4.4	5.0	104.8
1985	33.4	12.2	2.3	10.3	5.7	5.5	110.2
USA							
1970	119.3	53.5	1.4	29.4	30.2	18.1	262.3
1985	139.8	65.4	1.4	26.7	41.9	17.5	282.3
Asia							
1970	4.6	1.9	1.0	4.6	1.6	1.7	35.9
1985	6.5	2.5	1.3	6.2	2.3	2.3	44.1
Africa							
1970	12.6	6.8	2.8	0.8	1.3	1.3	44.6
1985	14.9	8.0	3.4	0.9	1.7	1.5	48.6
Latin America							
1970	36.7	21.2	1.5	6.8	2.9	4.6	87.3
1985	39.7	21.2	1.8	8.0	4.1	5.6	100.1

Source: University of California Food Task Force, A Hungry World: The Challenge to Agriculture (University of California, 1974).

Table 22. Estimated Total Consumption of Ruminant
Livestock Products, 1970 and 2000.

Region, Year	Dairy Products (milk equivalent	Beef & Veal (carcass weight)	Mutton & Lamb (carcass weight)
Developed Regions			
1970	286.2	27.6	3.7
2000	403.7	43.7	5.6
Developing (India)			
1970	38.8	0.2	0.4
2000	91.2	0.4	1.0
Developing (others)			
1970	77.5	10.9	2.7
2000	205.5	26.3	6.3
World			
1970	402.5	38.7	6.8
2000	700.3	70.4	12.9

Source: Fitzhugh, H.A., H.J. Hodgson, O.J. Scoville,
T.D. Nguyen and T.C. Byerly, The Role of Ruminants in
Support of Man (Morrilton, Arkansas: Winrock Inter-
national Livestock Research and Training Center, 1978).

COST OF PROTEIN

This section attempts to establish a cost comparison between various food items based on protein content. The values in Table 23 represent the cost of 29
grams of high quality protein (the daily requirement)
based on the retail cost of each food product and the
adjusted protein content (see page 192). In other
words, these values indicate what would be the cost to
the consumer if he/she were to derive his/her daily
protein requirement entirely from one food item. The
formula used to calculate the cost of protein is:

$$\text{cost of 29 g of protein} = \frac{\text{cost per kg of food} \times 29 \text{ g}}{\text{adjusted protein content}(\%) \times 10}$$

This methodology is not ideal since it does not
include the value or contribution of other nutrients,
but it is useful in establishing relative protein costs
to the consumer.

With the exception of wheat flour, the cheapest source of protein in the United States is poultry meat followed by eggs. The daily protein requirement costs less from poultry meat, eggs, cheese, and milk than from bread, rice, or potatoes. Similar relationships appear to be true in Kenya and the United Kingdom although the data are incomplete.

The generalization that animal products are more expensive sources of protein than staples such as potatoes, rice, and wheat (bread) does not seem to hold true. Starchy foods (cereals, grains, and tubers) are basically high energy sources. It seems that appropriate mixtures of food items would be in the long run more economical and more nutritious.

Table 23. Relative Cost of 29 g of Protein from Various Food Sources at Four Locations in US$

Food Item	United States	United Kingdom	Kenya	Jamaica
Beef				
Hamburger	0.670			
Rump Steak			0.480	
Stewing Steak		0.490		
Pork				
Leg of Pork			0.599	
Pork Chops	0.930			
Ham	1.010			
Sausage		0.654		
Lamb				
Leg of Lamb	0.852		0.590	
Lamb Chops		1.305		
Shoulder			0.540	
Poultry				
Whole Chicken, dressed		0.248		0.552
Eggs	0.312	0.312	0.381	
Cheese, cheddar	0.580	0.272		
Milk, fluid	0.423	0.362	0.305	0.405
Wheat				
Flour	0.222		0.252	0.159
Bread	0.678			
Corn, flour	0.400		0.132	
Rice, white polished	0.616			0.415
Peanuts, shelled			0.326	
Peas, dry		0.714	0.609	
Broad Beans			0.141	
Potatoes	0.959		1.053	
Tomatoes	7.816		7.250	
Oranges	6.853		6.440	
Turnips			4.500	

CONCLUSION

Over the past ten thousand years, man has evolved systems for the production of food in which empirical methods have given way to systematic procedures based upon better understanding of the behavior of plants and animals and their interaction with environmental factors. Research has given man the tools through which he can maximize yields, prevent losses, avoid disease, ameliorate climatic stresses, and improve the nutritional quality of his food resources.

We have in relatively recent times learned that our capacity to grow, work, develop mentally, and resist diseases depends upon the consumption of various substances called "nutrients" which are found distributed among a wide array of foodstuffs. We can satisfy our needs for nutrients by utilizing an almost infinite number of combinations of foodstuffs. Theoretically, we can calculate perfect diets which would assure a person of getting precisely all of the nutrients he/she requires. In the United States, producers of pigs and poultry and dairy and beef cattle use precision formulation in feeding their livestock, thus assuring good health, maximum reproductive performance, and minimum wastage of expensive feedstuffs. Although it is possible to formulate such diets for people, their use is generally limited in practice. The point is that each individual is left with the task of "balancing his own ration." In doing so, many factors determine the degree of success in achieving a balanced nutrient intake; for example, the kinds of foodstuffs available, the capacity to acquire food, food preferences, and the nutritional value of foods. The more limited the access to a variety of food materials, the more difficult it becomes to assure the consumption of all of the necessary nutrients. Premium foods, therefore, become important in filling the nutritional gap since they tend to assure the ingestion of the most necessary nutritional substances. Ironically, because people do not conscientiously eat to satisfy nutrient requirements, serious imbalances in nutrient intake can and do occur in our own society in spite of and/or because of the wide range of foods available.

The survival and growth of any society is dependent upon the health of its people. Leaders must recognize the importance of the provision of adequate food supplies to the development process. Political and economic independence rest on the ability of a nation to satisfy its food requirements. Yet the investment in agriculture in many nations is pitifully low. The United States, unfortunately, still has not

assumed the world leadership role in agriculture which
its preeminence in agriculture would suggest. Developing countries must not only be exhorted to solve
their food problems, they must be assisted more effectively and forcefully in that direction. Specifically,
their national policies should be designed to encourage
the optimum utilization of the resource base in the
production of a wide range of food commodities whether
they are grains, fruits, tubers, animals, or vegetables. Furthermore, national policies must give
attention to the collection, transport, and processing
for storage and/or redistribution of foodstuffs from
the production sector to the consuming public. Agricultural research and development programs can be
strengthened to accomplish these goals.

In the United States it is reasonably certain that
we can maintain a level of food supply far in excess of
our needs. Nevertheless, we confront a series of
important issues related to food supply which will
affect not only our own welfare but that of other
nations as well. These issues are:

The use of land and other resources for food
production. As a nation we have not yet established
any forward looking policies which will safeguard the
productive capacity of our lands, assure the provision
of water resources, and guarantee the supply of energy,
fertilizer, and labor for food production.

The disposition of farm products. Consumers will
have to pay more for food products. The form and cost
of nutrients will become increasingly important.
Economics will determine how much grain is converted to
animal products. World demand for United States grains
is likely to intensify.

The need to sustain our agricultural research
capability. Our agricultural research establishment
currently suffers from several constraints including
limited resources, aging facilities, parochial orientation and political pressures which seriously impede
the process of aggressively addressing important problems facing the nation.

The necessity of adequately addressing international food and agriculture issues. The United
States lacks a professional career cadre whose principal function is to maintain a capacity to understand,
interpret, and interact in a total array of food and
agriculturally related problems.

The <u>urgency to crystallize United States food and agriculture policies.</u> While avoiding over regulation and bureaucratization there is an urgent need to develop total strategies and appropriate policies which do affect United States agriculture.

Signaling problem areas is not the same as providing solutions. The issues are complex as may be their resolution; however, need we wait for another world food crisis before appropriate solutions are forthcoming?

APPENDIX

CONTRIBUTION OF ANIMALS TO MANKIND

In addition to food, animals provide man with a variety of products and services, which in many instances are difficult to accurately quantify. This appendix summarizes the contribution of animals not presented in the text.

Fiber, Skins, and Fats

A less accurately quantifiable contribution of animals is in the form of materials such as fibers (wool and hair), hides, skins and pelts, and animal oils, greases, and fats. Wool is the most important product among the animal fibers. In 1977, some 2.59 million metric tons were produced with an estimated market value exceeding US$5,569 million. Although wool production has decreased in the last decade, it is expected that the rapidly increasing costs of raw materials for manufacture of synthetic fibers will favorably affect the profitability of wool production. Mohair from Angora goats is the main hair in world trade. The United States produces about 50 percent of the world supply or about 3000 tons annually.(41) Hair from camels, buffaloes, sheep, yak, reindeer, pigs, and other animals is used in the manufacture of a variety of products such as brushes, felt, and stuffing materials.

Hides and skins are obtained from a variety of animals both wild and domestic, but the most important sources of supply are cattle, sheep, and goats. Mc-Dowell (42) reports that hide exports rank in the first ten commodities earning foreign exchange for about thirty countries. The principal use of hides is in the production of leather for the manufacture of footwear, apparel, and various industrial products.

In 1977, around 9.74 million metric tons of lard, tallow, and grease were produced throughout the world (43) with an estimated market value of well over US $4,000 million. The United States produces roughly one-half of the world's tallow and grease and exports about 1.3 million metric tons annually which represents nearly three-fourths of the total world trade.(44) In many developing countries most of the animal fats are consumed locally or used as fuel or in candle and soap making.

Waste Products

Fuel, feed, fertilizer, and building materials are also important contributions from animals throughout the Middle East, India, Northern Africa, and other developing regions. Cattle, buffalo, camel, sheep, and goat dung is used extensively as fuel due to the scarcity of wood or suitable plant material for heating and cooking. Makhijani and Poole (45) estimate that more than 200 million tons of manure are used yearly as fuel in the developing countries. To replace dung with coal and oil in India alone, it would cost more than US$3,000 million annually, exclusive of distribution costs.(46)

The use of animal manures as a substrate in the production of biogas (methane) is attracting a great deal of attention. Countries presently utilizing this technology include Korea (29,000 plants), India (20,000 plants), and Taiwan.(47) Research is currently underway to solve some of the limitations in the applicability of this technology.

Inedible tissues, such as blood, bones, inedible offal, trimmings, and poultry heads, feet and feathers from processing plants, are commonly used in developed countries as raw materials in the manufacture of highly nutritious feeds. Recycling of manure (feeding processed manure to livestock) is a subject of intensive research; cattle and poultry manure have been successfully fed, as part of the diet, to ruminant animals. As processing technology increases, undoubtedly so will the use of manures in the rations of ruminants and possibly nonruminant animals.

Animal feces, including night soil (human excreta), have been used for centuries as fertilizers to replenish soil nutrients removed through harvesting of crops and to improve soil texture. McDowell (48) reports that approximately 40 percent of the farmers of the world depend wholly or in part on animal wastes to improve soil fertility. The rising cost of chemical fertilizers is likely to force the poorer farmers in developing areas to depend more and more on the use of animal manures. Conversely, manure disposal has become a problem to commercial beef, dairy, poultry, and swine producers in developed countries. Innovative means of handling, conserving, and effectively using these large quantities of manure are being actively sought.

Manure and hair are extensively used in some developing areas for construction of dwellings. Mixtures of animal feces, mud, and/or clay are made into

234

bricks or used as cement to plaster frameworks made of
twigs or heavier pieces of wood.(49) Nomads in the
Middle East incorporate goat and camel hair in the
material used to build their tents.

Traction

Power and transport are two important, but often
overlooked, contributions of certain species of animals
to people in developing regions. Estimates of the
animal power used in agriculture in selected regions
are impressive; of the total animal and tractor power
used in Africa, the Far East (excluding China), the
Near East, and Latin America, 82, 92, 88, and 75 percent, respectively, is derived from animals.(50) In
contrast, animals provide only 1 percent of the farm
power in the United States and 24 percent in France.
McDowell's (51) assessments further indicate that close
to 20 percent of the world's population largely or
wholly depends on the transport of goods by animal
cartage or packing. In view of the rising costs of
fossil fuels (and probable shortages), it is unlikely
that major changes will occur in the use of animal
power in the near future.

Pest Control

The value of animals in pest and weed control is
difficult to assess. It is well-known, for example,
that some breeds of dogs were developed to protect
sheep from predators; mongoose are used to hunt and
control snakes in Asia; and chickens, geese, and other
fowl kept by farmers in developing areas help control
the proliferation of weeds, bugs, and insects. Ruminant animals (cattle, buffaloes, sheep, and goats) play
a significant role throughout the developing world in
maintaining roadsides, irrigation canals, and urban
fringe areas clear of weeds, grasses, brush, and waste
material. Goats have been successfully used in biological control of brush in Texas, California, and Australia. A farmer in north central Mexico reports that
he has significantly reduced his insecticide application and successfully controlled boll weevil in cotton
by intensive grazing of his fields after harvest with
goats.

Religion

Throughout history various animals have been the
object of a diversity of religious beliefs. Worship of
sacred animals or sacrifices of animals to appease the
gods have been components of several important reli-
gions.(52) Present day Hindus, for example, consider

the cow a holy animal. Harris (53) points out that the economic importance of the cow to the Indian farmer is at the root of "cow love."

Research

The contribution of animals to man's understanding of biological processes is often underestimated. In the medical field, animals are used to develop new surgical techniques, to test new drugs, or as sources of drugs (e.g., hormones). Most of the basic knowledge in the areas of nutrition and physiology has been obtained through animal experiments. Byerly (54) quotes data from the Institute of Laboratory Animal Resources which show that some forty-five million mammals and birds were used for research purposes in 1971.

FOOTNOTES

(1) Bell, H. B., L. N. Davidson and H. Scarborough. Physiology and Biochemistry. 6th ed. Baltimore: Williams and Wilkins Co., 1965.

(2) FAO. The Fourth World Food Survey. Food and Nutrition Series, No. 10. Rome, 1977.

(3) Ibid.

(4) FAO. Amino Acid Content of Foods and Biological Data on Proteins. Nutritional Studies No. 24. Rome, 1970.

(5) FAO/WHO. Energy and Protein Requirements. FAO Nutrition Meetings Report Series No. 52. Rome, 1973.

(6) Ibid.

(7) Block, R. J. and H. H. Mitchell. "Correlation of the Amino Acid Composition of Proteins with Their Nutritive Value." Nutrition Abstracts and Reviews. 1946.

(8) Schmitt, B. A. Protein, Calories and Development: Nutritional Variables in the Economics of Developing Countries. Boulder, Colorado: Westview Press, 1979.

(9) FAO/WHO, Energy and Protein Requirements.

(10) University of California Food Task Force. A Hungry World: The Challenge to Agriculture. University of California: Division of Agricultural Sciences, 1974.

(11) FAO/WHO, Energy and Protein Requirements.

(12) FAO, The Fourth World Food Survey.

(13) Lappe, F. M. and J. Collings. "When More Food Means More Hunger." In R. B. Talbot, ed. The World Food Problem and the U.S. Food Politics and Policies: 1977. Ames, Iowa: Iowa State University Press, 1978.

(14) Anonymous. 1977. "Caprina e Ovinos, Criacoes Ideais Para o Poligono Das Secas." Dirigente Rural xvi-9 (1977): pp. 6-18 (Brazil).

(15) Svedberg, P. "World Food Sufficiency and Meat Consumption." American Journal of Agricultural Economics 60 (1978): p. 661.

(16) Bowman, J. C. Animals for Man. London: E. Arnold Publishers Ltd., 1977.

(17) McDowell, R. E. Ruminant Products: More Than Meat and Milk. Morrilton, Arkansas: Winrock International Livestock Research and Training Center, 1976.

(18) FAO, The Fourth World Food Survey.

(19) FAO. Production Yearbook, 1977. Rome, 1978.

(20) Szczepanik, E. F. "Recent Changes in World Livestock/Feed Price Ratios." Monthly Bulletin of Agricultural Economics and Statistics 25 (1976): p. 1.

(21) Turner, H. N. "Extensive Production Systems." Workshop Proceedings on the Role of Sheep and Goats in Agricultural Development. Morrilton, Arkansas: Winrock International Livestock Research and Training Center, 1976.

(22) Fitzhugh, H. A., H. J. Hodgson, O. J. Scoville, T. D. Nguyen and T. C. Byerly. The Role of Ruminants in Support of Man. Morrilton, Arkansas: Winrock International Livestock Research and Training Center, 1978.

(23) Bula, R. L., V. L. Lechtenberg and D. A. Holt. Potential of the World's Forages for Ruminant Production. Morrilton, Arkansas: Winrock International Research and Training Center, 1977.

(24) Hodgson, H. J. "Forages, Ruminant Livestock, and Food." BioScience 2 (1976): p. 625.

(25) Fitzhugh, et. al., The Role of Ruminants in Support of Man.

(26) Ibid.

(27) Barat, S. K. "Abattoir By-products: Potential for Increased Production in Developing Countries." Proceedings Conference on Animal Feeds of Tropical and Subtropical Origin. London: Tropical Products Institute, 1975.

(28) FAO, Production Yearbook, 1977.

(29) Freivalds, J. "Cassava: an Old Crop Captures New Markets." Feedstuffs, January 1979, p. 7.

(30) Martinez, A. Protein Development Program -- Iran. Report to the Ministry of Agriculture and National Resources. Tehran, Iran: FMC Corporation, 1975.

(31) Bula, et. al., Potential of World's Forages.

(32) Lebdosoekojo, S. "Mineral Supplementation of Grazing Beef Cattle in Eastern Plains of Colombia." Ph.D. thesis, University of Florida, Gainsville, 1977.

(33) Committee on Foreign Animal Diseases. Foreign Animal Diseases. Their Prevention, Diagnosis and Control. Richmond, Virginia: United States Animal Health Association, 1975.

(34) Martinez, A. Personal communication, 1978.

(35) University of California, A Hungry World.

(36) Allonby, E. W. "Disease and Parasite Constraints to Production." Proceedings Workshop on the Role of Sheep and Goats in Agricultural Development. Morrilton, Arkansas: Winrock International Livestock Research and Training Center, 1976.

(37) Stonaker, H. H. "Increased Animal Production Through Breeding." A. M. Altschul and H. L. Wilcke, eds. New Protein Foods, Vol. 3. New York: Academic Press, 1978.

238

(38) Reid, J. T. and O. T. White. "Energy Cost of Food Production by Animals." A. M. Altschul and H. L. Wilcke, eds. New Protein Foods, Vol. 3. New York: Academic Press, 1978.

(39) U.S.D.A. Alternative Futures for World Food in 1985. World GOL Model Analytical Report Vol. 1. Economics, Statistics and Cooperatives Service. Foreign Agricultural Economic Report No. 146. Washington, D.C., 1978b.

(40) Ibid.

(41) McDowell, Ruminant Products.

(42) Ibid.

(43) U.S.D.A. Agricultural Statistics 1978. Washington, D.C., 1978a.

(44) Makhijani, A. and A. Poole. Energy and Agriculture in the Third World. Cambridge, Massachusetts: Ballinger Publishing Co., 1975.

(45) Ibid.

(46) McDowell, Ruminant Products.

(47) Ibid.

(48) McDowell, R. E. "Inventory and Capacity for Expanding Protein Available From Animal Sources." Symposium Protein and the World Food Situation. North Carolina State University, Raleigh, 1974.

(49) Bowman, Animals for Man.

(50) FAO. Smaller Farmlands Can Yield More. World Food Problems, No. 8. Rome, 1969.

(51) McDowell, Symposium.

(52) Bowman, Animals for Man.

(53) Harris, M. Cows, Pigs, Wars and Witches: The Riddles of Culture. London: Hutchison and Co. Ltd., 1974.

(54) Byerly, T. C. "Competition Between Animals and Man for Agricultural Resources." A. M. Altschul and H. L. Wilcke, eds. New Protein Foods Vol. 3. New York: Academic Press, 1978.

6
AQUATIC ANIMAL PROTEIN FOOD
RESOURCES—ACTUAL AND POTENTIAL

Carl J. Sindermann

Whenever the interrelated subjects of world population and world protein production are discussed, the
conversation ultimately gets around to the role that
aquatic animals can, and undoubtedly will, play in
meeting future human nutritional needs. It is important, therefore, to assess as objectively and candidly
as possible, the present and potential contributions
from the aquatic environment to the protein food supply -- to sort out the myths from the realities. The
sorting process can be a sobering experience, particularly for those who feel that somehow, if famine stalks
the land, the seas will provide.

Since over 70 percent of the earth's surface is
covered by water, and since an enormous amount of plant
and animal production goes on in the world's oceans and
fresh water, it should be reasonable to expect a
greater food yield than we have had so far. The amount
of increase that is possible is a subject of much
discussion and extensive misunderstanding. It seems
evident, though, that any increased yield from the
oceans will not be food of plant origin, since the
predominant plant production there is in the form of
microscopic algae not easily harvested or utilized by
humans. The importance of the ocean now and in the
foreseeable future is and will be as a source of animal
protein (at present about 13 percent of the world's
animal protein is derived from aquatic sources).

Realistic appraisals of the productive potential
of the oceans have been made recently. It has become
clear that in the sea, just as on land, there are great
areas of very low productivity. Most of the really
productive areas huddle close to shore, over the continental shelves and in the estuaries, or follow the
great current systems of the oceans. Ninety percent of
the world's catch of fish is taken on the continental

shelves, but less than 10 percent of the ocean surface is over the shelves. Furthermore, we still do not "harvest" most species in the real sense, because we have no control over planting or survival, nor do we fully understand the extent of the standing crop or the dynamics of its production. The marine fisheries still exist largely at a hunting economy level (or at best a primitive range-management level), except for slow inroads being made by legally constituted national and international fisheries management bodies, and by the halting emergence of aquaculture as a significant factor in aquatic food production.

There is a large amount of current activity in the United States and elsewhere concerned with management of natural populations of fish and shellfish, as well as aquaculture development. With the enactment of the Fisheries Conservation and Management Act of 1976, the United States assumed fisheries jurisdiction on its continental shelves and is now in the early and painful throes of attempting rational conservative management of fish stocks, some of which had been heavily ex-ploited and depleted by efficient foreign distant-water fleets since the early 1960s. Other nations have extended their fisheries jurisdictions, and there are numerous efforts -- national and international -- to manage fish stocks to ensure sustained yields of "tra-ditional" species. There are also continuing efforts to expand fisheries to underutilized or unutilized "nontraditional" species wherever they exist in ade-quate concentrations in the world oceans.

Concerning aquaculture, a federal interagency committee is at present drafting a national aquaculture plan for the United States, and aquaculture development bills have been introduced yearly (without success so far) since 1977. The National Oceanic and Atmospheric Administration (NOAA) supports aquaculture research and development programs at several of its National Marine Fisheries Service laboratories and through grants to universities from its Office of Sea Grants. The De-partment of Interior is responsible for freshwater aquaculture, and other federal agencies are involved in lesser ways. The Department of Agriculture was named lead agency for aquaculture in the Food and Mariculture Act of 1977. Internationally, the Food and Agriculture Organization (FAO) has an active program of fishery development in many nations, with aquaculture one of the foci. The International Council for the Explora-tion of the Sea has taken new interest in marine aqua-culture since 1977 and has a number of working groups involved in areas such as pathology, introductions of nonindigenous species, and genetics. Additionally, there are numerous bilateral activities around the world, such as the United States-Japan Natural Resources Panels on Aquaculture (UJNR), encouraging joint projects and exchange of technology.

This paper attempts to look broadly at food production from aquatic sources, with emphasis on marine contributions -- present and potential. The emerging role of aquaculture, despite significant constraints, is also considered against a background of continued production from fisheries on natural stocks.

FISHERIES ON NATURAL STOCKS

Despite the limitations of our vision, there is cause for reasoned optimism when considering food production from the world's oceans. During the period 1950 to 1970 harvests from fisheries on natural populations of marine animals tripled, with an average annual increase in landings of about 6 percent, from eighteen million tons in 1950 to sixty million tons in 1970. (The catch from fresh water in 1970 was estimated at just over ten million tons, making a grand total from aquatic sources of about seventy million tons). Beginning in the 1960s, fisheries on natural stocks were dominated by highly efficient fleets of modern fishing vessels, principally from the European socialist countries. Production peaked in 1970, then leveled off and even declined slightly during the current decade, due in part to a reduction in herring and Peruvian anchovy catches and to the very recent restriction of fisheries resulting from extension of fisheries jurisdiction out on the continental shelves by a number of nations, including the United States.

Constraints on increases in food production from marine-fishery sources are severe. A number of stocks of "traditional" food species, such as cod, haddock, certain flounders, lobsters and redfish, are fully exploited, and have been in some instances overexploited; landings in recent years have been sustained by development of fisheries for nontraditional species. Costs associated with fishing -- particularly fuel costs -- have escalated alarmingly, and concern has been expressed about impacts of coastal pollution on fisheries.

Despite these constraints, estimated production from United States commercial fisheries in 1978 was a record three million tons, with an exvessel value of $1.8 billion (the previous record was 1962 with 2.7 million tons). Leading species in 1978 were (in quantity) menhaden, crabs, shrimp, and salmon, and (in value) shrimp, crabs, salmon, and tuna. These are all food species with high unit value, except for menhaden which is processed for fish meal and oil. This record catch still constitutes only about 5 percent of the world total, despite the fact that the continental shelves of the United States are among the most productive in the entire world. (By comparison, the estimated harvests by Japan and USSR were each in excess of ten million tons).

Many knowledgeable fishery scientists have accepted an annual global production of roughly one hundred million tons of traditional fish species as a maximum expected yield from the oceans, with an uncomfortably large range -- from twenty-one million tons to two billion tons (and with one estimate by Graham and Edwards (1) at 60 million tons). More recently this figure of one hundred million tons has been revised downward (2), and the present catch of sixty million tons of traditional species seems more plausible on a sustained basis. (It should be recognized, though, that this estimate would be considered very conservative by some authorities, who cling to the one hundred million ton or higher sustained-yield estimate. Also, the estimate does not take into account changes in marine climate, which may affect future yields of important species).

Further increase in world food production from the sea is feasible, but it will depend more and more on utilization of natural stocks of nontraditional species now underharvested (such as squids, Antarctic krill, and some of the small herring-like fishes) or not harvested (such as lantern fishes); on technological advances that increase the efficiency of capture; on better utilization of fish that are harvested (a greater percentage of the catch used for direct human consumption, rather than for fish meal and oil); and on solution of economic, environmental, and institutional problems which now suppress expansion of marine aquaculture. World food production from aquaculture now constitutes about 10 percent of total fisheries production, and there have been substantial advances in the technological base for aquaculture, for marine as well as freshwater species.

In examining world fish catch statistics, it is important to note that over one third of the total is not used directly as human food, but is processed into fish meal and oil, and the meal is fed to domesticated animals in those few countries that can afford such expensive animal protein. Thus the fish become food

for humans, but in a somewhat inefficient way, considering the loss during conversion of fish protein to
pig or chicken protein (a loss of 60 to 70 percent).
The inefficiency becomes much less significant, however, when fish which are considered inedible or
undesirable as food for people are turned into pork or
chicken. Despite this, there is still a challenge to
find ways in which fish proteins now fed to domestic
animals can be prepared in forms acceptable for direct
human consumption, to avoid the inevitable loss in
conversion.

AQUACULTURE

Aquaculture, defined as the culture or husbandry
of aquatic organisms in fresh or salt water, yielded an
estimated six million metric tons of food in 1975 --
less than 10 percent of the world production of fishery
products (3). Yields from aquaculture doubled in the
period 1970 to 1975, according to FAO statistics; much
of the increase was in high-unit-value species in
developed countries. Some countries now depend on
aquaculture for a significant part of fish and shellfish production. Japanese aquaculture production
increased fivefold (to 500,000 metric tons) in the
period 1970 to 1975, while Israel now derives almost
half its finfish production from aquaculture. Additionally, there has been expansion of traditional
culture practices for carp, milkfish, tilapia, and
other species in many countries, particularly in Asia.

Aquaculture in the United States

United States aquaculture production in 1975 was
estimated at only 65,000 metric tons -- slightly more
than 2 percent of U.S. fish and shellfish landings and
about one-hundredth of estimated world aquaculture
production. Even this limited amount still constituted
(in 1975) about a quarter of our salmon production,
about two-fifths of our oyster production, and about
half of our catfish and crawfish production.

Aquaculture in the United States must look to
fresh water for its past successes. Aquaculture of
trout, salmon, and catfish in fresh water is well
established. Trout and salmon culture began early in
the century to be an important factor in sport fishery
production and stream rehabilitation; catfish culture
in the southeastern states has been a viable and expanding industry for over a decade. Recent exciting
developments in fresh water include the successful
expansion of introduced populations of salmon in the

Great Lakes, pilot-scale polyculture of fish in ponds, and the commercial success of crayfish and minnow culture in the southeastern states.

Future expansion of United States aquaculture should be in estuarine and coastal waters, with salmon, shrimps, and bivalve mollusks leading candidates (4). The history of salmon production on the west coast of the United States includes a number of dramatic changes -- high levels of natural production in the early days of the 20th century; drastic decline due to overfishing, building of dams and increasing indus- trialization; partial rebuilding of stocks with the development of a system of public hatcheries; and most recently, the exploration of pen culture in sea water and the beginnings of private ocean ranching of salmon in the Pacific Northwest.

Shrimp production in the United States has long been dominated by the Gulf of Mexico fishery on penaeid species. Recent developments include expansion of shrimp fisheries elsewhere in the United States, gradual but significant annual increases in imports of shrimp, and initial attempts at commercial culture of penaeids as well as freshwater shrimps of the genus Macrobrachium.

Molluscan shellfish production is still dominated by landings from fisheries on natural stocks, particu- larly from Atlantic surf clam populations. Oyster production methods have long employed simple culture techniques, but the development of commercial hatch- eries as a source of seed is becoming an important method of augmenting production.

Aquaculture is a logical aquatic counterpart of agriculture. A reasonable assumption would be that the development of the animal husbandry component of aqua- culture in the United States and elsewhere could follow the same sequence of steps as did modern agriculture (selective breeding, disease control, diet formulation, automation, etc.). Unfortunately, there are funda- mental differences which emerge when comparing agricul- ture and aquaculture. Agriculture development met a specific 20th century need for increased food produc- tion; it was built on a core of already-domesticated species; and it was supported by relatively generous funding for both applied and fundamental research. None of these factors have existed or now exist for aquaculture.

Attempts to find parallels with agriculture must be carefully circumscribed, since it is highly unlikely that aquaculture will ever replace fisheries on natural populations to the extent that terrestrial animal husbandry has replaced hunting as a protein food source. Aquaculture production in estuarine and near-shore waters could conceivably surpass production from wild stocks of the continental shelves and open ocean, but such offshore areas are unlikely candidates for the effective environmental control required for aquaculture and will probably remain for the foreseeable future as a source of food only from fishing. Only estuaries and protected coastal waters are amenable to the manipulations required for aquaculture, and even these environments are often resistant to man's attempts at control. Despite such resistance, these margins of the sea are where significant protein food should be produced by aquaculture in the decades ahead. The extent of potential production can be perceived only dimly, but is substantial.

A recent well-researched report (5) by the National Academy of Sciences titled "Aquaculture in the United States" offered a number of perceptions about the future of aquaculture in the United States. Significant among the conclusions reached were these:

"...in the United States, aquaculture will have only a minor impact on food production in the near term, in comparison with other food production systems."

"...in the long term, aquaculture will be a means of increasing protein supplies."

"...aquaculture has the potential to contribute to increased food production. If this potential is to be tested, expenditures for current programs and for research and development must be increased."

"Constraints on orderly development of aquaculture tend to be political and administrative, rather than scientific and technological."

"Aquaculture in the United States has lacked coherent support and direction from the federal government. Poor coordination, lack of leadership, and inadequate financial support have traditionally characterized programs relating to aquaculture."

World Aquaculture -- Projection

From an international perspective, there is cause for reasoned optimism when considering increased food production from aquaculture. Despite institutional, economic, environmental, and technological constraints, global yields are increasing. Intensive culture of high-unit-value species, such as pen-rearing of yellowtail in Japan, is now the basis for a large and economically-viable industry; saltwater rearing of salmon is approaching the point of economic feasibility; and pond and raceway culture of shrimp is now in pilot-scale production. Additionally, extensive culture of animals which utilize very short food chains -- such as oysters, mussels and mullet -- has the potential for enormous expansion with existing technology. The recent FAO Technical Conference on Aquaculture reported encouraging progress in aquaculture in the past decade; realistic estimates place future yields at twice the 1976 level (six million tons) by 1986, and five times the 1976 level by the year 2000 "if the necessary scientific, financial, and organizational support becomes available (6)."

This optimistic report must be tempered by the observations that the recent increases in aquaculture production may reflect better statistical collection rather than any real increases in production, and that further increases will be determined largely by the kind of support provided. The overriding force in development of modern aquaculture is clearly a per-ceived national economic need. Those countries which have recognized such a need and developed a national aquaculture policy (Japan and Israel, for example), have moved furthest toward significant production, while other countries (such as the United States), without a recognized need or policy, have made little progress, except to increase the amount of available technical information.

Development of energy-intensive, high-technology culture of species requiring high-protein diets will undoubtedly continue in the next two decades, especially in industrialized countries, but substantial production of herbivorous or omnivorous species in natural waters -- designed to yield relatively low-cost animal protein -- should expand even more rapidly, particularly in developing countries, and particularly in tropical and subtropical areas with year-round growing seasons. An important role for the industrialized countries (probably functioning through FAO) will be to improve and promote the use of the technology required for extensive culture production of inexpensive animal protein in less-developed parts of the world (by such methods as selecting genetically for high food-conversion efficiency and rapid growth, testing of low-cost diets from natural products, training of technicians, etc.). Additionally, there is a significant educational role beyond training for production -- a role in encouraging changes in diets and in encouraging acceptance of aquaculture as a major occupation. The role of aquaculture in integrated rural development, through provision of better diets, jobs, and cash crops, can be significant in developing countries. Aquaculture there would be primarily in the form of small-scale, low-technology, labor-intensive operations, conducted in lakes and ponds or in coastal waters.

The potential of ocean ranching -- not only of anadromous species, but also of coastal-migratory species -- will be exploited within the next two decades, and substantial increases in yields (as well as augmentation of fished stocks) can be expected in proportion to public and private investment in this exciting new approach to fish production which involves rearing and release of juveniles to forage in natural habitats. An important qualifying comment here would be the need for consideration of impacts of introduced populations on natural stocks and the need to determine and consider the total carrying capacity of the ocean areas involved.

Expansion of food production through aquaculture must be a matter of national policy and national priority -- much as the expansion of distant-water fishing fleets was in many countries (particularly the European socialist countries) during the decade of the 1960s. Included in such policy would be improvement in the technological base, development of legal protection for aquaculture enterprises, control of coastal/estuarine pollution in grow-out areas, and encouragement of capital investment. With increasing restriction on harvests from fish stocks in continental shelf waters of other nations brought about by extended fisheries jurisdictions, the aquaculture option should become much more attractive and compelling on a worldwide basis as a protein food source.

EFFECTS OF POLLUTION

We must, if we are to realize the potential food production of inshore waters, reduce the massive and increasing pollution load that has already had significant local impacts on a number of commercial fish and shellfish species. Destruction and degradation of estuaries is of particular importance, since many of the fishes of the continental shelf are dependent on these inshore waters, particularly during the early part of their lives, and most shellfish are produced in estuaries or close to shore. In these very important estuaries, environmental degradation includes physical modification by diverting freshwater outflow, dredging channels, and filling marshlands -- in addition to chemical and biological alterations caused by domestic and industrial pollution. Estuarine populations of commercial species decline and disappear as industrial pollution makes conditions for life untenable -- or the survivors of these species are legally excluded from markets because they are contaminated and represent a danger to the health of human consumers. This process continues and accelerates at the present time. For example, each year there is a net loss nationally of about 1.2 percent of United States shellfish growing areas due to legal closure because of increasing coastal/estuarine pollution. As human populations on the rims of the oceans (the bays and estuaries particularly) increase, pressures increase proportionately to remove water areas from both production of food and from use as recreational areas, and instead sacrifice them to so-called "industrial progress." We cannot talk sensibly about potential food production from areas that have been abandoned in this way.

In terms of pollution impacts on abundance of natural populations of fish, it is important to make as realistic an assessment as possible. This is difficult because pollution is only one of many environmental factors that affect survival and well-being of marine organisms. At present it is possible to identify severe localized effects of pollutants on fish and shellfish in bays and estuaries, and it is possible to demonstrate experimentally that contaminants such as heavy metals, petroleum hydrocarbons, and synthetic chlorinated hydrocarbons, can kill or injure individual animals, but it is almost impossible to demonstrate general effects of environmental pollution on the abundance of resource species. It may be that such effects are occurring, but our baseline data and our monitoring programs do not yet provide adequate data to separate pollution effects from the "background noise" of effects of natural factors on changes in abundance of marine species.

Severe localized pollution problems exist in many bays and estuaries, which are, of course, prime aquaculture areas. Use of inshore waters for food production is absolutely incompatible with their use as waste dumping and discharge sites for an expanding human population. There can be no question of multiple use of these waters; we must make a firm and permanent commitment of certain water areas to food production if marine aquaculture is to have any future in the industrialized nations. Water quality is an overriding consideration.

An alternative might be culture of marine species in complete artificial environments -- totally withdrawing from dependence on the natural environment at any stage in the life cycle of the cultivated species. This may be feasible, especially for shellfish, where brood stocks could be maintained in trays, where larvae could be fed with cultured algae in artificial sea water, and where growth to market size could occur on racks in fertilized artificial ponds. At present these closed-cycle artificial systems are well outside any cost-effective level, and, somehow, this retreat to artificial, energy-demanding systems seems like an admission of defeat. Surely we should be intelligent enough to devise ways to take advantage of the tremendous productivity of unfouled inshore waters as a principal source of protein food for the human species.

There has been much talk of cleaning up the aquatic environment, and much publicity has been given to a few fish reappearing in rivers that until recently were too foul for their survival. Unfortunately, much of the gain in rivers and canals has been at the expense of the estuaries and coastal waters -- we have simply moved much of the pollution problem seaward.

Decisions made now about the extent to which degradation of estuarine/coastal waters should be allowed to continue can have a very important bearing on aquaculture in the future. If the edges of the sea are considered important to the nutrition of future world populations (and I believe they should be) then actions must be taken now nationally and internationally to ensure that such production areas will be available to meet the developing need for protein food. We cannot afford to delay facing realities in food production (as we have with world petroleum consumption) until a crisis is imminent.

As a footnote, there is one positive aspect of what we have termed "pollutants" -- which is that domestic sewage wastes are made up principally of organic nutrients which can enhance natural productivity of coastal/estuarine waters. If very carefully controlled in amounts per unit area of water surface, and if free of toxic pollutant chemicals, such domestic

wastes can serve as fertilizers. Growth rates of molluscan shellfish and certain other marine species can
be increased dramatically by such limited additions of
organics. If proper attention is given to protection
of public health (possibly by depuration procedures),
there is no reason why the organic nutrient residues
cannot become a positive factor in nearshore productivity.

PROPOSALS FOR THE FUTURE

A concise statement of national goals for the
United States in fish and shellfish production might be
(a) to understand and manage effectively the renewable
natural resource base, and (b) to supplement this base
with aquaculture production where feasible.

Fisheries on Natural Populations

The continental shelves of the world will continue
to produce fish and shellfish in amounts governed by
biological principles of sustainable yields, by actions
of fisheries management bodies sensitive to those
principles, and by fishing power of nations.

Obviously, to fully realize the potential productivity of the oceans we need greater knowledge of the
resources and their dynamics, and we need better
methods of locating and catching marine animals, especially those in the lower links of the food chains. We
must keep in mind that we are dealing with a renewable
resource that is highly mobile, vertically and horizon-
tally (except for shellfish), that can change in concentration and location daily or seasonally, and that
reacts to variations in a number of environmental
factors such as salinity, temperature, and availability
of food. We must understand the population dynamics --
birth, growth, reproduction, longevity, death -- of
exploited species to permit proper management, and we
must also understand how marine populations interact.
Only with such knowledge can we manage the entire
ecological complex effectively and predict distribution, abundance, and population responses to human
predation. The ultimate management form must be total
ecosystem management, which includes the habitat and
the populations which act as prey, predators, or competitors. This level of resource management is attainable, but the level of understanding required will be
difficult and expensive to achieve.

Within the concept of total ecosystem management,
marine range management can be effective in the near

term in areas of high productivity on the continental
shelves. The degree of control is largely dependent on
depth of water and the configuration of the shoreline.
Coves and bays may be fenced or diked, predators or
unwanted competitors may be selectively removed, sa-
linities may be manipulated, and fertilizer may be
applied. In open waters, management might take the
form of crop rotation; particularly, fishing grounds
might be exploited for a certain number of months or
years, then fishing pressure could be sharply reduced
for a subsequent period. There is some concern that we
are depleting stocks of valuable species on some fishing grounds with present selective fishing methods and
thereby permitting expansion of populations of less
desirable species. Assuming that some market can be
found for all species -- either as food, as fish meal,
or as fish protein concentrate -- a possible management
plan could be evolved that would encourage retention
and use of all species and sizes taken in trawls, but
would restrict fishing to a rather rigid pattern, that
is, to certain squares of a checkerboard overlay of the
fishing grounds, or to certain longitudinal tracts
through the grounds. Such areas could be shifted
annually or in some longer time sequence. This form of
management involves the entire productive ecosystem, of
which the exploited species are only a part.

Aquaculture

Despite some successes in production of a fish
species in freshwater aquaculture (principally trout
and catfish) in the United States, we cannot claim that
marine aquaculture has yet reached a remotely comparable stage. At present, marine culture is a high-risk
venture with a number of uncontrolled variables. The
necessary technology is being developed, but there is
still a substantial amount of "art" involved in rearing
marine animals. With a few notable exceptions, which
involve substantial financial commitments by a few
large companies, much of the aquaculture research and
development in this country has been done by small
underfinanced private ventures, or by underfinanced
government programs. Large-scale research and development programs, adequately funded for a number of years,
and representing joint industry-university-government
action, are needed for each of the species which seem
most amenable to culture. Emphasis must be placed on
development of inexpensive, chemically-defined, probably pelletized food; on genetic selection for rapid
growth, disease resistance, and suitable market quali-
ties (flavor, texture, color); and on automated production systems. The present methods of poultry production and marketing in the United States provide an

excellent model and illustration of what might be accomplished, despite continuing economic problems of the industry.

An interesting possibility is that marine aquaculture may be developed in the United States primarily to provide recreational saltwater fishing, much as trout hatcheries were developed early in this century to provide freshwater angling. Cultivated marine fish such as snappers and groupers could be used to stock artificial reefs; cultivated crabs and lobsters could be used to stock skin diving areas; and cultivated clams could be seeded in inshore recreational areas -- just as examples of the possibilities. It may be that, as with trout hatcheries, a significant amount of the needed marine research and development work could be accomplished for the initial purpose of supplying recreational fishing, and that in time (again as with trout hatcheries) the culture operations would be economically feasible in themselves.

There are, of course, other channels for aquaculture production. Commercial catches of some species may be augmented by "ocean ranching," an activity in which young animals are reared beyond the most vulnerable early-life history stages and then released into coastal waters. Some beginnings are being made with ocean ranching of salmon in the Pacific Northwest, and the Japanese are attempting similar augmentation of coastal migratory species such as shrimp and red sea bream. Such methods could be used as well in the future with certain endangered or severely depleted species.

Another interesting possibility is that marine aquaculture on a large scale may be vigorously supported as a policy by certain countries other than the United States which are interested in large quantities of animal protein. Through deliberate programs of price support, subsidy, or massive government research, development, and production, some of the marine and estuarine animals that feed directly on plant plankton (oysters, clams, some herring-like fishes, and others) could be produced in great quantity. Nations with tightly controlled economies might well travel this route if fishing on natural stocks decreases in productivity, or if vessels are excluded from major fishing grounds because of extended national jurisdiction.

CONCLUSIONS

Keeping in mind the many qualifications and obstacles discussed in previous sections, it seems that a
number of general statements about food production from
aquatic sources can be made.

1. The oceans will continue to be a very important source of high quality protein essential
for human existence and well-being. At present, fisheries (fresh water and marine) provide an estimated 13 percent of the animal
protein consumed by man.

2. Production of protein food from the oceans
tripled during the two decades from 1950 to
1970 and has stabilized since then at about
sixty million tons. There are still stocks of
underutilized or unutilized marine animals --
particularly smaller forms (under six inches)
and especially the herring-like fishes -- that
must form the basis for any substantial increase in total ocean food production from
natural populations.

3. Annual production of protein food from freshwater has been estimated at about ten million
tons, which may be a very conservative figure.
Although some production in industrialized
nations is derived from high technology culture at salmonids, most of freshwater production is from carps, tilapia, and other herbivorous or omnivorous species, particularly
in Asia. Expansion of production depends on
available water supply.

4. Aquaculture offers exciting potential avenues
for increased production. Though now an
insignificant contributor (less than 10 percent) to total food production from aquatic
sources, if properly developed it could easily
provide a much larger percentage of the total.
Coastal and estuarine areas seem most suitable
for expanded efforts. For the forseeable
future, however, marine aquaculture will
produce limited quantities of high-priced
seafood, and will become a major source of
inexpensive animal protein only if national
policies so indicate.

5. Multiple uses by man of fresh and salt waters
have led to habitat degradation that has
adverse effects on living resources and is a

serious deterrent to aquaculture. Because of
pollution abatement measures, some improvements have been noted in fresh waters, but
pollution of coastal and estuarine waters is
increasing. Serious steps toward environmental management must be taken if food production from such waters is to be maintained
or increased.

6. Despite decades of research, the problems of
understanding and manipulating the dynamics of
food production in the sea are still enormous.
At present our knowledge is superficial, and
much of it may be based on misconceptions. As
a noted marine biologist (7) observed two
decades ago, the oceans truly represent a
frontier, not only in the literal sense, but
also as a frontier in the minds of men -- as
the boundary between knowledge and ignorance.

FOOTNOTES

(1) Graham, H. W. and R. L. Edwards, "The World Biomass of Marine Fishes," In Fish in Nutrition
(London: Fishing News (Books) Ltd., 1962), pp.
3-8.

(2) Hennemuth, R. C., "Marine Fisheries: Food for the
Future?" Oceanus 22 (1979), pp. 2-12.

(3) This may be an underestimate in that estimated
aquaculture production from fresh waters of the
People's Republic of China has recently been
revised upward by some informed observers to be in
excess of ten million tons. See Ryther, J. A.,
"Aquaculture in China," Oceanus 22 (1979), pp.
21-28.

(4) Kaul, P. N. and C. J. Sindermann, eds., Drugs and
Food from the Sea (Norman: University Oklahoma
Press, 1978).

(5) National Academy of Sciences, Aquaculture in the
United States (Washington, D.C.: U.S. Govt.
Printing Office, 1978).

(6) Food and Agriculture Organization, Report of the
FAO Technical Conference on Aquaculture, FAO
Fishery Report 188 (Kyoto, Japan, 1976).

(7) Walford, L. A., *Living Resources of the Sea: Opportunities for Research and Expansion* (New York: Ronald Press, 1958).

THE PROCESS OF
AGRICULTURAL DEVELOPMENT

INTRODUCTION

If feeding the world's people requires increased capacity to produce and distribute food on the part of the developing world, then it is crucial to understand the process of agricultural development in order to influence it effectively. Whether one approaches the task from the perspective of the low-income, developing country or from the point of view of the aid donor, it becomes important to know how to structure cooperative activity intended to improve the nutrition of the neediest people. Fortunately, experience with agricultural development abroad has been accumulating so that it is possible to draw some general conclusions about the most effective developmental avenues to pursue.

The secretary-general of the 1974 United Nations World Food Conference and a former president of the World Food Council, Sayed Marei provides a wealth of suggestions. Stressing the pervasive role of agriculture in the national economies of the developing countries, he notes that the transition of developing-country agriculture is both an internal and an external problem, of necessity linked to the ac·:ions of the developed countries. He sees food distribution as being connected with land tenure, income distribution, and employment opportunities, and he believes that a central problem is the development of integrated rural development systems.

Six major, necessary conditions for increasing the rate of agricultural growth in the developing countries occupy the attention of D. Gale Johnson. Deficiencies in one or more of these conditions can substantially retard improvement, he believes. It is his view that fulfillment of these six conditions depends largely upon the policies followed by the developing countries, but he points out that they do not always have access to the needed resources and therefore require contributions from the high-income countries. Gaining support from the developed world means overcoming three diffi- culties, one of which is the ambivalent and often conflicting views concerning the objective of rapid agricultural development. Unlike Marei, D. Gale John- son is doubtful that it is wise to attempt the redis- tribution of income through programs of agricultural development.

Arthur T. Mosher cites five essential conditions for agricultural growth, similar in some respects to those of D. Gale Johnson. In addition, Mosher des- cribes four accelerators of production which are useful where the five essentials are already fulfilled. Against this background of the nature of agricultural growth, Mosher specifies six strategies for rapid agricultural growth in the developing countries. A concluding concern is expressed about the need to reorganize some public agrisupport programs within developing countries and to change operating procedures to make them more effective. Mosher also points to the need for aid agencies, both international and bi- lateral, to initiate some changes individually and in cooperation with each other.

These papers stress the actions which can be taken, with reasonable assurance of success, to meet the current and projected challenge of building satis- factory food systems within the poorer countries. While there are serious impediments to progress, such as political insufficiency, lack of research, and international trade barriers, it is clear that suste- nance for the world's people can move ahead if men will simply use their intelligence and invest their re- sources wisely.

7

AGRICULTURE IN TRANSITION

Sayed A. Marei

WORLD INTERDEPENDENCE

The invitation to prepare this paper prompted a host of memories -- memories of 1974 when, as Secretary General of the United Nations World Food Conference, I was busily preparing for its November meeting and the preliminary ones. It seemed to me then, and so it does now, that since 1970 events have left us gasping for breath and looking for explanations.

Food crises! Was the cause the huge Russian purchase of wheat from the U.S. in the summer of 1972? In 1973 there was a shortage of meat. In 1975 there was a glut in cattle feedlots and meat-storage facilities. Supply of dairy products in 1973 was limited. In 1974 there was an abundance.

One thing is clear. There is a food problem. But it would be erroneous to attribute it to the sporadic phenomena of a big purchase or bad weather. Our researches and deliberations in 1974 made it clear that there are deep-seated problems that have been accumulating, and those intermittent happenings have acted as a trigger to the loaded powder keg of food problems.

In many of my speeches in 1974 I emphasized that the issue is no longer one of a famine here and another there. We hear of a famine in some part of the world in one year and bumper crops with storage problems the next. The heart of the issue is the agricultural sector. And without an analysis of that sector our vision of the issue becomes distorted.

My subject here is "Agriculture in Transition." Regrettably, it would be far from the truth to say that agriculture is already in transition. Accordingly, much of what I shall say represents a hope -- a hope which we all cherish if the world is to extricate itself from the grips of the most pressing problems.

The topic, "Agriculture in Transition," is one which first and foremost concerns countries of the Third World. An analysis of this subject must cover several dimensions which, in practice, are interwoven all the time, but for analytical purposes can only be presented separately. We do so, however, in full realization that they are intertwined and interrelated. It seems to me that we should try to trace what is happening to these countries along the path to development, be that agricultural or industrial. Many aspects are involved here, such as demography and growth in production as well as rather common features which one finds in third world countries.

However, were I to restrict the subject to the Third World, I believe that an incomplete picture would emerge. Accordingly, while focusing on these countries, I have endeavored to include the economically advanced countries. These countries are also producers of agricultural goods, after all. More important for my approach here, though, are the relations between the developed countries and the developing ones. We cannot ignore such relations in our increasingly interdependent world.

THE IMPORTANCE OF AGRICULTURE (1)

I must get the perspective straight before I enter into details. For most of the developing countries the word "agriculture" does not simply bring to mind food, but also foreign exchange. This is so because a major component of their exports is agricultural goods. The concept of "foreign exchange" may not be all that important for a developed economy. But it is certainly of pivotal importance to a developing country where a foreign exchange gap is an all too familiar problem.

In addition, in the majority of these developing countries, agriculture is the biggest contributing activity to the gross domestic product. This simply means that its <u>weight</u> in total productive activity is greatest.

Also of importance is the agricultural labor force in the total labor force of a country. Taking 1975 figures, in Egypt this percentage was 52 percent; in the Sudan it was as high as 80 percent. It is important here to draw attention to a point of fact which is often not realized: in most of the big oil countries

Table 1. Indices of World and Regional Food and Agricultural Production

FOOD PRODUCTION

	1971	1972	1973	1974	1975	1976	Annual rate of change 1961-70 %	1970-76 %
	1961-65 average = 100							
Developing market economies (1)	125	125	129	132	140	145	3.0	2.7
Africa	119	117	113	120	123	127	2.6	1.2
Far East	126	122	131	131	143	143	2.7	2.6
Latin America	128	128	132	139	144	154	3.5	3.2
Near East	127	137	130	141	151	158	3.2	4.2
Asian centrally planned economies	126	125	130	134	138	140	2.9	2.4
TOTAL DEVELOPING COUNTRIES	125	125	130	133	139	143	2.9	2.6
Developed market economies (1)	123	122	125	129	133	135	2.4	2.5
Western Europe	120	119	123	129	128	126	2.3	1.6
North America	124	122	124	126	135	141	2.4	3.1
Oceania	127	126	139	132	141	150	3.4	3.2
Eastern Europe and U.S.S.R.	125	122	141	136	130	140	3.1	2.0
TOTAL DEVELOPED COUNTRIES	124	122	129	131	132	136	2.6	2.3
WORLD	124	123	129	131	135	139	2.8	2.5

AGRICULTURAL PRODUCTION

	1971	1972	1973	1974	1975	1976	Annual rate of change 1961-70 %	1970-76 %
	1961-65 average = 100							
Developing market economies (1)	124	124	128	131	137	141	2.8	2.5
Africa	119	119	115	122	123	128	2.7	1.1
Far East	125	122	133	131	141	141	2.7	2.4
Latin America	123	124	126	134	136	144	2.9	2.8
Near East	128	138	131	142	148	156	3.3	3.8
Asian centrally planned economies	127	126	132	135	139	142	3.0	2.5
TOTAL DEVELOPING COUNTRIES	125	125	129	133	138	142	2.9	2.5
Developed market economies (1)	120	119	122	125	128	131	2.1	2.3
Western Europe	120	118	123	129	127	126	2.2	1.7
North America	119	118	120	121	128	134	1.7	2.9
Oceania	124	122	126	120	129	135	3.1	1.4
Eastern Europe and U.S.S.R.	125	123	140	135	131	140	3.1	2.1
TOTAL DEVELOPED COUNTRIES	121	120	127	128	129	133	2.4	2.2
WORLD	123	122	128	130	133	137	2.6	2.3

Note: Crops and livestock only. In addition to other nonfood products, the index numbers of food production now also exclude coffee, tea, linseed, and hempseed, and are therefore not completely comparable with those published in earlier years.
(1)   Including countries in other regions not specified.
Source: FAO, State of Food and Agriculture, 1977 (Rome, 1978), pp. 1-5.

in the Near East, agriculture is the biggest user of manpower because the petroleum industry does not generate a large volume of employment. For example, in Saudi Arabia, which is the biggest oil-exporting country in the Near East, some 63 percent of its labor force is in agriculture.

The foregoing contrasts sharply with the prevailing state of affairs in developed countries. Even if we take the example of the United States of America, where agricultural output and exports are huge, we find that agricultural activity belongs to only 6 percent of the labor force, and agricultural output contributes 3.9 percent to GDP.

All considered, agriculture has a pervasive role in the national economies of most of the developing countries. Consequently, what happens in the agricultural sector has multiplier effects in the rest of the economy as well as an impact on the level and pattern of external trade

Performance in Agriculture

Having established the importance of agriculture to developing countries, it is fitting to examine their performance in this regard. Such an exercise is necessary before looking into the future.

Table 1 gives indices of food production and agricultural production for developing countries and developed countries according to region.

Some of the salient figures are reproduced below:

Average Annual Growth Rate of Food and Agricultural Production for Developing and Developed Countries

	1961-70	1970-76
Developing countries		
Food	2.9%	2.6%
Agriculture	2.9%	2.5%
Developed countries		
Food	2.6%	2.3%
Agriculture	2.4%	2.2%
World		
Food	2.8%	2.5%
Agriculture	2.6%	2.3%

It is evident that, first, both agriculture and
food production performances seem to have slowed down
in the 1970s compared with the 1960s. This applies to
developing and developed countries alike. Second,
contrary to popular belief, the developing countries as
a group, despite their difficulties, were able to
expand their agricultural and food output faster than
the developed countries. Third, for the world as a
whole, it appears that food production has been growing
at a rate higher than that of agriculture.

Of all the regions in the developing world, the
Near East shows the most impressive production growth
rate. Not only does it have the highest growth rate
for both agriculture and food, but also it is the only
region where the growth rate figure for 1970-76 is
higher than that for the decade of the 1960s.

While maintaining this highly aggregative picture,
let us see what will happen if the factor of population
growth is added. In order not to burden the text with
tables, Table 2 is given by regional breakdowns only
for food. The aggregation for both food and agriculture is given below, on a per capita basis:

Per Capita Growth Rates

	1961-70 %	1970-76 %
Developing countries		
Food	0.6	0.3
Agriculture	0.6	0.2
Developed countries		
Food	1.6	1.5
Agriculture	1.4	1.4
World		
Food	0.8	0.5
Agriculture	0.6	0.3

Source: Computed from data in FAO, State of Food and
Agriculture, 1977, (Rome, 1978).

A comparison to the above figures with the previous respective growth rates immediately reveals that
the introduction of population growth changes the
previous findings. In spite of the lower rate of
production in the developed countries, their per capita
agricultural and food production still rose by 1.4
percent and 1.5 percent, respectively, against 0.2

Table 2. Indices of World and Regional Food Production
Per Capita

	1971	1972	1973	1974	1975	1976	Annual rate of change 1961-70	1970-76
	1961-65 average = 100						%	%
Developing market economies	102	99	100	100	103	104	0.4	0.1
Africa	98	94	88	91	91	91	-	-1.5
Far East	103	98	104	99	106	103	0.2	0.2
Latin America	103	101	101	103	104	109	0.8	0.4
Near East	103	108	100	105	109	111	0.5	1.3
Asian centrally planned economies	111	108	111	112	113	114	1.1	0.6
TOTAL DEVELOPING COUNTRIES	105	102	104	104	107	107	0.6	0.3
Developed market economies	113	111	113	115	118	119	1.4	1.6
Western Europe	114	111	115	120	118	116	1.5	1.1
North America	113	110	111	112	119	123	1.1	2.3
Oceania	110	107	116	109	115	121	1.5	1.7
Eastern Europe and the USSR	116	112	128	123	116	124	2.5	1.2
TOTAL DEVELOPED COUNTRIES	114	112	118	118	118	121	1.6	1.5
WORLD	107	104	108	107	108	109	0.8	0.5

Source: Ibid., p. 1-8

See footnotes to Table 1.

percent and 0.3 percent for developing countries. Two
additional remarks are worth making: The per capita
production-growth rates of developing countries are
very low -- close to zero percent; these rates are
noticeably lower in the 1970s as compared with the
1960s, unlike the developed countries. A simple
reflection will show that if these figures were to
represent a trend extending into the future, the
emerging picture would be frightening.

The Emerging Picture

A rather complex picture emerges. I have shown
that, while agricultural growth in developing countries
per se is higher than that in developed countries, the
picture is reversed when the population factor is
included.

Significant, and disquieting, is the discrepancy
in the per capita production figures of developing and
developed countries. Alarming also is the perceivable
slowdown in production in recent years. Given the
continual increase in demand, mainly as a result of
population growth, there is concern about the increasing food imports by developing countries. Their
net cereal imports rose from three million tons in
1950-51 to thirty-three million tons in 1969-71 to
fifty tons in 1975-76. It has been estimated that the
cumulative excess costs of these imports in the 1974-76
period, because of higher food prices, are $12 billion.
Given the increasing trend of urbanization in developing countries, it is reasonable to expect that their
food demand will expand at higher rates in the future.

More seriously, these gaps between food supply and
demand correspond to existing nutritional levels. It
does not mean that fifty million tons of cereal imports
have satisfied requirements. The World Food Conference
estimate of roughly 460 million seriously undernourished people has now been surpassed. The need for
upgrading the dietary position in these countries is
huge. According to UN figures, of 128 developing
countries, seventy-one had average dietary energy
supplies below requirements. In particularly critical
situations are populous countries such as Bangladesh,
India, the Phillipines, and as many as eighteen countries in Africa. Moreover, in many of the poorest
countries, average dietary energy supply appeared to
deteriorate in the 1970s.

It may be cautioned here that these are average
figures and, accordingly, gross inequalities at the
national level are masked. Even if the average energy

supplies appear adequate, this does not necessarily mean that undernourished groups are nonexistent. It is unfortunate that data on food intakes by socioeconomic groups of the population are lacking for most developing countries. This is surely an area which is well worth researching.

The persistent problem of hunger and malnutrition is the real food problem. It has become more serious, despite the rather good harvests since 1974 and increased awareness by the international community. The figure of seriously undernourished people I alluded to earlier represents about 20 percent of total population in Africa and Asia and 13 percent in Latin America. Even if these percentage figures have not increased, absolute numbers have gone up and the world is moving away from the declared objective of abolishing hunger and malnutrition. Perhaps we should be more wary in using percentage figures when talking of food where the absolute number of mouths to feed is what matters.

Of course, domestic agricultural production is the main determinant of the level of available food supplies in most developing countries. While for most of these countries agricultural output has failed to keep pace with population growth, one should not forget that there are some success stories. For example, from the 1950s to the early 1970s eleven developing countries have achieved a food production increase averaging 5 percent or better.(2) By any standard, such a growth rate figure in the agricultural sector is high.

The broad picture that emerges from what I have been saying so far is one where agriculture has not been growing fast in developing countries, as a whole. Indeed, agricultural exports as a percentage of total exports have been declining for most developing countries. We have seen that the role of agriculture is more crucial in developing countries than it is in developed countries. Not only is agriculture the main determinant of available food supplies, but also it has important roles to play in economic and social development. It is the main earner of foreign exchange. It creates savings which are needed for investment. It provides employment for the majority of the labor force.

Given these considerations, it is imperative to ask why agricultural development has faltered in many instances. We turn our attention to this subject in the next section.

TRANSITION OR TRANSFORMATION

There is always a danger that a few success stories on the Green Revolution or "miracle seeds" may lead us to conclusions that blind the eye to modest or low achievements. We read that Pakistan's 1971 wheat production was up 76 percent from the 1961-65 average. Latin American corn production was up more than 50 percent and the Indian wheat crop of 1971 was almost double that of six years earlier. In recent years, India has been having difficulties in storing wheat. Pakistan's 1974 rice crop set an all-time record.

Decidedly these are achievements. I am not going to engage here in questions raised about the Green Revolution: for example, how in some instances it reduced the nutritional level of people in farming areas because more emphasis was placed on cereals rather than on food legumes. My main concern has to do with where the Green Revolution is heading. Can it be extended to the majority of other developing countries?

It seems clear that, broadly speaking, the reasons for the relative neglect of agriculture are both internal and external. Since World War II, a hundred or so developing countries have gained independence. Once this was achieved, there developed a craze for industrialization. In fact, economic growth became synonymous with industrialization. Politically, industrialization fitted well with the aims of politicians currying public favor. Industrial projects can be completed fairly quickly and they afford a certain glamor. On the other hand, agricultural projects are neither dramatic nor rapidly completed.

The magic charm of industrialization blinded logical minds, and agriculture appeared to be an unproductive activity in view of the wonderful prospects of industrialization. In the course of this enthusiasm, the leaders of the developing countries got encouragement from everywhere: from economic advisors and from the developed world as well. The advisors seemed to join in the fervor, and the industrialized world responded eagerly to the desires of the developing countries. After all, the developed countries are the source of manufacturing equipment and of the expertise needed for industrialization. Thus, what happened externally reinforced the internal bias. Even aid channeled from the developed world to the developing one favored industry, the returns of which are more predictable than those of agriculture and the gestation period of which is shorter.

The foregoing analysis should not be construed as an argument against industrialization. To industrialize for industrialization's sake is one thing, and to industrialize on sound economic grounds is another. What matters in the end are cost structures and efficiency considerations. Moreover, an often neglected fact is the preponderant weight of agriculture in the GDP of the national economies of the developing countries. As a result of this, the growth rate of GDP (which is a weighted average of the growth rates of all sectors) tends to be closer to the agricultural growth figure. It is, therefore, no wonder that even when the industrial sector grew healthily, the overall growth of the economy was disappointingly slow.

Clear cut evidence which shows that most developing countries did not get their priorities right can be obtained by reference to investment. In most of the countries agriculture did not receive its due share in investment. Studies carried out by the United Nations show that the share of capital that went to agriculture was not commensurate with the contribution of agriculture to the national economy. It is curious that even the currently planned allocations of investment show this bias against agriculture. Table 3 gives the share of agriculture in total planned investment for several developing countries. If these rates are compared with the contribution of agriculture to GDP it will readily be found that it is considerably lower. This situation becomes worse when it is seen that often plan implementations fall short of the stated figures -- a trend that seems to go on unabated.

After the decade of the 1970s, the cold light of realism does not show that governments have grasped the magnitude of the problem. It is an increasing magnitude at that, and a Malthusian apocalypse is indeed in sight. In 1900, the _annual_ increase in the world demand for grains stood at 4 million tons. In 1950 it was 12 million tons. Currently it is over 30 million tons. I repeat, these are _annual_ figures and they are not the only alarming figures. If we just turn our attention to the use of chemical fertilizers, it is worth pointing out that one should not be deceived by statistics. One is told that the use of chemical fertilizers has increased sixfold since 1950, which is an impressive percentage. However, it is estimated that no less than 80 percent of small farmers in the developing countries are still without any fertilizer. While some of the developed countries use more than 300 pounds per arable acre, Bangladesh uses eight and many countries in Africa only one. Thus, even the average figures are woefully unrepresentative.

Table 3. Share of Agriculture in Total Planned
Investment in Developed Countries (*)

Country	Duration & scope of plan	Share of agricul- ture %	Country	Duration & scope of plan	Share of agricul- ture %
Bolivia	1976-80C	9.6	Turkey	1973-77C	11.7
Brazil	1975-79C	6.0	Cameroon	1976-81C	17.3
Costa			Congo	1975-77C	15.0
Rica	1974-78C	15.0	Morocco	1973-77C	15.8
Ecuador	1973-77C	18.9	Niger	1976-78C	21.7
Haiti	1976-81C/As	15.0	Nigeria	1975-80C	8.3
Bangla-			Senegal	1973-77C	23.3
desh	1973-78C	23.8	Sierra		
Malaysia	1976-80C	10.7	Leone	1975-79C	15.5
Thailand	1977-81C	15.5	Tunisia	1977-81C	15.8
Iran	1973-78C	11.4	Upper		
Sudan	1977-83C	26.0	Volta	1972-76C	19.9

(*) C = comprehensive, As = Agriculture sector

Source: FAO, The State of Food and Agriculture, 1976,
(Rome), pp. 156-157.

Notwithstanding these remarks, agricultural development is not by any means synonymous with increased
fertilizer use. Agricultural development is far much
broader, and it also includes economic, social, and
political dimensions. Nor is agricultural transformation a matter of high-yielding seed varieties, especially wheat, rice, and maize that have come to be
associated with the term Green Revolution. One cannot
question the productivity of the new seeds when adequate water, fertilizers, pesticides, capital, and
credit are available. But the precise question at hand
is whether these complementary resources are available.
Their availability in developing countries is more
often the exception than the rule. The Green Revolution shows how to become richer when resource availability is secured. But it does not address itself to
the core problem: what should we do in order to increase agricultural production and improve nutritional
standards? What should we do to overcome the social
polarization of the Green Revolution in the countryside
as well as some undesirable ecological effects? And
who are "we?" Is it developing countries or is it the
international community at large? If so, what are its
stakes?

It seems that, although output may increase, agricultural production only passes through a transition phase without a transformation. And we see the Green Revolution faltering even in the few areas where it once flourished.

I am inclined to feel that, despite the gains in output by a few developing countries, the current supply-demand picture is bleak. In many instances, more for some means less for others. An approach that tries to go into the intricacies of the many issues involved seems necessary. I shall outline my ideas on this subject.

The Potential

Can the "food gap" be closed? One determinant of this is the potential to increase production. This potential itself is a misleading notion. It should be gauged against the exponential increase of population and increased demand as a result of rising affluence.

Paradoxically, hope for the Third World lies in the degree of underdevelopment in much of its agriculture. If India matched the United States' phenomenal advance in productivity per acre in the last twenty years, its grain production would now be 130 million tons higher than the current annual figures. If Japanese standards of productivity prevailed in Bangladesh, its rice output would quadruple. As mentioned earlier, in virtually the whole of the Third World fertilizer consumption is so low that this input alone could produce a surge of growth. This contrasts with many developed countries where the point of diminishing returns has been approached.

A word of warning is necessary here. It is not a matter of more application of one input that will transform agriculture. Nor would I like to list here a number of "requirements" or "requisites" for agricultural growth, such as capital, technology, inputs, and so forth. Such an approach has been exhaustively repeated in the past decades to the point of becoming a platitude.

It seems to me that a more fruitful approach is one which highlights interrelated issues and analyzes them. Let us leave the achievements of the Green Revolution for those areas where it can be implemented and emphasize the variety of situations and ecosystems. Let us learn first from the local people what they already know about their own systems. We very often use the term "food" as synonymous with "grains." This

is misleading. In many areas of tropical Africa no
less than 30 percent of the food is in the form of
vegetables -- not tomatoes, not cabbages, but indigenous vegetables. Let us undertake original research in
the search of the potential value of unexploited root
crops such as cassava and plantains. Surely, this is
an important area for research.

What is needed is a technology that is relevant to
the agroecological conditions of a country. In the
short run the most profitable returns can be obtained
by investing in the improvement of existing schemes.
Without hardheaded knowledge of farming systems, this
cannot be accomplished. In the long run investments in
augmenting land resources can be undertaken. While in
some developing countries land is a major constraint,
in a large part there are vast land resources which are
either unutilized or are utilized in productive processes of very low return. The largest land reserves
of the developing world can be found in South America,
Africa, and in parts of Southeast Asia. About 3.2
billion hectares of land in the world could be used to
grow food crops and raise livestock, but only 1.4
billion hectares are presently being used. Selective
step-by-step development of this potential is necessary
because many of these regions suffer from specific
limitations often connected with high rainfall and high
temperatures.

There are two points which run through what I have
just been saying. First, while the population/food
equation is on our minds, and rightly so, we should
give almost equal weight to a population/employment
equation. I think it is fitting to bring together
simultaneously a Malthusian type model and a Keynesian
one. The reason for this is that the malnutrition
problem can essentially be stated in terms of the
man-hours of jobs and not just in terms of shortage of
grain. Food may be there, but poverty may prevent
individuals from buying it. This has been happening on
a much bigger scale than one tends to think. Accordingly, it is the requirements which we should emphasize
rather than effective demand. When talking of employment generation, I would like to avoid being narrow and
stress labor-intensive techniques throughout. I should
think that while labor availability cannot be ignored,
very often fewer labor-intensive techniques may in the
long run generate employment along lines of the old
notion of a wage-fund theory. Agribusinesses have an
important role to play in the countryside in this
regard.

My second point aims at clearing up an often-heard misconception about the small farm versus the small farmer. The two should never be confused. After all, the eight-acre farm (a small one) in Japan is one of the most productive in the whole world. On the other hand, the concept of a small farmer is a euphemistic expression for a poor one.

The Challenge

As was agreed in the World Food Conference, in the long run there is no other solution but to produce more food in the food-deficit countries, particularly the poorest.

It is clear that the performance of agriculture in developing countries is wanting. But it is also clear that agriculture is not going to develop in the next year or two. It would be tragic if the rich nations think that the present food problem is caused solely by population growth in the poor countries. True, in the long run, the readiness of the poor countries to reform their farm systems will be the key to increasing world food supplies and stabilizing world population. But here and now the issue of justice turns on the responsivness of the rich nations. It is not justice alone. At the heart of today's world there are conflicts. It is a world which is poised between an epoch of technological strides and advances and one of poverty and despair. Three-quarters of the world population is now poor. In just about 20 years' time, at the close of this century, the ratio may rise to four-fifths.

The interdependence which colors our world of today will characterize that of tomorrow. Economic growth in the industrialized world goes hand in hand with growth in the poor countries. Economic stagnation breeds instability, and an expanding world economy is necessary for development. In an increasingly interdependent world, lopsided expansion cannot continue for long. Agricultural development, flow of raw materials, industrial output and all productive activities are interrelated at the world level. Accordingly, an expanding world economy would not be compatible with stagnating economies in the Third World.

The world system now faces a challenge. Agriculture and food provide a challenge to all nations for increased economic cooperation in order to share the benefits as well as the responsibilities of a growing economy at the world level.

With a minimum of vision, it is clear that the
future outlook of agriculture will depend both on the
efforts of developing countries to increase production
and the willingness of developed countries to assist
these efforts. It will depend on the agricultural and
trade policies of the main food-exporting countries.
One of the pillars for solving the food problem is food
security. Is the international community genuinely
interested in food security? If so, let us not forget
that food security is very different from market security.

It was estimated in 1975 that by the year 2000 the
developing countries should produce some 550 million
tons of cereals over and above their current production
of 400 million tons. Pessimists say that this is an
impossibility because already many of the major food
production systems have been undermined. A more objective viewpoint, however, would argue that mass
starvation can be averted if the problem is tackled at
the national and international levels. After all, the
international community at the World Food Conference in
November 1974 undertook to abolish hunger and malnutrition within a decade. The World Food Council has been
working on the implications of this objective in terms
of requirements for investment and inputs as well as
postharvest losses.

Notwithstanding the importance of production, we
must not forget the central role of distribution,
especially when talking at the international level.
This aspect was emphasized equally with production in
the World Food Conference. Regrettably, it has not
received adequate attention.

It is undeniable that at the international level a
solution is linked to the policies of the main cereal
exporting countries. The United States exports 60
percent of world grain exports, and the dependence of
the world on U.S. grain exports is increasing. Accordingly, a lot depends on the kind of agricultural policies which the United States wishes to follow.

At the national level the problem of food distribution is not an issue of distributing free or subsidized food. Instead, it is intimately linked with land
tenure and income distribution systems as well as
employment opportunities. Conceivably, all resources
may be available (finances, inputs, technology, and so
on), but if, for instance, the poorest 20 percent or 30
percent of the population has little or no land and no
employment opportunities, it will not benefit and will
remain hungry.

Central to the agricultural and food issue is the evolution of integrated rural-development systems. In the last few years there has been an increasing tendency to view rural development as a whole, instead of merely using mechanistic concepts of increasing food and agricultural production. This realization stems from the observed tendency of increasing rural poverty. As yet there is little awareness of the obstacles involved in the implementation of integrated rural-development policies. This is a fertile area for future research, in the full recognition that rural development does not mean there is only one course of action. In any case, it appears that developing countries need to give more importance to pricing policies in agriculture and rural credit. These two issues have proved to be serious obstacles to increased production and should be entered as integral elements in agricultural production.

My subject would be incomplete if I made no reference to economic cooperation among developing countries. I have already spoken about cooperation between developed and developing countries at the global level. Since World War II, there has been a worldwide tendency to form regional groupings. This trend is noticeable also in the Third World. In this connection I would like to make two points.

First, economic cooperation (or integration among developing countries) is not incompatible with cooperation between advanced countries on the one hand and Third World countries on the other.

My second remark is concerned with the specific place of agriculture in regional integration. The reasons for the formation of an economic regional grouping are well-known. Apart from trade liberalization, an important aim for developing countries is increased output within the grouping or the economic union. After all, the goal is development, and a regional grouping that results in increases in production is certainly welcome. The problem that arises in practice is that emphasis is put on the industrial sector or merely trade in general.

It is clear from the foregoing that regional economic groupings among developing countries have met difficulties and failures. Nevertheless, taking a longer perspective of development, it seems certain that regional groupings will be a permanent and important part of the institutions by which poor countries seek to improve their economic performance. These groupings are needed because some aspects of development issues need to be tackled in a wider context than a national economy, but at the same time it would be difficult to do so at the full global level or only at that level.

Integration schemes among developing countries have so far neglected agriculture.(3) If, for example, we look at the Central American Common Market (CACM), we find that, despite the importance of agriculture in the economies of these countries, the institutional and legal mechanisms of the economic integration process were concentrated exclusively on the development of industrial policy. The sole exception is the Special Protocol on Basic Grains under the General Treaty (Limon Protocol), which came into force in 1968. The Caribbean Free Trade Association (CARTFIA), established in 1958, met with operational difficulties in regard to the Agricultural Marketing Protocol which came into force in 1969. The main reasons were: (a) inadequacy of transport facilities and market information and (b) lack of efficient production plans permitting agricultural specialization. The Latin American Free Trade Association (LAFTA) represents a tenuous compromise between the short-term subregional objectives. Economic development appears as a secondary objective. Although a little over 50 percent of the intra-area trade consists of commodities, agricultural cooperation shows no signs of progressing.

In Africa, the agreements establishing regional groupings seem to emphasize the importance of food and agriculture in economic development, but in practice no practical steps have been carried out. There has been more success in the Near East where the Arab Common Market has in the last few years turned its attention to food and agriculture, and the Arab Authority for Agricultural Development and Investment has been established with an initial capital of more than half a billion U.S. dollars with the specific aim of increasing food production in the region. Regional integration can help put these resources to work and raise productivity, thus accelerating food production. This can be done by enlarging and making more assured the agricultural markets within the region and by common action regarding the financing and supply of production inputs -- both those imported and those produced within the region.

In addition to what has been said above, if a large part of the population has no identifiable role in integration, they cannot be expected to support it. Regional integration is a demanding process, politically as well as economically, which needs continuing mass support to be successful. A scheme of regional integration which does not have the support of the most populous sector in developing countries may fail completely, or, even if it continues to exist, can never be taken to the point where potential benefits can be fully realized. Thus it is to the long run advantage of the more immediately benefitting industrial sectors to ensure that agriculture is included.

Our recognition of the necessity for inclusion of the food and agricultural sector in regional groupings should not blind us to the difficulties and problems involved in so doing. There is inadequate understanding of the process of integrating agricultural sectors, and there are difficulties in modifying agricultural structures by national action within a regional framework. The nature of the structure of agriculture is such that changes are not made easily. There is no quick response in agriculture when attempts are made to stimulate or to free trade within the region, partly because of the weight of the large subsistence sectors and partly because of limitations put on the actual extent of freeing trade by governments which largely control trade in agricultural products.

A set of problems accompanying the inclusion of agriculture in regional integration is that governments are often unwilling in practice to envisage changes which might increase their dependence on other countries, even those within the regional integration scheme, for supplies of food. It takes time for governments to be persuaded that an appropriate degree of regional self-sufficiency is a better and more feasible economic goal than the pursuit of a higher degree of self-sufficiency primarily at the national level.

Another general difficulty with including agriculture in regional integration among developing countries is the overall backwardness of the sector and its infrastructure. Transport is lacking or high in cost, communications are poor, market information is scanty and unreliable, and a large part of farming is outside the money economy and therefore outside any immediate integration influence.

Nevertheless, these obstacles do not invalidate the very strong arguments for the inclusion of agriculture. If anything, these problems underlie our inadequate understanding and the desperate need for research in this subject.

CONCLUSION

What emerges is the realization that agricultural development in poor countries is not only desirable but necessary. Virtually all these countries are poised between the industrialized countries (which account for 65 percent of world's output and 70 percent of its trade) and the oil countries (suppliers of the main sources of energy). Unfortunately, each of the two sides reinforces the other in generating inflationary pressures and an unstable economic order which is showing signs of increasing stress.

In mentioning the foregoing I am not oblivious to the responsibilities of the developing countries. After all that has been said, the issue of agricultural development in poor countries is still here. Progress will take place only if there is the political will to provide funds and resources for agriculture and for appropriate technology. The developed countries can help not only in financial aid but also in the provision of technology. The latter can be provided in line with complementary inputs and must be done as an evolving process. After all, the technology of today in the advanced countries was not born full-grown. Although the goal of agricultural development is a shared goal, what the developing countries need may differ from the needs of the developed world.

As apparent in the unusually large number of international meetings of the last few years, the international community has become increasingly aware of a need for a new international economic order. Economic issues have decidedly turned into central political issues. Several international institutions have been set up, but practical steps are not yet in sight.

The central test of statesmanship and vision requires realism and a spirit of accommodation in approaching the problems of the world and needs of the people.

Ultimately, the economic well-being of any nation will depend on international cooperation in which the developing countries are participants. Inflation, depression, high energy costs, and food shortages hit every nation, but they hit hardest the poor and poorest countries. In the context of the evolution in the diffusion of power since World War II, competitive tests of economic and political strengths will only increase the plight of the majority of the peoples of this earth. The end result will be an international system whose benefits are even less widely shared -- a system which will be considered as unjust -- leading to a total crumbling of institutions.

The realization that in the end we all inhabit ONE WORLD is a victory for common sense.

FOOTNOTES

(1) The word agriculture is used here to cover also hunting, forestry, and fishing -- this being the commonly accepted usage of the term internationally.

(2) The eleven countries are: Bolivia, Costa Rica, Cyprus, Ecuador, Lebanon, Libya, Malaysia, Mexico, Thailand, Togo, and Venezuela.

(3) Unlike the initial situation in regional integration among developed countries, the distribution of benefits among developing countries from the integration of industry can be achieved largely in the course of assigning new manufacturing processes among members. Thus, contrary to the situation in the European Economic Community (EEC), for example, there is usually no need to bring agriculture into an integration scheme among developing countries in order to achieve a politically acceptable initial distribution of benefits from freeing trade in manufactures. This reduces the political necessity of including agriculture.

8

CONDITIONS FOR MORE RAPID AGRICULTURAL DEVELOPMENT

D. Gale Johnson

By some comparative standards it could be said that the past two decades have witnessed rapid agricultural development or growth in the developing countries.(1) Agricultural output for 1976-78 was 50 percent greater in the developing countries than in 1961-65. This compares to a 37 percent increase over the same time period for the developed countries. For the developing countries the output growth rate was nearly 3 percent per annum; for the developed countries, 2.5 percent.

It is true, of course, that most of the increase in food production in the developing countries was offset by population growth. Per capita food production in 1976-78 was only 9 percent above that of the earlier period, while in the developed countries per capita food output increased by 23 percent. The slower growth of per capita food production in the developing countries should not lead to the conclusion that agriculture has stagnated. We should remember that at no time in the history of the world have population growth rates approached those that have occurred in the developing countries since the end of World War II.

The rate of growth of agricultural output, of course, has not been uniform among the developing countries or regions. East Asia has had the largest increase (70 percent), followed by West Asia (69 percent), Latin America (51 percent), South Asia (45 percent) and Africa (28 percent). Africa has seen a decline in per capita food production between 1961-65 and 1976-78 of almost 10 percent. Increases in cereal and other food imports have prevented a decline in per capita caloric consumption. Clearly some developing regions have had substantially greater success in mobilizing their agricultural resources than have others.

I shall address my remarks to conditions for more rapid development of agriculture in the developing countries. By more rapid agricultural development I mean nothing more than a further increase over the rates recently achieved in the growth of agricultural output.

It is not possible, except in quite general terms, to specify the conditions that will assure more rapid development of agriculture in the low-income countries. It is fairly easy to indicate a number of the necessary conditions -- conditions that are required to some degree if growth is to be more rapid -- but they will not assure achieving it. It is far more difficult and risky to indicate what changes would be sufficient to assure more rapid growth, either in general or in particular.

Obviously, it is not my intention to imply that nothing can be done to increase the rate of agricultural development. I do wish to note, however, that there is considerable uncertainty about the effects of any program that may be promulgated and even greater uncertainty about the effect of any one action of policy change. Furthermore, and this is very important in any consideration of actions and initiatives taken by the United States, the effects of many actions may be apparent only after several years. A farmer knows that when he plants a seed or tree that he will not have a harvest tomorrow but only after an appropriate time, which for some tree crops may be a decade or more. All too often policymakers, and even many who are not policymakers, expect relatively quick results and are then disappointed when no observable effects are apparent in a year or two or three.

A Brief Perspective

Some further perspectives on the performance of agriculture and rural development during the past three decades may be helpful in our efforts to better prepare for the future. As noted, the group of countries that we call developing countries have achieved a rate of output growth greater than that of the industrial countries, including the United States. When you take into account that most of the world's agricultural research is done in the industrial countries and that these countries subsidize their agricultures in numerous ways -- high support prices, credit subsidies, tax advantages, and import restrictions -- and that in many cases the developing countries penalize agriculture in a variety of ways, the performance of agriculture in the developing countries can only be applauded.

Modern agriculture is of recent origin, hardly a half century old. The superiority of yield of grains in the industrial compared to the developing countries is of equally recent origin. During the latter half of the 1930s, grain yields in the two groups of countries were identical; today, there is a difference of 60 percent. The yield difference that has emerged is not due to our greater intelligence, initiative, and energy nor, probably, to better natural resources, but is due primarily to the more extensive application of the results of the modern sciences of genetics, biology, and chemistry. These results occurred due to substantial investment in research in both the public and private sectors and in development efforts that produced literally hundreds of modern farm inputs on which the agriculture of the industrial countries are now dependent.

We now have a much better understanding of the motivations, adaptability, and capacities of the farmers, all too often illiterate, of the developing countries than we did only two decades ago.(2) We now know that these farmers are no different than the rest of us, except that they are much poorer in material goods and resources and have had many fewer opportunities to use their human resources than we have had. There now exists an enormous amount of evidence for all to see that poor and illiterate farmers everywhere are quick to adopt new varieties and methods of production that are economically superior to the old. The rapid rates of adoption of new high-yielding varieties of grains, the expanded use of fertilizer, and the use of herbicides and insecticides indicate significant responses to new and superior alternatives.

One more point is worth making: Few of us who talk and write about world food problems could survive if we traded places with any of the hundreds of millions of poor farmers of the developing world. We simply would not have the knowledge and skills that would permit us to feed, clothe, and house ourselves and our families with the meager resources under the command of a large fraction of the world's farmers.(3)

Increasing Agricultural Productivity

There are two categories of responses to the question implicit in the subject "Conditions for More Rapid Agricultural Development": One relates to the general conditions and policies that would facilitate a more rapid rate of growth of agricultural productivity and output. The other relates to the contributions that the high-income countries could make to achieve the objective of more rapid agricultural development in the low-income countries.

I shall deal primarily with the second category of responses, though I shall comment briefly upon the first. Keeping the comment expressed at the beginning of this paper in mind, the major necessary conditions for increasing the rate of agricultural growth are:

1. Adequate incentives are needed for farmers through prices for products and inputs that have a reasonable relationship to the real alternative values in the international market. This does not mean that any departure from international price relationships is fatal -- if it were, we wouldn't have witnessed so much growth of agricultural output during the past three decades. But a cost is incurred when the prices of modern inputs are kept high by protection of domestic sources, or prices of major crops are kept substantially below international prices. Product prices significantly above international prices (after appropriate adjustments for the overevaluation of currencies and high prices for inputs) also have a resource cost, though such product prices may increase the rate of output growth for the particular product if not necessarily for all agricultural output.

2. The availability of useful knowledge and technology is a condition that will make higher levels of productivity of agricultural resources possible both from land and labor. For most of the developing countries there do not exist crop varieties or methods of production that would result in significantly higher productivity of land and labor and at the same time would result in higher incomes for farm families. Yields are not lower in the developing than in the developed countries due to the failure of the former to adopt the technology of the latter. The transfer of knowledge and technology is a highly complex process in agriculture. A great deal of research, both basic and applied, will be required if the farmers in developing countries are to have the same productive opportunities as farmers elsewhere. At the present time at least three-fourths of the world's agricultural research is undertaken in the developed countries. There is more than adequate evidence to indicate that research can produce results that will increase agricultural productivity in the developing countries.

3. Farmers must have adequate and reliable supplies of modern inputs. Increasing agricultural productivity means that farming becomes more dependent upon the rest of the economy. It is a false hope to

expect that a doubling of grain yields by the end of the century could be achieved without the development of sophisticated new crop varieties, available supplies of high-quality seeds that quickly reflect recent breeding improvements, sufficient supplies of fertilizers to provide the nutrients for the higher yields, supplies of the required plant protection materials, and the equipment and facilities to minimize harvest and postharvest losses. One of the great advantages of American agriculture is that it is so well served by the nonfarm sector of the economy. As agricultural productivity increases farmers require more credit, both for longer term investments, such as tube wells, and for current production activities.

4. If farmers are to produce significantly more than they can consume or trade in local markets, they must have access to national and international markets to permit their acquiring the goods and services that they desire. This condition is related to the provision of adequate incentives, but it goes beyond questions of governmental policies that directly affect the prices farmers receive and pay. More than two centuries ago, Adam Smith showed clearly the relationship between the extent of the market and economic development. This basically simple proposition remains equally valid today for developing countries. The greater the confidence that farmers have that their production will find a dependable market in which their prices will not be significantly influenced by the level of output in their immediate area, the more willing they will be to accept the risks involved in production expansion or in shifting to potentially more profitable commodities. Roads, railroads, storage facilities, competitive processing and marketing systems, and reliable and pervasive communications through radio and other modern means contribute to the creation of broad and dependable markets.

5. In many agricultural areas improvement and expansion of irrigation are required if the recent rate of rural development is to be significantly increased. Irrigation improvement and expansion are desirable not only to facilitate increased production but also to reduce the instability of output. The new high-yielding varieties of grain require more effective water management than the traditional varieties. It is not only that the new varieties are shorter but also that their other requirements -- fertilizer and plant protection materials -- are much more productive when the irrigation water is effectively controlled. A rather small percentage of the land irrigated from surface water sources in South and Southeast Asia has

sufficient control of water at the field level to permit effective use of the high-yielding varieties. Water storage is required in most cases if irrigation is to be provided during the rainy season. Unfortunately, both the improvement of water management in existing systems and the development of new irrigation systems involve large capital expenditures.(4)

6. The economic, social, and political systems must permit rural people, including all segments of the farm population, to share in economic growth of their countries. Farm people should have the choice of not being farmers. As economic growth occurs, agriculture provides a decreasing proportion of total employment opportunities. This is a universal law of economic development, contradicted only during periods of enormous stress such as war, famine, or deep depression. Educational opportunities between rural and urban areas should be more equal than they now are in the developing countries, though none of the current industrial economies ever achieved such equality until their farm populations were a minor fraction of the total. Equal access to education for farm and urban areas is at least as important, if not more so, to facilitate occupational choice by farm young men and women as it is for the direct contribution to agricultural productivity. As long as there are significant differences in access to education, farm incomes will lag behind nonfarm incomes.

The six conditions that have been briefly noted, if achieved simultaneously, would contribute to more rapid agricultural development than has been achieved in the majority of the developing countries. However, if one or more of the conditions are not met to a sufficient degree, the remaining conditions might well not be sufficient to achieve more rapid agricultural development. For example, where new seed varieties and methods of production that result in increased yields are not available, more rapid agricultural growth can hardly be expected. When new varieties are available but which require substantial cash outlays to achieve increased yields, low farm prices and/or high input prices may make the new seed varieties unprofitable. Even if new seed varieties are available and farmers are not exploited by low prices for their input but the production inputs required to achieve higher yields such as fertilizer, pesticides, and herbicides are not made available, there will be no basis for more rapid agricultural development.

The fulfillment of the six conditions depends largely upon the policies followed by the developing countries and by the resources they are willing to devote to agriculture and its development. However, the developing countries are not all on the same footing. While all developing countries can adopt price policies that provide adequate incentives for their farmers, many may lack sufficient resources to invest enough in research, to improve and expand their irrigation facilities or to expand education in rural areas. Further, access to international markets is in many cases controlled by the high-income countries. Consequently, there are contributions that the high-income countries can make to achieve more rapid agricultural development.

Contributions of High-Income Countries

The extent of the contribution to more rapid agricultural development in the developing countries that can come from outside should not be exaggerated. Most of the differences that one can see among countries in the rate of their agricultural development reflect the effects of internal policies, events, and programs. But this does not mean that the contribution from external sources would not be significant in many cases.

There is not, in my opinion, all that much mystery or disagreement about the actions and efforts that the high-income countries could take to assist the low-income countries to achieve more rapid agricultural development. The major difficulties are of three sorts. One is the unwillingness of the high-income countries to permit the adjustments in their own economies to facilitate increased exports from the developing countries. A second is the reluctance, one might say inability, of governments to make long-term commitments to support activities that are likely to provide benefits only after several years. The third is that in recent years we and other countries have held ambivalent and often conflicting views concerning the objective of rapid agricultural development. In many cases we have put greater or at least equal emphasis upon other objectives, such as assisting the poorest of the poor or changing the distribution of income and power within the countries being assisted. Such objectives may be laudable, but I suspect that we would express the strongest possible objections if external forces attempted to impose such requirements upon us. Furthermore, such objectives, if pursued, may well be inconsistent with more rapid agricultural development.

If we wish to assist other countries to modify their distribution of income, we should be more forth-

right about it and not cover such efforts under the
guise of agricultural development. I do not want to be
interpreted as denying that concern about the distribution of income and resources is unwarranted, or that
efforts to provide the poorest with more resources will
not make some contribution to more rapid agricultural
development, but only to emphasize that we must not
confuse either ourselves or the recipients of our
assistance concerning the trade-offs that may be involved. We also have the responsibility to make our
decisions on the basis of reasonable evidence and not
by accepting cliches and dogmatic positions.

Let me now turn to a small number of suggestions
of the elements of a program that could be supported by
the high-income countries and international agencies
and that would contribute to more rapid agricultural
development. There is nothing particularly new about
the elements of the program, nor would there be revolu-
tionary effects on the growth of agricultural output,
productivity, and incomes. What might be achieved by a
concerted and well-designed program to which firm commitments were made in terms of a decade or more might
be an increase in the rate of growth of agricultural
output by 0.5 percent per annum. This seems small but
in twenty years this would mean 10 percent more agricultural production annually than might otherwise have
been produced. This would have noticeable effects upon
the well-being of billions by the year 2000, both rural
and urban people.

Investment in Agricultural Research

The past two decades have seen an important innovation in agricultural research institutions -- the
international agricultural research institutes. There
are now nine institutes located in developing regions.
The combined budgets have now grown to an annual rate
in excess of $110 million. These institutes have made
important contributions through the development of
high-yielding varieties, mainly wheat and rice.(5)
Without these new varieties food production in many
developing countries would be significantly lower than
is now the case.

The substantial successes that have been achieved
by the institutes should not gloss over two important
limitations of our agricultural research institutions.
One is the relative paucity of quality agricultural
research institutions in the developing nations. With
a few notable exceptions, and in spite of recent increased levels of financial support, most of the developing nations lack the capacity to take advantage of

the agricultural research undertaken in the rest of the world, including the work done at the international institutes. There is a high location specificity in agricultural production, especially in crop production. Each agricultural and climatic region has its own peculiar problems that require adaptations of crop varieties to take advantage of local conditions and to solve problems that may be local in nature, such as controlling diseases and insects.

The other limitation is the decline in the support of basic agricultural research that has occurred over the past two decades. The international institutes are primarily engaged in adaptive and applied research. Any expectation that these institutes will be responsible for major research discoveries has little likelihood of being fulfilled. This is not said in criticism of the international centers. Due to the heavy concentration of both basic and applied agricultural research in the temperate zones, much that has been learned can be adapted to tropical and semitropical conditions of the developing countries. And given the current relationship between food supply and nutritional requirements in many parts of the developing world, it is appropriate that emphasis should be given to research that makes use of what is already largely known. However, such exploitation of current and past knowledge cannot go on indefinitely with equally fruitful results. The basic research in the sciences that permitted the creation of modern agriculture must receive increased support if the remarkable advances of the past half century are to be sustained into the next century.

I do not think it necessary to dwell at any length upon how the United States and other high-income countries could mobilize their resources to increase the contribution of agricultural research to more rapid agricultural development. The National Academy of Science World Food and Nutrition Study: The Potential Contributions of Research makes the case for additional investment in agricultural research, both in the United States and in cooperation with the developing countries.(6) In addition, the report identifies twenty-two high-priority research areas, selected from several score possibilities.(7) It also dwells at some length on the limitations of our efforts, primarily through the Agency for International Development (AID), to support productive international collaboration, to assist the development of national agricultural research systems and to further the education and training of scientists from the developing countries. The report was, in my opinion, unduly mild in its criticisms of the performance of the United States in assisting agricultural development during the 1970s.

The World Food and Nutrition Study deserves far more attention than it has received since its publication several years ago. It should not be ignored by the Executive, the Congress, or the public as simply another effort of the agricultural research establishment to obtain more financial support for the land grant colleges and universities, the agricultural experiment stations or the U.S. Department of Agriculture. Only two members of the fourteen-member steering committee could be tagged with significant claim of self-interest and one of those two was emeritus. Of course, all of the members of the steering committee had an interest in and concern for the state of agricultural research in the United States and elsewhere.

There is little in our recent experience that gives one much hope that our support of research relevant to the agricultural problems of the developing countries will make significant contributions over the next decade. We no longer seem to have any sense of direction. Our institutional arrangements are marred by bureaucratic entanglements, short-term commitments and instability of leadership. In terms of our support of the development of agricultural research institutions in the developing countries, much more could be accomplished with even our present level of expenditures if firm commitments could be made for a decade, subject to review only for progress in meeting stated objectives.

But such a pessimistic view need not prevail. While one should always be reluctant to assume that new institutional arrangements will make any difference in the case of programs to assist in international agricultural research, including the development of viable national research institutions, the Agency for International Development seems to be incapable of significant improvement. Its mission is complicated by too many deficiencies -- annual appropriations, conflicting directions concerning its actions, and high turnover among its personnel -- to provide any hope of reform and revitalization.

Nor will a new institutional arrangement such as the Institute for Technological Cooperation suggested by President Carter do the job of supporting more rapid agricultural development through agricultural research and technological transfer appreciably better than AID unless it is appropriately financed on a long-term basis, its objectives are limited and simply and clearly stated, it is not encumbered with a variety of political restraints, and it can attract administrative and scientific personnel of high quality.

The "new directions" embodied in the 1973 amendments to our aid legislation and their continuation and amplification in the International Development and Food Assistance Act of 1975 cannot be effectively applied to decisions on agricultural research, except perhaps on the most applied and developmental research. No one can predict in advance what will be the outcome, if any, of a specific basic research undertaking. Even the effects of developing a new plant variety upon the distribution of income cannot be predicted with any certainty. The shrill predictions of the social disruption and the further impoverishment of the poor in rural areas that would result from the introduction of the new high-yielding varieties of rice and wheat have largely been answered by events. This is not to say that no one was disadvantaged by the new varieties, but millions of other poor have had higher incomes and more adequate food than would otherwise have been the case.

Contrary to some expectations, the new varieties were scale neutral -- after a brief learning period small farmers were quite capable of adopting the rather complicated technologies that were involved. Also contrary to many dire predictions, the new varieties resulted in the employment of more farm labor. It was true that the amount of labor used per ton of grain declined but the increase in production was large enough to increase the total amount of labor used. It is perhaps worth noting that the government of India almost reached the decision in the mid-1960s to prevent the introduction of the high-yielding varieties of rice and wheat because of the expected effects on the distribution of income among farm people. It was felt that new varieties would favor the already most developed and progressive areas and thus primarily the higher income farmers. It would be difficult to imagine the harm that would have been done to poor people in India if the Indian government had decided to prohibit the introduction of the new seed varieties.

Our longer-run objective in assisting the expansion of agricultural research in the developing countries should be to create the capacity in most developing countries to do most of their own agricultural research and to educate their own scientists. In other words, we should assist in the creation of viable national universities and research institutions that would largely eliminate the need to rely on external assistance. As a part of that effort, we should signi-

ficantly expand our fellowship programs for the education of scholars and scientists from the developing
countries in the numerous fields of study required for
the building of high-quality national agricultural
research and educational institutions. This suggestion
is not designed to increase the dependence of the
developing countries upon the United States, but rather
to make the institutions in the development country of
such quality that they can provide education and
training that approaches what can be obtained in the
industrial countries. We should not forget that the
United States achieved its scientific preeminence only
after a long period of heavy dependence upon Western
Europe for the advanced education of its scientists.

Investment in Irrigation and Other Agricultural Facilities

The amount of economic development assistance
provided by the United States is low by almost any
standard that you might choose by comparison with our
past performance or other industrial countries. If we
were really serious about assisting the world's poorest
people, we would provide substantially higher levels of
assistance for investment in rural areas. Most of the
world's poor people -- 80 percent according to World
Bank estimates -- live in rural areas. In fiscal 1979
we budgeted less than $750 million for all AID programs
for agriculture.

The expansion of irrigation in developing countries, especially in South and Southeast Asia, would
contribute to both increased food output and greater
stability of that output. A recent study of The Trilateral Commission, in which I participated, indicated
that, with a total capital expenditure of approximately
$54 billion for irrigation expansion and improvement,
it would be possible to double rice production in South
and Southeast Asia in about twenty years.(8) If such a
program were to be realized, it would require approximately a doubling of the current annual investment in
irrigation in that region. Of the investment required
over the period, about 30 percent would come from the
developing countries and the remainder from loans and
grants from the high-income and, possibly, from the
OPEC countries. The achievement of the stated level of
investment in irrigation in South and Southeast Asia
would require a rather small increase in the percentage
of the gross national product of the high-income,
private-market economies provided as economic aid from
0.35 percent in 1975 to 0.55 percent by the early 1990s
and a modest increase in the share of such assistance
going to the region and to irrigation development in

that region. Obviously, the program is not one that would impose great additional burdens upon the industrial countries.

There are many other potentials for investments that would assist agricultural development -- roads, storage facilities, schools, and medical facilities, for example. At the present time only about a fourth of all development assistance is devoted to agriculture. While this is an improvement over a few years ago, it is not obvious that assistance to agriculture has reached an appropriate share of the total.

Increasing the Stability of Prices and Food Supplies

Instability in prices and supplies has adverse effects upon both producers and consumers as well as upon governmental programs and policies. International price instability is at least as much a function of the policies followed by governments as it is of variations in production. Numerous governments, including those of many developing countries, protect their domestic producers and consumers from variations in world demand and supply by deliberately manipulating their imports and exports in order to maintain domestic prices at the desired stable level. These policies, which are followed by the European Community, most other countries in Western Europe, Japan, and the Soviet Union, impose substantial price instability upon international markets and for those countries whose domestic prices vary with international prices. With only a few exceptions, the United States does not follow such policies but is affected by the actions of other countries. The recently concluded multilateral trade negotiations made no progress in modifying these policies.

Food aid could be used to assure a significant degree of stability of supplies to the developing countries. An important use of food aid, especially in the form of grain, would be for the United States alone or in cooperation with other high-income countries to offer to supply to every developing country any short-fall in grain production below the trend level of production in excess of a given percentage of the trend or average production level. By holding quite modest levels of reserves or with small changes in net international trade, each developing country could hold consumption shortfalls to 3 percent or less of the trend level of consumption.(9)

The contribution of the food insurance proposal to more rapid agricultural development would be modest. However, three or four benefits may be noted. One is that somewhat greater risks of production variability can be accepted in programs to increase the rate of growth of agricultural output. Another is that the costs of offsetting production variability would be substantially reduced, whether these costs would have been incurred through storage or the accumulation of foreign exchange reserves, and more resources would be available for investment to increase agricultural production. A third benefit would be a modest transfer of food resources to the developing countries, not every year for any one developing country but averaging about five to six million tons annually for all developing countries.

A fourth benefit is somewhat more intangible, but perhaps as important as any of the others. If the food insurance program became the major form of food aid, the disincentive to agricultural production in the recipient countries would be much smaller than under the existing procedures for distributing such aid. If there is to be a quite constant flow of food aid, as has been the objective of FAO and recent international negotiations on a grain agreement, or if the amount of food aid responds to the needs of the major suppliers, there is little correspondence between the flow of food aid and the potential benefits that the recipient countries can derive from the aid. In the past, we have distributed food aid with little concern about the effects of the aid upon the prices received by farmers in the recipient countries. Under the food insurance program, food-aid shipments would enter a country in response to a production shortfall. There would be some disincentive effect in the sense that during such a year food prices would be lower than they would have been if no food aid had been received.

I do not mean to argue that there should be no other form of food aid. But the other types of food aid should be carefully evaluated in terms of their potential benefits to the recipient country. Food aid can have a value nearly equivalent to an unrestricted cash transfer under a number of circumstances. Examples might be special feeding programs for pregnant women, mothers, and children; school lunch programs; providing the food component for rural development projects and for new settlements during the first year or two. We should never again use food aid as a means of solving our domestic agricultural problems through providing an outlet for products that we find inconvenient and costly to hold. We should remember that while most developing countries are net importers of grain, there are some developing countries that are grain exporters and such surplus-disposal operations have an adverse effect upon prices in international markets.

I have tried to highlight the important conditions for more rapid agricultural development and three important contributions that the industrial countries could make. I have not attempted to provide an exhaustive list or a complete set of recipes. Some of my exclusions have been deliberate. I have not included land tenure reform as one of the necessary conditions for more rapid agricultural development because I know of no substantial evidence that it contributes to that end. There may be other arguments for land tenure reform that are considered to be important, such as increased equity in the distribution of income and political power. I certainly have not included land tenure reform as one of the contributions that industrial countries could make to assist in more rapid agricultural development in the developing countries. Until we have far more knowledge than we now have of the political, social, and economic costs of land tenure reform and a much better understanding of the consequences, I can find no rational basis for our attempting to induce developing countries to undertake such reforms.

I have not included the reductions of the numerous and onerous trade barriers imposed by industrial countries upon the entry of both the agricultural and manufactured products of the developing countries as one of the important contributions that could be made to more rapid agricultural development. Nor have I discussed the benefits to agricultural development of slowing down the rate of population growth and of reductions in morbidity and mortality. A slower rate of population growth will reduce the need to bring more land under cultivation and will increase the percentage of the population that is of working age. These changes, and others that result from slowing the rate of population growth, contribute to higher per capita incomes in rural areas. Reductions in morbidity and mortality have the effect of increasing the productivity of farm people and thus contributing to more rapid agricultural development.

A modest increase in the rate of growth of agricultural output and of the incomes of rural people is attainable in the years ahead. Natural resource limitations are not a major barrier to the achievement of these objectives. There is an enormous potential for increasing the productivity of land and labor in the developing countries. The crop-yield potentials for tropical and semitropical areas are at least as great as the potentials in the temperate zones. This is no longer a supposition or an assumption; it has been demonstrated by the yields actually realized from the new varieties grown in tropical areas. Nevertheless, much remains to be done -- the yield potentials for numerous other crops that are important to the developing countries must be expanded and the actual farm yields must be brought closer to the yield potentials. These results can be achieved if the appropriate research effort is undertaken and sustained over a long period of time.

More than additional productive research is required if more rapid agricultural growth is to be achieved. Agricultural and food policies must provide farmers with adequate incentives; farmers must have access to production inputs and credit, and industrial countries must provide assistance for major capital requirements. If the farmers of the developing countries are given improved opportunities, we can be confident that they will effectively utilize them. If there is a failure to achieve more rapid agricultural development, it will not be the farmers who must accept responsibility. Rather it will be the policymakers of both the developing and developed countries who have not taken the actions that are both appropriate and possible.

FOOTNOTES

(1)   I have used the terms "developing" and "developed" as more or less synonymous with "low income" and "high income," respectively. Neither set of descriptive terms is particularly satisfactory. A number of the countries that are classified as "developing" have very high per capita incomes and several others have reached income levels that should not be classified as low. The term "developing countries" has become primarily a geographic designation, including Latin America, Africa (except South Africa) and Asia (except Japan).

(2)   Theodore W. Schultz contributed substantially to our understanding of the skills, capacities, and intelligence of farmers in poor countries in his important book, Transforming Traditional Agriculture. (New Haven: Yale University Press, 1964; reprint edition, New York, Arno Press, 1976).

(3) A. H. Bunting of Reading University makes this same point rather more elegantly than I have: "We are all bound to be humble in the presence of the producers: they are not only the principal beneficiaries of all we are trying to do, but they, both men and women, are people of substance and achievement. How many of us, who are so wise in international gatherings about what other people should do, could emulate them in winning subsistence, survival, dignity, and fortitude in the face of calamity from the meager resources of traditional rural society in tropical environments." ("Science and technology for human needs, rural development and relief of poverty," iads Occasional paper, International Agricultural Development Service, 1979), p.8.

(4) The Trilateral Commission study Reducing Malnutrition in Developing Countries: Increasing Rice Production in South and Southeast Asia, page 31, estimated that the capital cost of improving and extending irrigation to support a doubling of rice production in 15 years would be approximately $53 billion in terms of 1975$.

(5) The high yielding varieties of wheat were developed before the formation of CYMMT (International Maize and Wheat Improvement Center in Mexico) through the research program jointly supported by the Rockefeller Foundation and the Government of Mexico.

(6) Published by the National Academy of Sciences, Washington, 1977. In addition to the report of the steering committee, five volumes of supporting papers were published. The supporting papers were prepared by the 14 study teams that were organized to support the efforts of the steering committee. Approximately 200 individuals participated in the study teams. Major efforts were made to consult with qualified and responsible individuals in the developing countries. I should note that I was a member of the steering committee.

(7) The high-priority research areas indicated are: nutrition-performance relations, role of dietary components, policies affecting nutrition, nutrition intervention programs, plant breeding and genetic manipulation, biological nitrogen fixation, photosynthesis, resistance to environmental stresses, pest management, weather and climate, management of tropical soils, irrigation and water management, fertilizer sources, ruminant live-

stock, aquatic food sources, farm production
systems, postharvest losses, market expansion,
national food policies and organizations, trade
policy, food reserves and information systems.

(8) Umberto Colombo, D. Gale Johnson and Toshio
Shishido, Reducing Malnutrition in Developing
Countries: Increasing Rice Production in South
and Southeast Asia (New York: The Trilateral
Commission, 1978).

(9) This proposal for a food or grain insurance program was first made by me in late 1974 and was
included in an FAO publication Food Reserve Policies for World Food Security: A Consultant Study
on Alternative Approaches (ESC:CSP:75/2, January
1975) by Jimmy Hillman, D. Gale Johnson and Roger
Gray. Further work on the proposal is presented
in my paper "Estimating Appropriate Levels of
Grain Reserves for the United States: A Research
Report," Office of Agricultural Economic Research,
The University of Chicago, Paper No. 77:26, Revised, February 10, 1978. Grain could be provided
under the proposal without any charge or on the
basis of a charge related to the level of per
capita income in the recipient country.

9
STRATEGIES FOR RAPID AGRICULTURAL GROWTH

A. T. Mosher

SPURTS OF AGRICULTURAL GROWTH

The word "rapid" belongs in this topic because it emphasizes the pressing needs for sustenance in the world. Widespread undernourishment and malnutrition, coupled with current rates of population growth, make rapid increase in the world's food supply important.

However, in addition to food, agriculture produces fibers, industrial raw materials, and other products useful to man; in many instances more than one of these agricultural products are produced in combination on the same farms. For this reason, as well as the rising rural incomes, it is important to emphasize agricultural growth rather than expanded food production alone, regardless of the nature of the products produced.

Moreover, as we seek more rapid agricultural growth, it is important to recognize that not all of the activities needed to accelerate agricultural growth can be hurried. Some can, but some cannot.

That spurts of agricultural growth, each lasting several years, can be achieved has been demonstrated repeatedly in recent decades. The most publicized instance was the so-called Green Revolution, involving wheat and rice in the late 1960s. A similar spurt occurred with respect to corn in Thailand a few years earlier. Another, involving wheat and cotton, occurred in undivided India fifty to eighty years ago. Finally, there was the agricultural spurt in this country based on the introduction of hybrid corn about fifty years ago.

What brings on these spurts of agricultural growth? In the case of the Green Revolution it was the availability of new fertilizer-responsive varieties of wheat and rice. In the case of the rapid expansion of corn production in Thailand it was a combination of the completion of a new highway (begun several years earlier) that connected the northeast of Thailand to Bangkok and the seaports of southern Thailand, and the search by Japan for sources of feed grains to supply a growing livestock industry. In the case of India at the turn of the century it was construction of a major irrigation system that increased both the area under cultivation and the productivity of land already being cultivated. In the U.S. in the 1930s and 1940s of this century it was hybrid corn.

These four instances (and there are many others) illustrate that spurts of agricultural growth are not all triggered or set-off by the same immediate "cause." That immediate "cause" may be a new variety, a new highway, a new market, a new farm input, a new irrigation system, a new set of price relationships, or some other. These are very dissimilar. Is there any characteristic that they have in common?

The Five Essentials

There are five essential conditions for agricultural growth:

1. a market demand for at least one product that can be produced in a particular region and established market channels through which trade can take place;

2. a technology which, if adopted, could increase production in the area;

3. convenient availability, locally, of the farm supplies and/or equipment that use of the improved technology requires;

4. adequate incentives to increase production;

5. transportation facilities to bring supplies and equipment to farms and to take farm products to market.(1)

These requirements are essential. They are like the five spokes and parts of the rim of a wheel. None is useful without the other four.

Markets for farm products
provides one:

New farm technology
adds a second:

The local availability of
farm supplies and equipment
is a third:

Adequate incentives
for farmers provides
the fourth:

Transportation facilities complete the wheel.

What spurts in agricultural growth in the instances cited above have in common is that <u>the spurt in
growth was triggered by the last essential condition to
be achieved in particular geographic regions.</u> In the
case of the Green Revolution and of hybrid corn in the
U.S., the spurt occurred where a market demand already
existed, where needed supplies and equipment were
available, where price relationships provided adequate
production incentives, and where transportation facilities were adequate. When improved varieties became
available, a spurt of increased production followed.
In Thailand, good varieties of corn and the needed farm
supplies and equipment were already available. When
the Japanese market demand led to attractive corn
prices, and transportation access to the outside world
was added almost simultaneously, a spurt in production
occurred. In India at the turn of the century a strong
market demand in Great Britain for wheat and cotton at

adequate prices provided the incentive. When irrigation and road construction were introduced and market towns grew up along the roads, the remaining essential conditons were fulfilled and production increased rapidly.(2)

What do these spurts tell us about strategies for achieving rapid agricultural growth?

First, they tell us that attention must be paid to all five of these essential conditions.(3) Since they are all essential, one is no more important than the others. Any one of them can be the triggering mechanism in situations where all of the others have been fulfilled.(4)

Second, they tell us to expect spurts to occur at different times and at different rates in different agricultural regions. Different regions have different resource endowments of soil fertility, rainfall distribution, and/or irrigation. Also, different crops are grown in different regions, and more productive technologies are frequently (but not always) specific to particular crops. Moreover, different geographic areas have different development histories. Some are already highly commercialized; others are strongly related to subsistence farming.

Third, they tell us that each spurt results from prior activities conducted over a considerable period of years, perhaps without any spurt occurring, as much as from the particular occurrence that met the last unfulfilled essential condition and led immediately to a spurt.

We should note that the time when some essential conditions can be met is predictable, in other cases it is quite uncertain. Creation of adequate marketing facilities is a relatively mechanical process of establishing an adequate number of collection points, and of wholesaling and storage facilities. The same is true for retail outlets for farm supplies and equipment.

Similarly, establishing an adequate network of farm-to-market roads or installing an irrigation system are tasks of calculated magnitude in any particular region as well as the estimated date of completion.(5) What is required with respect to such essential conditions is simply to design each system, get started on the task of creating it, and then keep at it steadily until it is in place.

By contrast, developing higher yielding crops or farming systems through research is a much less predictable process which usually requires persistent effort over a number of years, without knowing with much precision when (or whether!) a substantial payoff will occur. Similarly, what will be required in the way of production incentives is subject to changes over time due to unpredictable (or unpredicted) changes in world prices for farm products and/or changes in the cost of production of farm supplies and equipment, especially of fertilizer. Unlike providing marketing and irrigation facilities and building an adequate transportation network, research needs to be a continuous process with the dates of payoff uncertain; and providing adequate production incentives has to be a matter of continuous attention and adjustment.(6)

What this adds up to is that, if attention is given to all essential conditions, there will, from time to time, be spurts of increasing production but we cannot predict exactly when they will occur.

CONTINUOUS AGRICULTURAL GROWTH

Up to this point the emphasis has been on "spurts" of agricultural growth. What we need to be concerned about, however, is continuous agricultural growth over an extended period of time.

A distinctive feature of agriculture as an industry is that it involves a particular production process, but not the production of a particular commodity. Agriculture uses solar energy and biological growth processes to produce many human foods, several fibers, and other products useful to man. Each spurt of increasing agricultural production is usually achieved with respect to a particular commodity. And just as each spurt begins in a particular region when the last of the five essential conditions is met, it ends when the production potential of the region using the new technology or facilitating service has been fully exploited. Thereafter, there may, or may not, be slight annual increases, but for the most part the amount of each commodity that can be produced in a particular region is on a kind of plateau, with the plateau being lifted, from time to time, by a new spurt. (Figure 1)

If a region can profitably grow only one crop, or one crop each season, Figure 1 represents total agricultural growth as well as the profile of a spurt which shows that the only way to secure constantly increasing

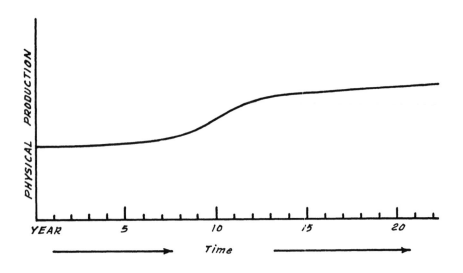

Figure 1

A SPURT IN PRODUCTION OF COMMODITY `A`

production is to keep feeding in new technologies, or improved agrisupport services. But most regions can grow more than one crop. Where that is true, an appropriate research program can provide new technologies that trigger spurts with respect to several different farm commodities from time to time and the cumulative effect of these can, as illustrated by Figure 2, be steadily rising aggregate agricultural growth.(7) In Figure 2, commodity #1 (which might be wheat) experiences a spurt in production (at point A) due to a new technology that sets the process for more rapid growth in motion and ushers in Stage 2. For a few years production of commodity #1 increases rapidly as it spreads throughout the area where it can be grown and where the other four essential factors are present, but then it begins to level off (at point B). From that time on the rate of increase in production of commodity #1 may continue to rise more rapidly than it did before point A either due to land improvement or to making up deficiencies in one or more of the other four essentials, or to the adoption of a series of new and gradually improving varieties.

At point C a spurt in production of another commodity -- commodity #2 (perhaps maize) -- is brought on by the addition of the last essential element for agricultural development that can, again, be any of the five listed on page 301. The rapid increase in commodity #2 continues until point D when both the trend line for commodity #2 and that for total value-added in agriculture again subside to a less dramatic rate of increase.

Additional spurts in total value-added are brought on by developments with respect to commodities #3, #4, and #5 at points E, F, and G.

It should be understood that Figure 2 is schematic only, its purpose being to illustrate two things. One is that even though the rate of increase with respect to each commodity during the period of each spurt cannot be maintained, the rate of growth in total value-added (considering all commodities together) can continue to increase over a considerable period of time. The other purpose is that since most of the essential elements require preplanning and investment some years before they will pay off in a spurt of increased production, it is important to launch efforts to meet the required conditions for increases in additional commodities without waiting for the potential for any one commodity to be fully exploited.

306

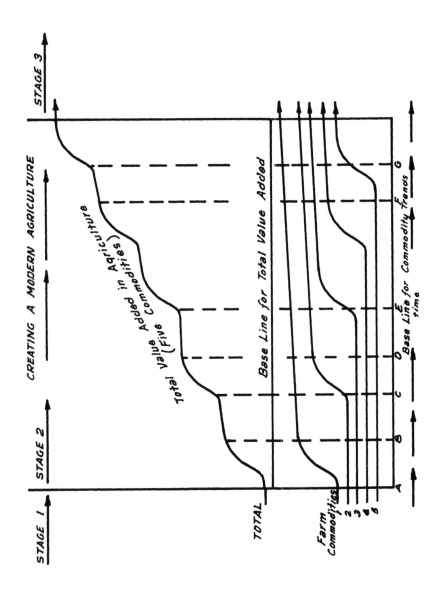

FIGURE · 2

SCHEMATIC DIAGRAM
OF THE NATURE OF AGRICULTURAL GROWTH

To many people, the way to achieve rapid agricultural growth is to concentrate on the accelerators
(footnote #4, page 314) -- especially extension education and production credit -- in order to fully
exploit the potential of presently available technical
possibilities as quickly as possible.(8)

Such efforts are extremely important and should be
part of any effort to achieve rapid agricultural
growth. What they take as given, however, are the
present limits to increased production that are imposed
by (a) the present resource endowment of each agricultural region and/or (b) failure to have met the existing shortfalls of the five essential conditions for
agricultural growth. Concentration on extension and
credit activities and other accelerators makes no
contribution toward extending irrigation and/or drainage, or transportation facilities. It takes present
production incentives as given. It does not expand
research to find new technologies that can form the
basis for additional spurts in production or for
smaller incremental increases of production during
"plateau" periods (Figures 1 and 2). It ignores the
stubborn fact that each spurt must be preceded by
long-term attention to several of the essential conditions for agricultural growth. The spectacular spurt
in wheat production in Mexico could not have been so
big or so quickly accomplished if Mexico had not made
major investments in irrigation and roads -- especially
in the northwest -- in previous decades. The relatively rapid creation of fertilizer-responsive rice
varieties by the International Rice Research Institute
built on previous decades of research in rice breeding,
especially in Japan and Taiwan. Progress in both
places would have been much less rapid were it not for
the revolution in fertilizer production technology in
the 1950s that greatly reduced the cost of fertilizer.

STRATEGIES FOR RAPID AGRICULTURAL GROWTH

Against this background of the nature of agricultural growth, and the requirements for achieving it,
what can be said about strategies for rapid agricultural growth?

First, the "farming district" should be adopted as
the normal geographic unit for agricultural planning
and for public agrisupport action programs.

It is on individual farms that agricultural production takes place. However, it is only those farmers
who have easy access to a set of agrisupport services

who are in a position to take advantage of improved methods of farming. As a bare minimum they need access to seeds, fertilizers, and farm equipment, and they need markets for what they produce. It helps if they have, in addition, the help of extension workers, farm credit, local verification trials, and farm-to-farm roads.

All of these facilities need to be readily available nearby, and where farmers must travel on foot or by bullock cart, that means within not more than three or four miles. Such an area can be called a "farming locality."

If the services available in farming localities are to function effectively, they need ready access to wholesaling and supervisory activities servicing a larger area, and it is that larger area which is meant here by a "farming district."(9)

It is these farming districts that constitute the optimum geographic units for agricultural planning and for public agrisupport action programs. Consequently, two considerations need to be honored in settling on the size of a farming district. One is that the farming localities within it can be effectively served. The other is that it be of an acceptable size for units of governmental administration.(10)

Second, activities to spur agricultural growth should be tailored to the most pressing current needs in each farming district, especially taking into account which areas have an immediate, a future, or a low potential for agricultural growth.

There is a strong tendency, whenever a new activity to spur agricultural growth is launched, to try to extend it uniformly to the whole country. That is wasteful because soils vary in productivity, some lands are irrigated while others are not, at any one time more productive technologies are available for some commodities than for others, and previous activities have fulfilled some of the essential requirements for agricultural growth but not others.

The variations in programs that should be introduced to take into account these differences in local circumstances should correspond particularly to whether each part of a district has an immediate, a future, or a low potential for agricultural growth.

Areas of immediate growth potential are those in which the soils are good, rainfall is adequate or the area is efficiently served by irrigation facilities, and completed research has already produced technologies that provide the basis for substantial agricultural growth with respect to at least one major crop or type of livestock for which there is an adequate market demand.

Areas of _future_ growth potential are those in which soils and climate are favorable but where irrigation, or research to develop pertinent high productive technology, or (if the region is remote) physical access are currently lacking.

Areas of _low_ growth potential are those where circumstances are such that farming can never be a highly productive occupation without major technological changes that cannot now be foreseen.(11)

Even _within_ individual districts conditions are usually not uniform. Many districts embrace some areas having an immediate growth potential while other areas in the same district have only a future growth potential and some may have a low growth potential. Consequently, it is important not only to vary programs as between districts but within individual districts as well.(12)

It is only in areas of immediate growth potential that quick agricultural growth is possible.(13) Consequently, it is there that immediate attention should be given to making sure that needed supplies and equipment are locally available, that farmers' incentives are adequate, that numerous local verification trials are carried out, and that extension and credit facilities are available.

Meanwhile, activities of a different type should be undertaken promptly in areas of future growth potential. There the most urgent need is to engage in the research and/or to make the investment in irrigation or in roads that will convert them into areas of immediate growth potential a few years hence.

Third, in deciding when and where to undertake particular activities to spur agricultural growth, special attention should be given to complementarities among, and gestation periods of, various activities to spur agricultural growth.

As for complementarities, these are strongest among programs to distribute inputs, to provide extension and credit services, to strengthen rural marketing facilities, to carry out local verification trials, and

to establish and maintain farm-to-farm roads.(14) Too
frequently a single one of these services is established all over a country without simultaneously assuring the adequacy of the other five. However, the
effectiveness of each of the six is enhanced or limited
by the quality of the others. They should all be
tackled together in order to take full advantage of the
high complementarities among them.

As for gestation periods, it is important to
recognize that considerable time must pass between the
time when certain activities to spur agricultural
growth are launched and the time when they can begin to
make higher production possible. This is true especially of large irrigation projects, research activities, and establishing schools and colleges to educate
and train agricultural technicians. Especially where
rapid agricultural growth is a prominent objective,
there is a tendency to set activities having long
gestation periods aside "for the time being because
they take too much time." That is always a shortsighted policy from the national standpoint, although
it may be to the advantage of individual politicians.(15)

Fourth, "commodity campaigns" should be undertaken
within immediate growth-potential areas for those
commodities for which more productive technologies are
currently available.

In the discussion of spurts of progress and of the
way these spurts can contribute to continuing agricultural growth over a period of years, it was pointed out
that progress is achieved commodity by commodity.
Whenever research has resulted in profitable new technologies for producing a particular commodity, it is of
the essence of agricultural growth to get those new
technologies into widespread use as quickly as possible.

Such efforts were given the name "commodity campaigns" by CIMMYT, the International Institute for the
Improvement of Wheat and Maize. They consist of bringing together in a single, coordinated effort all of the
activities that bear directly and importantly on rapid
diffusion of new technologies with respect to a particular commodity. Those activities should always include regional adaptive research, on-farm testing (16),
and training extension workers as commodity production
specialists. In addition, depending on what has been
accomplished previously, a commodity campaign may need
also to give attention to making the needed farm inputs
widely and conveniently available, providing farm

credit, and perhaps adopting price policies that will give farmers adequate incentives to increase production of that particular commodity.(17)

Fifth, "farming-district programs" should be launched to demonstrate a pattern of public agrisupport programs that can be replicated to achieve a nationwide pattern of programs that is effective and efficient.(18)

The key difference between a commodity campaign and a farming-district program is that the former seeks to exploit current opportunities with respect to a particular commodity while the latter seeks to make maximum use of all of the agricultural production possibilities of particular farming districts. A related difference between the two is that whereas commodity campaigns are relevant only in areas of immediate growth potential, each farming district program should include those activities that are especially needed in areas of future growth potential, such as: irrigation, drainage, land-shaping, research to find improved technologies, and improved transportation facilities.

Farming-district programs are not commodity specific. They seek to find new farming systems, usually involving two or more crops and frequently livestock as well, that will make the best use of all of the physical, biological and human resources in each district. They take advantage of present (commodity specific) opportunities to increase production and simultaneously seek to increase the resources of the district to support still greater production in the future.

Sixth, the organization and operating procedures of ministries of agriculture, individual public agrisupport services, and international aid agencies should be reviewed and modified to make them more efficient and effective.

It is important that we in the developed countries not underestimate the competence of officials in most of the developing countries with respect to accelerating agriculture growth. Hundreds, if not thousands, of them have been as well-trained as our own technicians. They are constantly in touch with persons facing similar problems in other countries and with the international agricultural research institutes and international and regional banks. They are much more aware of peculiar factors in their own local situations that need to be taken into account.

At the same time, most of the agricultural technicians in lesser-developed countries are saddled with patterns of governmental organization and operating procedures that stand in the way of effective action. Many of those procedures are a legacy of colonialism. Some have been uncritically borrowed from other countries. Some have been urged on them by representatives of aid agencies with inadequate knowledge of local circumstances.

Rapid agricultural growth requires the effective functioning of a wide variety of organized activities, some private and some public. It frequently could be expedited by extensive reorganization of public ministries and bureaus. This is not something which can be accomplished by outsiders; it has to be done from within. However, outsiders can urge the importance of it and press for it to be done.

Some of the problems reside within aid agencies. Many of those in aid agencies do not themselves understand the dimensions of the problem. Where they do, specialization by functions within aid agencies precludes anyone from acting in accord with each over-all national situation. Moreover, international and bilateral agencies are now very numerous and each agency has its own procedures and philosophies of development. Representatives of all desire the attention of the same officials in each country. The result is a tendency for national officials to have to spend so much time trying to tailor aid requests to the idiosyncracies of various donors that it is difficult for them to sort out their own priorities and develop coherent national policies and programs.

Within each national government and within each aid agency, therefore, some reorganization and some modifications of operating procedures could substantially accelerate agricultural growth.

Obviously, this list of strategies to spur agricultural production could be considerably extended. One is tempted to stress the importance of giving special attention to each country's agricultural research program, for that is the source of most new technologies. Similarly, it is important that continuous attention be paid to farmers' incentives to increase production. Such constant review should take into account the effect of relative prices on shifting land use among commodities. It also must give due attention to the impact of farm prices on the urban cost of living for nonagricultural workers.

SUMMARY

Agricultural growth occurs commodity by commodity. Some of this growth occurs gradually and simultaneously in several commodities, as when the use of fertilizers becomes more and more widespread or when irrigation is extended in an area where several different crops are profitable.

Agricultural growth also, and more spectacularly, occurs in dramatic spurts whenever the last essential element for production expansion is made available in a particular region. Such spurts usually last several years, but eventually the growth in production ceases and a new plateau of production is reached.

To achieve steadily expanding agricultural production it is necessary simultaneously to exploit as rapidly as possible the potential for increased production that already exists and to undertake now the activities that can increase the production potential several years hence.

"Commodity campaigns" are an effective means of exploiting present opportunities to increase production in places where a commodity for which improved technologies already exist is or could be profitably grown. They can result in spurts of agricultural growth.

"Farming-district programs" are an effective way to tailor public programs to the current needs of different sets of present or potential local resources. They may include one or more commodity campaigns in appropriate places. They also should include activities that may convert areas of future growth potential into areas of immediate growth potential a few years hence. These are activities that lay the groundwork for spurts of increasing production. They honor the fact that achieving rapid agricultural growth depends in part on launching and persevering with several activities that have long gestation periods and, in the case of research, uncertain outcomes.

Successful implementation of programs to spur agricultural growth frequently requires some reorganization of public agrisupport programs and changes in operating procedures that can make those programs more effective. It also demands some changes in the operation of international and bilateral aid agencies, both individually and in cooperation with each other.

314

FOOTNOTES

(1) See Mosher, A.T., Getting Agriculture Moving, Agricultural Development Council, New York, 1966, pp. 63-120.

(2) In the beginning, the Punjab situation involved a major increase in the cultivated area without introduction of any new varieties that could increase production per acre. (I prefer to call this agricultural expansion.) Since Great Britain was eager to increase exports of wheat and cotton from India to Britain, the colonial government soon launched research programs that produced higher yielding varieties that gradually increased productivity per acre, thus considerably lengthening the spurts of increasing production of wheat and cotton well into the present century.

(3) There is some, but not great, substitutability among these essential elements. Profit incentives that are very strong can alleviate the influence of poor transportation; excellent and low cost transportation reduces the size of the economic incentives needed. Where a substantial increase in production can be achieved by introducing improved technology, it reduces the size of the economic incentives that are needed.

(4) There are additional activities that can accelerate the rate of growth in places where the five essential conditions have already been met. Those include education for development, production credit, group action by farmers, and improving and expanding agricultural land, especially through irrigation. Since these activities can accelerate the rate of growth, they are pertinent to strategies for rapid agricultural growth. However, growth can occur without them whereas it cannot occur except where the five essential conditions are met.

Of these four accelerators, irrigation and/or drainage is the most important wherever it is feasible, especially wherever in the tropics rainfall is low, highly seasonal, or undependable.

(5) Achieving an adequate network of farm to market roads is frequently retarded by insisting on too high a quality of roads in the beginning. What is required is simply improved facilities for moving heavy or bulky goods from one point to another.

That may, in the beginning, be only improved paths for beasts of burden or bicycles on carefully located rights-of-way. Frequently, however, engineers of public works departments insist that much more expensive widths and surfaces be incorporated in any roads built at public expense. Taiwan, South Korea, and the New Territories in Hongkong illustrate what can be done by starting with inexpensive paths that take little land out of cultivation.

(6) In this paper the attention is on agricultural growth. But governments must keep other considerations in mind in trying to manipulate production incentives for farmers, such as the cost of certain foods to urban consumers. Japan pays its farmers much more than the world price for rice. It is wealthy enough to do that. Poorer countries have to be more cautious in letting their domestic prices for farm products get out of line with world prices.

(7) It will be noted that, while Figure 1 expresses production in terms of physical production, Figure 2 expresses it in terms of "value-added." That shift is made in order to make aggregation possible. However, where the food-population race is the major concern it is physical quantities of food about which we must primarily be concerned.

(8) What such efforts accomplish, in terms of Figures 1 and 2, is to increase the slope of increasing production during each spurt and to shorten the time required to exploit fully the new production opportunities and to reach the new "plateau."

(9) Another way to describe a farming district is to say that it consists of the commercial hinterland of a major rural marketing center. In some countries it is about the size of existing governmental administrative units, such as the "district" in India or the kebupatan in Indonesia. It is usually about the same size as counties in the Corn Belt of the U.S.

(10) For a discussion of the factors involved in settling on the size and boundaries of farming districts see Mosher, A.T., Creating a Progressive Rural Structure, Agricultural Development Council, 1290 Avenue of the Americas, NYC, 1969, pp. 148-155.

(11) Areas of low agricultural-growth potential may be
ones in which topography is too rough for cultivation, or soils are too poor, or water sources for
irrigation are inadequate. The people now living
in such areas deserve attention, but it is useless
to devote resources to trying to increase agricultural production where that is virtually impossible. Instead, development efforts in such areas
should be used to try to increase nonagricultural
production and income, and/or provide educational
and health facilities, and provision of main roads
that can increase human mobility.

(12) In the 1960s the government of India launched an
"Intensive Districts Agricultural Program." The
intention was to select fourteen districts and
within them do everything that needed to be done
to increase agricultural production. Results were
very uneven among the districts selected. In
retrospect it became clear that one reason was
that circumstances within the selected districts
were dissimilar. For example, nearly all of the
area within certain districts was irrigated while
much of other districts was not. Previously
conducted research proved to be much more appli-
cable in some districts than in others.

(13) Areas of immediate growth potential are those
which Evenson has characterized as having "economic slack."

(14) It is because of the high complementarity among
these six activities that they have been jointly
characterized as constituting a Progressive Rural
Structure. (See Mosher, A.T., Creating a Progressive Rural Structure, Agricultural Development
Council, 1969.)

(15) It is important to recognize that two strong
impulses stand in the way of taking adequate
account of complementarities and of gestation
periods.

One arises from the short terms that most politicians in developing countries usually remain in
office. Because of that, they want activities to
spur agricultural growth that promise quick results. This leads them to give low priority to
activities that have longer gestation periods even
when those are of great importance to keeping
agricultural growth going three to ten years
hence.

The other obstacle arises from the strong preference of external funding agencies such as A.I.D.,
I.B.R.D., etc. to finance narrowly defined projects rather than more diverse programs. Most
funding agencies tend to prefer separate extension
projects, or credit projects, or transportation
projects, rather than coordinated projects that
take full advantage of the complementarities among
activities.

(16) On-farm testing is the same as conducting local
verification trials.

(17) Commodity campaigns aimed at rapid diffusion of
available technologies, although an important
strategy, usually do not include activities having
long gestation periods such as irrigation and
general programs of training agricultural technicians. Instead they concentrate on what can be
done quickly. Moreover, they seldom give attention to the impact on the production of other
commodities that may also be important nutritionally or in the national economy.

(18) Such programs are called "defined-area programs"
by Wortman and Cummings in To Feed This World,
Johns Hopkins University Press, 1979.

Section 4

THE INFLUENCE OF TRADE AND INVESTMENT

INTRODUCTION

As the world's people become more numerous, the significance of trade and investment to ensure an adequate and stable food supply becomes apparent. The emerging world food system is cemented into an interdependent network by bonds of trade and investment which must be dependable yet flexible enough to allow for change as the poorer countries develop their own productive and exchange capabilities. When viewed through an economic lens, agricultural development has a fundamental and pervasive influence upon the economic health of every country, and it is either poorly or well-served by trade and investment practices and policies adhered to by various nations.

Each country views agricultural trade and investment as activities which must fit its preferred political and social priorities, according to Ray Goldberg. These trade and investment activities are related to quite sensitive domestic and political considerations, such as food price stability and national security. Goldberg provides a conceptual framework for the understanding of both domestic and international agribusiness, then asks what U.S. trade policy should be in order to assist the expanding but impoverished market in the developing nations. He believes that a new public-private cooperative trade and investment policy is emerging which will better integrate U.S. participation than the present confused mixture of surplus disposal and aid programs. His view is that farm cooperatives will become increasingly significant participants in the domestic and international food system, and he favors linking these producer-initiated organizations with traditional foreign assistance efforts. He concludes with a series of suggestions for strengthening U.S. policy.

Alonzo McDonald reports the results of the Tokyo Round of the Multilateral Trade Negotiations. He describes the genesis and nature of agricultural trade barriers since the end of World War II and concludes that they result in the paradox of suboptimal food consumption in a world with surplus food production capacity. McDonald finds trade barriers to be not only destabilizing but also a source of intergovernmental confrontation. He describes the Tokyo Round of negotiations in some detail and concludes that the process established by the Tokyo Round promises long-term enhancement of world trade, provided that governments make it their business to continue working together in an effort to resolve trade problems.

These two papers give a sense of the importance of amicable trading relationships between countries as a significant element in the encouragement of agricultural development. Imperfections and instabilities in the world system of trade often work to the disadvantage of developing countries and provide another impetus for the generation of self-sufficiency in basic foodstuffs in the less-developed country.

10
INTERNATIONAL TRADE AND INVESTMENT POLICIES THAT INFLUENCE AGRICULTURAL AND ECONOMIC DEVELOPMENT

Ray A. Goldberg

INTERNATIONAL AGRICULTURAL AND ECONOMIC DEVELOPMENT

In order to understand agricultural and economic development in both developing and developed countries, it is necessary to explore the trends and structures affecting U.S. trade and investment policies. At the same time, one must remember that U.S. trade and investment policies are part and parcel of a complex set of U.S. and individual country and regional food-policy priorities, as enumerated in Figure 1. Each nation is concerned that agricultural trade and investment policies fit into its political and social priority system. These priorities include: (1) food-price stability for consumers and producers; (2) national security and avoidance of too much dependence on outside suppliers; (3) economic stimulation for the general economy and use of underutilized human resources; (4) avoidance of major change in food prices that then become part of a formal cost-of-living increase, which in turn fuels wage and other settlements adding to inflationary pressures; (5) policies that affect the nutrition of a country, such as shipping out needed basic food supplies while importing less essential food items; (6) foreign policy issues that give special barter arrangements to political allies, such as the exchange of sugar and oil at nonmarket prices between Cuba and the Soviet Union; (7) balance of payments considerations; (8) ecology requirements; (9) human welfare priorities, given the fact that food for most poor people in every country represents from 30 percent to 70 percent of their consumption expenditures; and, finally, (10) the social structure of life -- developing insulated policies of high commodity prices and protection to maintain small-scale farming agriculture as part of a way of life (as in the Common Market) or developing methods of reaching the small-scale producer in all countries.

324

FIGURE: 1

FOOD POLICY PRIORITIES

1. PRICE STABILITY FOR CONSUMERS & PRODUCERS
2. NATIONAL SECURITY
3. ECONOMIC DEVELOPMENT
4. COST OF LIVING & INFLATION
5. NUTRITIONAL STANDARDS
6. FOREIGN POLICY
7. BALANCE OF PAYMENTS
8. ECOLOGY
9. HUMAN WELFARE
10. SOCIAL STYLE OF LIFE

In order to keep these interrelated and complex priorities in perspective we have developed an agribusiness commodity systems approach at the Harvard Business School (see Figures 2, 3, 4 and 5). Agribusiness includes all the participants in a commodity vertical structure, from farm suppliers, farmers, assemblers, processors, and distributors to ultimate domestic and international consumers. These commodity systems include small-scale and subsistent producers and is nonpolitical in the sense that the approach may be and is used in a variety of political settings. The system also includes the coordinating machinery that holds it together: markets, futures markets, contractual integration, vertical integration, domestic and international farm cooperatives, governmental programs, marketing boards, trade associations, voluntary agency programs, and a variety of private, cooperative, and governmental joint ventures and long-term agreements and arrangements.

The agribusiness system exists for the ultimate purpose of satisfying the food and fiber needs of the consumer -- given the political, economic, and social priorities of the government. It has three levels of operation. The first involves the total macroenvironment and public policy. The government is not only a regulator and a facilitator of the agribusiness commodity system, but also a direct participant and joint-venture partner in many of the functions performed by the system. This brief paper has an overriding theme: how to improve the cooperation among the government, private, and farm cooperative segments of the U.S. food system in working out short- and long-term agricultural trade and development activities without creating either conflicts of interest or potential governmental takeovers. The second level of operation of the system involves the special commodity system itself in relation to its macroenvironment. The unique national and regional characteristics of each commodity system lead to the logical conclusion that each trade agreement and program should be developed on a commodity-by-commodity basis with no single pattern uniformly imposed. The third level of operation of the commodity system is the business firm, farmer, or governmental institution as discussed in specific terms of that entity in relation to the commodity system involved.

As we examine agricultural trade and investment policies, we must remember that exports represent the surplus supply above local demand for food at prevailing prices. In fact, all nations have an implicit amount of production reserves for themselves. Imports represent a residual demand for food products above

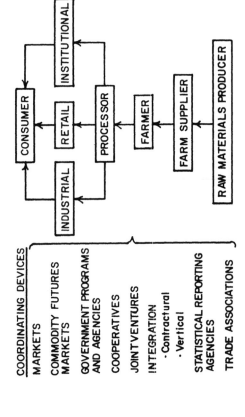

FIGURE : 2

DOMESTIC AGRIBUSINESS FLOW CHART

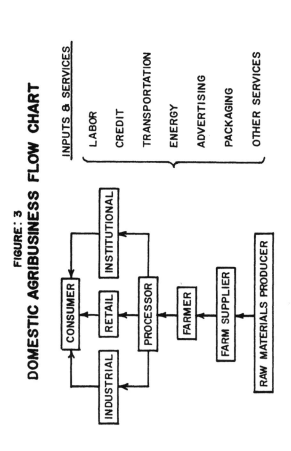

FIGURE: 3
DOMESTIC AGRIBUSINESS FLOW CHART

INPUTS & SERVICES

LABOR

CREDIT

TRANSPORTATION

ENERGY

ADVERTISING

PACKAGING

OTHER SERVICES

CONSUMER

INSTITUTIONAL

INDUSTRIAL

RETAIL

PROCESSOR

FARMER

FARM SUPPLIER

RAW MATERIALS PRODUCER

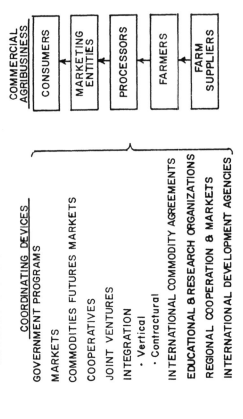

FIGURE: 4

INTERNATIONAL AGRIBUSINESS FLOW CHART

COORDINATING DEVICES

GOVERNMENT PROGRAMS

MARKETS

COMMODITIES FUTURES MARKETS

COOPERATIVES

JOINT VENTURES

INTEGRATION
· Vertical
· Contractural

INTERNATIONAL COMMODITY AGREEMENTS

EDUCATIONAL & RESEARCH ORGANIZATIONS

REGIONAL COOPERATION & MARKETS

INTERNATIONAL DEVELOPMENT AGENCIES

COMMERCIAL AGRIBUSINESS

CONSUMERS

MARKETING ENTITIES

PROCESSORS

FARMERS

FARM SUPPLIERS

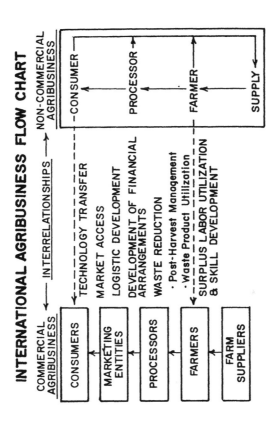

FIGURE: 5

INTERNATIONAL AGRIBUSINESS FLOW CHART

what is met by local production. Total food exports represent from 12 percent to 20 percent of world crop production. The export values are usually the most volatile because they represent "residuals" in the world food system and match one domestic governmental commodity and foreign exchange program against another. Several U.S. commodity and currency futures markets play an increasingly important role in the pricing of commodity trading arrangements between firms, coopera- tives, marketing boards, and governmental trading agencies as these commodity "residuals" utilize the futures market to lessen the risk of volatile price changes over time.

The Setting

The world food problem is summarized in simplistic terms in Figure 6. The developing countries with 70 percent of the world's population produce only 40 percent of the food, have only 30 percent of the world's income, and only 20 percent of the purchased agricultural inputs. In essence, the developing world does not produce enough food for itself nor enough income to purchase additional food or agricultural inputs for its population. What then should U.S. trade policy be in order to help develop this expanding but impoverished market in terms of relating governmental, private, and cooperative trading activities, and trans- fer of technology and assistance to this important and expanding part of the world's economy?

To answer this fundamental question one must first examine current economic relationships and future projections. From Figure 7 it is evident that most population growth will take place in developing socie- ties with three-fourths of the world being in the developing sector by 1985. It is also evident that over half of the population is engaged in the world's food system (agribusiness). In addition, three-fourths of the 145 million farm families in the world are noncommercial (produce primarily for themselves) and have farms whose average size is five hectares (eleven acres) or less. How to reach these subsistence pro- ducers and enable them to become part of a commercial food system is an important factor in our discussion on the role of international trade and investment policies as they affect agricultural and economic development. Adding to the food pressures of our expanding human population is the fact that, in addition to the 4.8 billion people in the world in 1985, we will also have by that year a livestock and poultry population of approximately 15.8 billion (see Figure 8).

FIGURE: 6

DEVELOPED vs DEVELOPING

	DEVELOPED	DEVELOPING
POPULATION	30%	70%
FOOD PRODUCTION	60%	40%
INCOME	70%	30%
INPUTS	80%	20%

332

FIGURE: 7

POPULATION STATISTICS

	1978	1985
POPULATION		
DEVELOPED	1.2 bil. (30%)	1.2 bil. (25%)
DEVELOPING	3.0 bil. (70%)	3.6 bil. (75%)
TOTAL	4.2 bil.	4.8 bil.
AGRIBUSINESS	2.5 bil. (60%)	2.4 bil. (50%)
FARM FAMILIES		
COMMERCIAL	40 mil.	38 mil.
NON-COMMERCIAL	100 mil.	107 mil.
TOTAL	140 mil.	145 mil.

Source: World Food Plan; World Bank.

333

FIGURE : 8

LIVESTOCK PRODUCTION
AND ESTIMATES FOR 1985

WORLD	NUMBERS			
	1950	1960	1970	1985
	NUMBERS x 10⁶			
BEEF CATTLE	763.3	920.6	1250.7	1936.8
SHEEP & GOATS	1012.0	1218.7	1457.3	2155.7
SWINE	227.7	343.1	626.9	1075.4
POULTRY	–	–	5560.2	9668.7

U.S. Response

The U.S. response to the agricultural economic
development needs of the developing world historically
has been a confused mixture of surplus disposal and aid
programs. Unique export and import concessional sales
and markets further confuse agricultural economic
development with their rapidly expanding export sales
in the 1970s, and of financial and technical support
directly from the United States or through the World
Bank, the Overseas Private Investment Company, or
financial and technical support through private groups
such as the Latin American Agribusiness Development
Company (LAAD), or individual private and/or cooperative firms. The encouragment of commodity agreements
for sugar, wheat, and feed grains adds to the confusion. The United States, under the leadership of
special trade representatives, also has encouraged the
freeing-up of markets by reducing trade restrictions on
over 150 products, thereby favorably affecting 25
percent of our imports and exports.

From this historical mixture of programs, a new
nonpartisan, creative, public-private-cooperative trade
investment policy is emerging. The implications of
this new policy and the changemakers' roles are best
understood by looking at the past, current, and future
U.S. positions in a changing world-food economy.

U.S. Position

Figure 9 indicates that North America has become
the only significant surplus-grain-producing area in
the world and, therefore, U.S. export policy is tied up
with the development, export, and import policies of
every nation. Figure 10 indicates that U.S. agricultural exports are forecast to be $38 billion and over
140 million metric tons in the 1979-1980 export year,
thus providing the United States with a positive trade
balance of $20 billion. At the same time, our imports
are estimated to be at an all-time high of $18 billion,
with the enormous responsibilities of providing market
access for both developing and developed countries to
the U.S. market. Figure 11 indicates that, although
developed countries are still the major market for U.S.
agricultural exports, sales to centrally planned economies, such as the Soviet Union, People's Republic of
China, and to developing economies are increasing at a
faster rate. In fact, Figure 12 indicates that in the
trading year ending September 1979, of the $32 billion
of U.S. agricultural exports, $16 billion went to
developing countries and centrally planned economies
and $16 billion to developed countries. These numbers

FIGURE: 9

World Grain Trade Balances by Major Geographic Regions
(Millions of Metric Tons)

REGION	1934-48	1960	1972	1976
NORTH AMERICA	+5	+39	+76.6	+94
LATIN AMERICA	+9	0	+0.7	-3
WESTERN EUROPE	-24	-25	-16.6	-17
EASTERN EUROPE (including USSR)	+5	0	-28.7	-25
AFRICA	+1	-2	-1.3	-10
ASIA	+2	-16	-40.0	-47
OCEANIA (Australia & New Zealand)	+3	+6	+9.3	+3

Key: +Exports   -Imports

FIGURE: 10

U.S. AGRICULTURAL TRADE BALANCE
1971-72 to 1979-80

YEAR BEGINNING OCT. 1

	71-72	72-73	73-74	74-75	75-76	76-77	77-78	78-79	79-80
				BILLIONS OF DOLLARS					
EXPORTS	8.2	15.0	21.6	21.9	22.8	24.0	27.3	32.0	38.0 (forcast)
IMPORTS	5.9	7.7	10.1	9.5	10.5	13.4	13.9	16.3	18.0
TRADE BALANCE	2.3	7.3	11.5	12.4	12.3	10.6	13.4	15.7	20.0
				MILLION METRIC TONS					
EXPORT VOLUME	60.7	99.7	93.4	87.2	106.7	102.2	121.7	127.9	140.0

FIGURE: II

CHANGING MARKETS
FOR U.S. AGRICULTURAL EXPORTS

	DESTINATION			
	DEVELOPED	DEVELOPING	CENTRALLY PLANNED	TOTAL
	BILLIONS OF DOLLARS			
1970	4.6	2.2	0.2	7.0
1977	14.9	7.4	1.8	24.1
% CHANGE 1970-1977	324%	336%	900%	344%

338

U.S. AGRICULTURAL EXPORTS
VALUE BY REGION
Oct. 1978 to Sept. 1979

REGION	BILLIONS OF DOLLARS
WESTERN EUROPE	9.8
EASTERN EUROPE	1.5
U.S.S.R.	2.3
JAPAN	4.9
CHINA	0.9
OTHER ASIA	6.0
CANADA	1.7
NORTH AFRICA	1.0
OTHER AFRICA	0.7
LATIN AMERICA	3.1
OCEANIA	0.1
TOTAL	32.0

are most important when one realizes that many of the procurement decisions are being made by governmental entities and that flexible long-term contractual relations will become more important in our future trade policies.

As one looks to the future as forecast in Figure 13, our relationships with both the developing world and centrally planned economies could become even more intertwined. Leontief noted in The Future of the World Economy that, as incomes and population rise by the year 2000, world agricultural production should increase by three to four times the 1970 level, and per capita output should double. The average total growth rate for all developing regions was projected to be 5.3 percent per year from 1970 to 2000 and 1.6 percent per year for the developed world. The U.S., starting with more than 50 percent of the international grain market of the 1970s, should increase its position despite a much needed surge in the developing world's productivity. The developing world is expected to move from 36 percent of total output in 1970 to a projected 62 percent of world agricultural output in 2000; its share of agricultural imports of the world to expand from 15 percent to a projected 40 percent in the year 2000. By that time, it is projected that one out of every four cultivated acres in the U.S. will be producing food for the developing world. As it is now, one out of every three cultivated acres in the U.S. is currently providing production for exports. From Figures 14 and 15 one notes that 64 percent of our wheat or thirty-two million metric tons are exported, as are 56 percent of our oilseeds or twenty million metric tons and 25 percent of our feed grains or sixty-one million metric tons.

Relation to Oil

Although the U.S. dominates agricultural exports, it is difficult to use this leadership export role as a bargaining factor with OPEC. As Figure 16 indicates, in 1978 U.S. agricultural exports were $28 billion, while OPEC's oil exports were $130 billion. Only $2.1 billion of OPEC's $11.0 billion for agricultural imports came from the United States, whereas $33 billion out of the $45 billion of oil the U.S. imported came from OPEC. Figure 17 makes the projection worse with $50 billion of oil imports projected for 1979 and $100 billion for 1985. The oil exporting countries' positive balance of payments went from $6 billion in 1978 to $43 billion in 1979. The developing oil-importing countries' negative balances went from a minus $31 billion in 1978 to a minus $43 billion in 1979.

FIGURE: 13

WORLD STATISTICS

	1970(%)	2000(%)
SHARE OF TOTAL AGRICULTURAL OUTPUT		
DEVELOPED	64	38
DEVELOPING	36	62
GRAIN CONSUMPTION		
DEVELOPED	45	33
DEVELOPING	55	67
ANIMAL PRODUCTS		
DEVELOPED	64	42
DEVELOPING	36	58

Source: World Food Plan

FIGURE : 14

U.S. AGRICULTURAL TRADE
1978 – 1979

COMMODITY	U.S. % OF WORLD TRADE	U.S. EXPORTS AS A % OF U.S.PRODUCTION
WHEAT	43 %	64 %
COARSE GRAINS	62 %	26 %
RICE	22 %	49 %
OILSEEDS	52 %	56 %
COTTON	29 %	54 %

FIGURE: 15
U. S. AGRICULTURAL EXPORTS
(MILLION METRIC TONS)

	1975/76	1976/77	1977/78	FORECAST 1978/79
WHEAT AND FLOUR	30.611	24.723	32.8	32.3
FEED GRAINS	49.855	50.602	55.5	60.6
RICE	1.953	2.229	2.1	2.4
SOYBEANS	15.050	15.156	20.0	20.5
VEGETABLE OILS	.868	1.142	1.5	1.6
OILCAKE & MEAL	4.870	4.336	5.8	6.3

FIGURE: 16

U.S. AND OPEC AGRICULTURAL AND OIL RELATIONSHIPS

1978

(in U.S. dollars)

U.S.	OPEC
TOTAL U.S AGRICULTURAL EXPORTS 28.0 bil.	TOTAL OPEC OIL EXPORTS 130 bil.
TOTAL U.S. AGRICULTURAL EXPORTS TO OPEC 2.1 bil.	TOTAL OPEC OIL EXPORTS TO U.S. 33 bil.
TOTAL OPEC AGRICULTURAL IMPORTS 11.0 bil.	TOTAL U.S. OIL IMPORTS 45 bil.

344

FIGURE: 17

U.S. IMPORTS OF OIL

YEAR	BILLIONS OF DOLLARS
1978	45
1979	50
1985	100

Impact on U.S.

U.S. agricultural trade has been the most significant feature in the agribusiness structure over the
last decade, with both exports and imports increasing
dramatically and affecting the well-being of the U.S.
farmer and consumers (see Figure 18). This also shows
up in changes in consumer food prices, as noted in
Figure 19. In 1977, two-thirds of U.S. food inflation
was due to rises of imported foods in that particular
year -- coffee, cocoa, and fish. Because food is such
an important item in the consumer price index, it is
critical to soften the impact of volatile domestic and
international commodity prices by developing long-term
trading relationships that do not cause a ratchet
effect in general price inflation and wage settlements.
This is especially necessary since labor is such an
important cost element in the total food bill, as noted
in Figure 20. Much of the increased value of U.S.
agricultural commodities has been capitalized back into
land values (as noted in Figures 21 through 26), which
raises questions about how newcomers can afford to
enter the farming sector of agribusiness and how small-
scale farmers can survive.

ACTION ALTERNATIVES

Role of Cooperatives

One method of involving the farmer more directly
in his domestic and international food system is
through farm cooperatives. Some 20 percent of the
world's population belongs to one or more farm cooperatives. Some 28 percent of U.S. agricultural production
is marketed by U.S. farm cooperatives, as is approximately 40 percent of all grains and soybeans (as noted
in Figure 27). In Figure 25, one notes that farmers
had close to $17 billion invested in their coopera-
tives. Some 23 percent of all farm supplies are produced and processed by farm cooperatives (see Figure
28).

Recently, several U.S. cooperatives along with
several European cooperatives joined together to purchase a German trading firm to become part of the
international grain business. Similarly, other cooperatives are coming to the United States, such as the
Japanese cooperative ZEN-NOH, which announced that it
will build a terminal elevator in the United States.
This is just the beginning of more farmer involvement
in agricultural trade. In reality, given the large
numbers of livestock and poultry in the world, the

FIGURE: 18

U.S. AGRIBUSINESS
(BILLIONS OF DOLLARS)

	1967	1972	1973	1977	1978	1979
FARM INPUTS	38	52	65	88	96	107
NET FARM INCOME	12	18(14)	32(25)	20(11)	28(14)	34(16)
AMOUNT IN SUBSIDIES & NON-MONEY INCOME	7	7	8	12	14	13
ALL FARM MARKETING	43	63	89	96	110	128
U.S. IMPORTS	4	6	9	13	14	16
U.S. EXPORTS	7	8	21	24	29	32
U.S. FOOD PROCESSING PURCHASES	40	61	77	85	95	112
ASSEMBLY, PROCESSING & DISTRIBUTION	54	62	67	132	144	153
CONSUMER FOOD EXPENSES	94	123	144	217	239	265

Note: Figures in parentheses net farm income in constant 1967 dollars.
Source: U.S.D.A.

FIGURE: 19

SOURCE OF INCREASES
IN GROCERY STORE FOOD PRICES

YEAR	TOTAL GROCERY FOOD PRICE INCREASE	CONTRIBUTIONS FROM		
		FARM VALUE	FARM-RETAIL SPREAD	FISH & IMPORTED FOODS
	PERCENT	AT	RETAIL	
1974	14.9	2.7	9.7	2.5
1975	8.3	1.8	4.6	1.9
1976	2.1	-1.6	2.5	1.2
1977	6.0	0.1	1.8	4.1
1978	10.5	5.1	4.1	1.3
1979	10.5	3.7	5.6	1.2
ALL FOOD 1978	11.3	16.4	8.2	7.2
ALL FOOD 1979	10.8	11.0	12.0	6.0

Source: U.S.D.A.

FIGURE: 20

Components of Retail Expenditures for Farm Foods

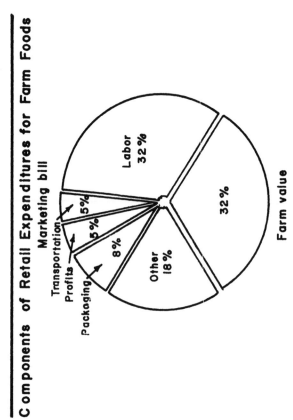

U.S.D.A.

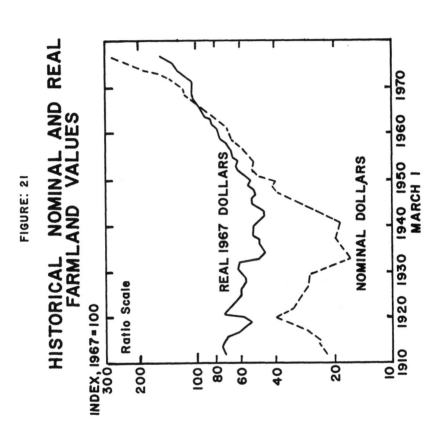

FIGURE: 21

HISTORICAL NOMINAL AND REAL
FARMLAND VALUES

INDEX, 1967 = 100

Ratio Scale

REAL 1967 DOLLARS

NOMINAL DOLLARS

MARCH 1

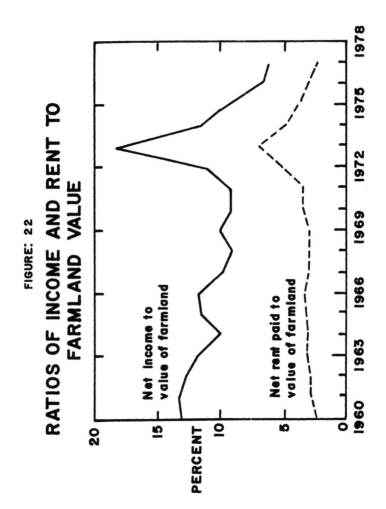

FIGURE: 22

RATIOS OF INCOME AND RENT TO FARMLAND VALUE

FIGURE: 23

RETURN TO FARM ASSET EQUITY
1960 – 1978

YEAR	RATIO OF EARNINGS TO ASSET EQUITY (%)	YEAR	RATIO OF EARNINGS TO ASSET EQUITY (%)
1960	2.6	1970	3.5
1961	3.3	1971	3.5
1962	3.5	1972	5.3
1963	3.3	1973	10.6
1964	2.6	1974	5.9
1965	4.5	1975	4.8
1966	4.7	1976	2.6
1967	3.3	1977	2.2
1968	3.0	1978	3.6
1969	3.7		

Source: Balance Sheet of the Farming Sector, 1978

FIGURE: 24

INCREASES IN PRICE OF LAND vs INCREASES IN DOW JONES INDUSTRIAL AVERAGE

YEAR	INCREASE IN PRICE OF FARMLAND (PERCENT)	INCREASE IN DOW JONES AVE. (PERCENT)	YEAR	INCREASE IN PRICE OF FARMLAND (PERCENT)	INCREASE IN DOW JONES AVE. (PERCENT)
1950	3	—	1966	7	-19
1951	14	15	1967	6	15
1952	9	9	1968	6	4
1953	2	-4	1969	5	-15
1954	-1	40	1970	3	5
1955	3	21	1971	3	6
1956	4	2	1972	7	15
1957	8	-13	1973	12	-17
1958	6	37	1974	25	-28
1959	8	17	1975	14	38
1960	3	-9	1976	12	17
1961	-1	19	1977	17	-17
1962	5	-11	1978	9	4
1963	5	17	1979	14	-4 (est.)
1964	6	15	1980	14 (est.)	—
1965	6	11			

FIGURE: 25

BALANCE SHEET OF FARMING SECTOR
AS OF JANUARY 1
(BILLIONS OF DOLLARS)

Balance Sheet of Agriculture

ASSETS	1940	1950	1960	1970	1975	1978	1979
REAL ESTATE	33.6	77.6	137.1	215.8	368.5	525.8	599.5
LIVESTOCK & POULTRY	5.1	12.9	15.2	23.5	24.6	32.0	51.3
MACHINERY & MOTOR VEHICLES	3.1	12.2	22.7	32.3	55.7	77.7	84.3
CROPS	2.7	7.6	7.7	10.9	23.2	24.9	27.4
HOUSEHOLD EQUIPMENT & FURNISHINGS	4.2	8.4	9.2	9.6	14.0	16.4	19.2
DEPOSITS & CURRENCY	3.2	9.1	9.2	11.9	15.1	16.3	16.8
U.S. SAVINGS BONDS	0.3	4.7	4.7	3.7	4.3	4.4	4.8
INVESTMENT IN CO-OPS	0.8	2.0	4.2	7.2	12.1	15.5	16.9
TOTAL	53.0	134.5	210.2	314.9	517.5	713.0	820.2

FIGURE: 26

BALANCE SHEET OF FARMING SECTOR AS OF JANUARY 1
(BILLIONS OF DOLLARS)

Balance Sheet of Agriculture

LIABILITIES	1940	1950	1960	1970	1975	1978	1979
REAL ESTATE DEBT	6.6	5.6	12.1	29.2	46.3	63.6	72.2
CCC DEBT	0.4	1.7	1.2	2.7	0.3	4.5	5.2
OTHER SHORT-TERM DEBT	3.0	5.1	11.5	21.2	35.2	51.1	60.1
TOTAL LIABILITIES	10.0	12.4	24.8	53.0	81.8	119.3	137.5
PROPRIETORS EQUITIES	43.0	122.1	185.4	261.9	435.7	593.7	682.7
TOTAL	53.0	134.5	210.2	314.9	517.5	713.0	820.2

FIGURE: 27

U.S. COOPERATIVES SHARE OF FARM MARKETING

	SIZE OF THE MARKET $ MILLIONS 1977	NUMBER OF CO-OPS 1977	CO-OPS MARKET SHARE % 1950	CO-OPS MARKET SHARE % 1977
COTTON	3,939	519	12	26
DAIRY PRODUCTS	11,776	579	53	69
FRUITS & VEGETABLES	9,923	436	20	27
GRAINS & SOYBEANS	26,697	2,599	29	40
CATTLE	20,230 ⎤			
HOGS	7,326 ⎬	654	16	9
OTHER MEAT ANIMALS	390 ⎦			
POULTRY & EGGS	7,219	151	7	8
OTHER	8,584	214	NA	20
TOTAL FARM MARKETING	96,084		NA	28

Source: U.S.D.A.

FIGURE: 28

U.S. COOPERATIVES SHARE OF FARM SUPPLIES

	SIZE OF THE MARKET $ MILLIONS	NUMBER OF CO-OPS	CO-OPS MARKET SHARE %	
	1977	1977	1950	1977
AGRICULTURAL CHEMICALS	2,008	3,597	11	33
EQUIPMENT	4,880	NA	5	11
FEED	13,840	3,819	19	19
FERTILIZER & LIME	6,089	3,949	15	36
PETROLEUM	4,050	2,903	19	28
SEED	2,856	3,526	18	15
TOTAL FARM SUPPLIES	33,723		16	23

Source: U.S.D.A.

biggest customer a farmer has for his grain is another
farmer.

U.S. cooperatives have also been active in the
developing world, helping to build fertilizer plants
and beginning to develop joint marketing arrangements.
The Japanese cooperatives have been more active than
U.S. cooperatives in Brazil, Thailand, and Argentina,
but I believe this will change as U.S. and developing-
country producers recognize their increasing interdependence.

Role of A.I.D.

Although over $25 billion has been spent on P.L.
480 shipments over the last twenty-five years, only in
recent years have we tried to combine this aid with the
changemaker's role in the developing countries. An
excellent example of this is the Amul Cooperative Dairy
in India. U.S. and European surplus dried skim milk
was blended into its dairy operations during the normal
shortage season, and the sales of these products provided the funds to expand the operations to include
small-scale producers and landless laborers without
destroying normal market prices. Currently, vegetable
oil shipments are being sent to India in an attempt to
provide expansion funds for farmer cooperative leadership in that industry. In a way, this cooperative
represents the change agent so necessary to fight the
"accommodation" to poverty described in John Kenneth
Galbraith's recent book, The Nature of Mass Poverty.

Role of Multinational Firms

Many U.S. multinational firms either collectively,
through organizations such as LAAD, or individually,
see the involvement of small-scale producers as very
important in their activities in developing countries.
One U.S. company in a joint venture in Southeast Asia
helped to develop a whole new corn economy in one area
of a country by taking on the role of an extension
service and providing seed, storage, and a guaranteed
price (where none existed before) to the small farmer.
In addition, U.S. multinational firms have taken on the
role of logistics experts for many developing countries' internal and export operations. For the most
part, though, European and Japanese firms have been
more active in this role, usually because their governments work more closely with their farmers and with the
developing countries than does the U.S. government.
New ways of working together should be explored for
both the private and cooperative segments of U.S.
agribusiness.

The Role of Government

--   The 1977 Farm Act, with a low loan price and a reasonable target price, with protection for the farmer in times of gluts and for the producer in times of surpluses, has enabled U.S. farm exports to be more competitive in world markets. This nonpartisan act should be continued.

--   Foreign assistance should be used in the developing countries in cooperation with local change agents and institutions, such as the Amul Dairy example.

--   Flexible long-term arrangements, such as the U.S. - U.S.S.R. grain agreement, should be developed so as to avoid unusual large-scale procurement changes. Financial credit arrangements should not be part of the budget process.

--   Import arrangements that involve major U.S. local producer areas also should develop long-term arrangements to provide market access for those supplying U.S. markets and for known market parameters for U.S. producers to avoid confrontations such as the current U.S. - Mexico tomato feud.

--   The government should encourage U.S. multinational and farmer cooperative involvement in the world food system.

--   Finally, the government also should encourage foreign investment in the U.S. food system as a valuable way to break down trade barriers, recognizing that there is a multiplicity of policy priorities to be served.

11
IMPLICATIONS OF THE
MULTILATERAL TRADE NEGOTIATIONS
FOR AGRICULTURAL TRADE

Alonzo L. McDonald

One of the most significant developments of the decade of the 1970s for world agriculture has been the Multilateral Trade Negotiations. The Tokyo Round of negotiations concluded in April of 1979. Although it was the seventh in a series of multilateral trade negotiations (MTN) since World War II, it was the largest and most significant, especially in terms of its impact on agriculture. Much of what will shape world agricultural trade for tomorrow was decided in those negotiations, so it would be difficult to minimize their potential impact on both importers and exporters.

INTERNATIONAL AGRICULTURAL RESTRICTIONS AND BARRIERS

Restrictions

All nations stand to gain from the Tokyo Round agricultural advances because all nations have been affected by the proliferation of trade barriers in the postwar period. These barriers stem from the adoption of domestic programs designed to support national agricultural economies through a variety of means. Most are based on domestic price and income-support mechanisms and include substantial barriers limiting the degree of price competition on imports.

Nations have adopted these programs for a variety of reasons. First, maintenance of national self-sufficiency in food production is generally a high national priority. With food the primary element for human survival, protecting the domestic agricultural economy from competition is seen by many nations as vital to their economy.

Second, price stability in domestic markets has traditionally been a major objective of governments since food occupies an important position in family budgets. Food price instability is economically disruptive and inflationary.

Third, high domestic price-support programs are politically attractive. Farmers are a major political force in many of the developed, as well as developing, countries. In our own country, recent agricultural strikes are indicative of the level of political involvement of farm groups. In Europe, Japan, and many other countries, farmers are able to use their political cohesiveness in support of specific farm programs.

Finally, there is a strong social rationale for the adoption of farm programs. The migration of farmers to the city in recent decades in search of industrial jobs and higher standards of living has caused many difficult dislocations around the world. Profiting from past experience, countries have sought to discourage urban migration by making rural existence more attractive. This tendency has been particularly pronounced in Europe and during periods of high unemployment in the cities.

Most of these programs operate through pricing objectives which frequently are higher than the cost of imports. As a result, barriers must be imposed to control the level of imported farm products. Consequently, consumers suffer, efficient agricultural producers suffer, and those countries in need of foodstuffs suffer.

Objectives may be achieved by means of tariff and nontariff barriers. Since tariffs have had only limited success in insulating domestic farm programs from the international market, there has been a proliferation of nontariff agricultural barriers in the postwar period.

Probably the most widely known of the national agricultural programs is the European Community's Common Agricultural Policy (CAP). The CAP incorporates two classic elements of domestic farm programs: high price supports and nearly invincible barriers to import competition. The European Community uses the ingenious "variable levy" to raise the duty-paid price of imported agricultural products slightly above domestic prices. Hence, import competition is eliminated and imports become a residual source of food.

The Japanese rely on various forms of protection, including state trading and quantitative restrictions on imports. In the case of wheat, for example, the Japanese food agency buys the product in international markets and resells it at domestic prices, which last year were one hundred percent higher than the imported price. The Japanese have quantitative restrictions or quotas on beef, citrus, leather, and many other products.

Even the less-developed world employs restrictive regimes to protect domestic agricultural producers. One example common to many developing countries is import licensing. Under this system only foreign products with a license issued by the government are allowed entry.

World trade in agricultural products is hindered by other nontariff mechanisms including standards, subsidies, and discriminatory government procurement practices. Standards may be issued on imported products designed to assure certain health and sanitary standards but which may in some cases provide unfair protection for domestic interests. One example would be the country which requires that live cattle be quarantined before they are allowed entry but does not provide any facilities to quarantine livestock. Imports are effectively prohibited.

Because of the special nature of agriculture, domestic farm programs are exempted from the discipline of the General Agreement on Tariffs and Trade (GATT). For example, under Article XI of the GATT, quantitative restrictions are generally prohibited; however, they are permitted for agricultural products under certain conditions. Another exemption for agriculture is the accepted use of export subsidies for agricultural products under Article XVI of the GATT, while such subsidies are prohibited for industrial products.

In addition to these blanket exemptions, there are also individual national exemptions for farm programs. The United States has its so-called Section 22 waiver, which enables it to impose fees or quotas on imported products that threaten to interfere with a domestic price-support program. The European Community's CAP is exempted as are all domestic agricultural programs in Switzerland.

Trade Barriers

The result of trade barriers is a paradox. Food consumption throughout the world is less than optimal, yet countries with surplus capacity cannot use it because access to export outlets is restricted. The situation is exemplified in U.S. trade with Japan and Europe. U.S. consumers use less than 20 percent of their spendable income for food, while the Japanese and the Europeans, with their very restricted import regimes for agriculture, must use 30 percent of their spendable income.

In addition to lowering the standard of food consumption worldwide, trade barriers also can have a destabilizing effect on the world food economy. It is a widely accepted notion that nations which stabilize their agricultural sectors through trade barriers transfer the instability of their farm sector to other nations. This transfer of price instability is particularly damaging to the food-deficit, less-developed countries. The imposition of an export tax by the European Community in the early 1970s to discourage wheat exports is an example of this destabilizing phenomenon.

Agricultural trade barriers also have led to increased governmental confrontation. Because of restrictions on trade, governments frequently have been at odds in the postwar period over bilateral agricultural trade problems.

THE TOKYO ROUND

Importance to World Agriculture Community

The Tokyo Round represented a major beginning, not an end, for adapting agricultural trade patterns to current and future needs and realities. It can be viewed from two perspectives: as a major step toward an improved, more rational, more humane world food production and distribution system, and as a source of major benefits for the United States. At this point, I would like to briefly review the specific accomplishments of the negotiation before considering its implications.

The Tokyo Round agricultural-package results fall into five general categories. The first are immediate benefits in the form of reciprocal reductions on specific tariff and nontariff barriers. During this aspect of the negotiations, countries agreed to reduce their own trade barriers on agricultural products in exchange for their trading partners' willingness to do the same. This "horse trading" is the traditional element of trade negotiations and the easiest one to understand and measure.

The second category constitutes potential benefits embodied in a series of new codes of conduct which were agreed to as a means of disciplining nontariff government intervention. Government involvement in the market place has been a growing problem overseas and has frequently placed U.S. exporters and others at a competitive disadvantage. The new codes establish rules for future international trade in a number of important areas, such as subsidies, product standards, government procurement, and others, where direct government action has been particularly onerous. Because of their pioneering character, the codes truly represent the major achievement of the Tokyo Round. For the first time, negotiators successfully assailed previously sacrosanct areas of government responsibility and in the process moved toward a more universally acknowledged role for market forces in agricultural trade.

To illustrate, increased discipline in the use of subsidies, especially export subsidies, will reduce the possibility of cutthroat competition between national treasuries. Although the code will not eliminate subsidy practices in agriculture, a step even the United States is not prepared to take, it will clarify their use and therefore reduce confrontation and needless misunderstandings. Improved international dispute settlement procedures under the subsidies code, and other codes as well, will encourage signatories to resolve disputes quickly and justly. As the world's most efficient agricultural commodity exporter, the United States stands to gain tremendously.

Third, the agricultural package includes agreement on an institutional framework for improved international cooperation in agricultural trade. Under this arrangement, countries recognize their mutual objectives in agriculture and agree to coordinate their efforts to achieve these objectives by means of regular consultations among agricultural policymakers. We hope in this way to reverse the trend in recent years toward greater conflict in agricultural trading practices.

Fourth, the package contains needed reforms and updating of the rules and institutions of international trade. This aspect will be increasingly significant as we seek to establish a firm foundation for responsible, mutually advantageous trade relations with developing countries. Without an improved framework for our relations with developing countries, the potential for serious conflict will grow rapidly in the future.

Finally, the agricultural package contains specific consultative arrangements in two important commodities: The International Dairy Arrangement and The Arrangement Regarding Bovine Meat. Both will result in a greater exchange of information concerning dairy and meat trade, production, and consumption. Regular consultations between the signatories will focus on problems before they become major issues.

Politicians and policymakers tend to view international trade negotiations in terms of their national interest in expanding exports. This is a natural and entirely appropriate approach which provided the necessary impetus to those of us who labored at the negotiating table. Viewed from that vantage point, it is gratifying to report that the United States did very well, indeed, in the agricultural negotiations, certainly better than in any previous round. This was reflected by the enthusiastic support of the agriculture community for the MTN implementing legislation.

Results for the United States

The United States placed continuing emphasis on agriculture throughout the negotiations. We demonstrated considerable flexibility in the initial procedural disputes which held up progress for several years in this area. It would be difficult to overemphasize the depth of feeling which existed domestically in the United States over the agricultural issue and the tremendous importance of making real progress in agriculture as an essential factor in obtaining congressional support for the results on the Tokyo Round. A solid agricultural result was also essential to retain the American farm community's long-standing support for liberal trade policies.

The results speak for themselves. The United States gave concessions covering approximately $2.7 billion in 1976 trade but received concessions on nearly $4 billion of U.S. exports, using 1976 figures. We anticipate that the concessions received by the United States will result in at least one-half billion dollars in additional U.S. agricultural exports, while the concessions we granted will result only in approximately a one-hundred-million-dollar increase in imports.

From the U.S. perspective, important as it is for us to look ahead to increased export opportunities across a wide product front, we should not lose sight of our stake in the current favorable export situation. Certainly, one of our primary objectives in the Tokyo

Round was to preserve our important position to the benefit of our own producers and to the world economy as a whole.

As the world's largest exporter of farm products, with one-third of our production going overseas, U.S. agriculture is truly international agriculture. American farmers are more exposed to developments in world markets than those of any other major country. Had the Tokyo Round faltered, American agriculture would have been one of the big losers. Moreover, the world agriculture system simply could not have tolerated such an added element of uncertainty.

Implications of the Tokyo Round for the Future

It is clear that we live in a climate of growing, sometimes exaggerated but, nonetheless, significant fears. We fear continuing inflation, chronic unemployment, slow economic growth and monetary instability. The recent flight to gold is clearly symptomatic. At times we seem to question our continuing ability as a nation to compete. These problems are serious and they will not disappear quickly. Our nation, and others, must learn to adjust to a situation of continuing uncertainty while maintaining our fundamental long-term objectives.

That is why the Tokyo Round is so important. We have been able to make necessary adaptations of our institutions to new and rapidly changing realities. We have been able to maintain the process of postwar trade liberalization as an essential motor for economic growth, for the creation of new jobs, and for the maintenance of investor confidence. We have been able to maintain sound working relationships between nations as well as to produce real economic benefits. We have, in short, been able to keep the fundamentals in view in a period of intense negative pressures.

However, the results of the Tokyo Round will not implement themselves. We have new tools; now we must use them. Three actions are necessary to achieve the full potential of the Tokyo Round. First, private sectors and governments must vigorously pursue their rights under the codes. Farmers and sellers of agricultural commodities must move aggressively in world markets where barriers have been reduced. In our country only the private sector can close the deals and make the sales that will turn potential benefits into solid economic gains.

Second, governments must continue to work together to resolve trade problems without resorting to confrontational tactics. Governments must adhere to new rules that they have agreed to in the codes, and they must be willing to submit themselves to the disciplines and dispute-settlement procedures they have accepted.

Finally, trade liberalization must not end with the Tokyo Round. Whether or not there will be a future round of such magnitude is subject to question. I personally think not. Nevertheless, we cannot simply put trade on the back burner as has been our traditional pattern after previous rounds.

What will world trade be like ten years from now, or twenty, or more? That is a very short but complicated question whose answer is still far from certain. Ultimately the answer rests with how well a tripartite working relationship evolves between the Administration, the Congress and the private sector. If these parties can work as well and as cooperatively together in the future as they did during the historic Tokyo Round negotiation, we can face with confidence and high expectations the continuing stream of challenges that await us in the field of trade.

Section 5

THE CONSEQUENCES FOR AMERICA

INTRODUCTION

Meeting the challenge of world nutrition as populations grow already involves the United States in significant ways. Thus far American contributions have been grain reserves for emergency needs, developmental assistance through bilateral and multilateral channels, and research and technical expertise for developing countries. The extent and nature of American involvement are matters which receive increasing study and attention as policymakers attempt to design a response which is appropriate and realistic. The rationale for American policy which finally emerges will undoubtedly be influenced by the expected impact which that policy will have upon the United States, as well as the impact it can be expected to have upon world needs for food.

Morton Sosland points to twenty five years' experience with P.L. 480, the Agricultural Trade Development and Assistance Act, as evidence that successful food production development in needy nations does not result in foreclosed American markets for agricultural exports. He also takes issue with those who would initiate measures to ensure price stability in the world food system, preferring instead a price-responsive system as the best mechanism for maintaining a healthy food production capability. Nations which pursue a goal of food self-reliance, as opposed to self-sufficiency, will continue to have needs for food imports at times of crop shortfalls and for the purpose of satisfying upgraded dietary preferences. These are conditions different from unhealthy dependence upon imports. Sosland believes that the United States should more actively encourage developing countries to pursue policies of free marketing of foodstuffs, not only for the United States but also for less-developed countries, because that is a policy of demonstrated effectiveness.

What will the impact of diminished energy supplies and increased energy cost be upon American agriculture? Emery Castle puts the American food system and its energy use in perspective; he finds no evidence that American agriculture is significantly more intensive than the remainder of the economy. He provides an analysis of the effects which more expensive energy will have upon the food system, including implications on the use of pesticides and fertilizers, irrigation, field machinery, and livestock. Castle also considers proposed adaptations, such as "gasohol," organic farming, and biomass conversion. Disruptions in supplies of energy would be more difficult than increased energy prices. American public policy, therefore, should be reassessed in terms of changed social values resulting from higher energy prices in order to make the food system less vulnerable to severe disruptions and heavy dependence on fossil fuels.

The world food problem, according to Steven Muller, should be viewed as inseparable from U.S. national security interests. As the unquestioned major actor in the world food arena, the United States cannot escape the consequences of its policy decisions or indecisions. Stability in world food supplies is an essential prerequisite to social and international stability. The upsurge in the incidence of political refugees and illegal immigrants is an early warning of the instability which can result if the food challenge is not met. What is lacking in the U.S. effort is priority, coherence, and a consistency of purpose. Muller provides several recommendations to sharpen the U.S. response.

12
THE IMPACT OF
INTERNATIONAL DEVELOPMENT
ON AMERICAN AGRICULTURE

Morton I. Sosland

Before dealing with the effect of overseas development upon U.S. agriculture, it is important to state in clear terms a few of my biases, particularly as they relate to the various sectors of agriculture in both the U.S. and abroad. Most essential is your understanding that I am a "grain person," which simply means that my major focus will be on defining development, trade and consumption, and their effects upon U.S. agriculture and the U.S. grain producer. My "grain" bias allows me to call the "Hamburger Revolution" in the food industry a more realistic "Bun Revolution." From my perspective the growth of the fast-food business in the U.S. is felt primarily in expanded sales of buns rather than in the things that go between the bun slices.

Such a viewpoint sometimes leads to much foolishness and even to a restoration of old-time disputes between the grain farmer and the cattleman. That is not my purpose. Instead, I mean to fashion a simple measure that is effective in describing developmental progress. Thus, my favorite way of signifying dietary well-being is derived from estimates of per capita consumption of grain in various countries around the world. In the U.S. we consume about 1,800 pounds of grain (wheat, coarse grains, and rice) per person per year. About 300 to 400 pounds of that consumption is in the form of grain foods -- those buns as well as bread, cereals, and other such products -- while the balance is consumed through grain-fed livestock and poultry.

Per capita grain-consumption numbers for other countries vary widely. In Western Europe the average is quite close to the U.S. at around 1,500 pounds per person per year. In the Soviet Union it has been officially stated by leaders that the official goal is to raise per capita consumption of grain from a current level almost on a par with the U.S. to an even higher rate to accommodate a large-scale increase in livestock feeding. You can see that the Soviet Union already has made major strides in improving its diet and raising the level of consumption of grains from a total well under the U.S. to near a parity, except for the lags between livestock and poultry consumption as compared to consumption of bread and similar products. It is the Soviet Union's striving in this regard that mainly accounts for that country's willingness to spend hundreds of millions of rubles for U.S. grain designed to forestall any setback in the upward trend of dietary levels.

In the developing countries, per capita consumption of all grains lags well behind the levels of the U.S., Europe, and the U.S.S.R. India, even after recent improvements in grain production from the famine-threatening period in the early 1970s, currently is at a consumption rate of 400 to 500 pounds. Many nations in other parts of Asia and in Africa and South America are not much better off, with Brazil, for instance, at 600 pounds. A fascinating example is China, where per capita consumption of all grain is currently at just around 500 pounds, which is at the low end of the scale. Yet, most recent travelers to China would indicate that they have seen few, if any, signs of hunger and malnutrition, particularly as compared to a nation like Chad in the center of Africa where per capita consumption averages practically the same as in China but where hungry people are an overwhelming part of the landscape. The conclusion from this apparent anomaly is that China has not been as successful in solving its food-supply problems as it has been in resolving its food - discipline problems. That is affirmed by China's stepped-up purchasing of food on world markets in the past several years and the increasing, but still rare, accounts coming from China of people evidencing dissatisfaction over the adequacy of their diets.

Another basic bias in this presentation is belief in the idea that the world does have the capacity to raise enough food to allow, and even to encourage, other nations to strive to attain a per capita grain-consumption level similar to that of the U.S. It is also my belief that there is very little that could be done to reverse that desire for a better diet, short of adopting by force a massive system of global food rationing. Indeed, this is the principal underlying force in the world food picture -- a constant striving, most frequently not well-defined, to match the U.S. consumption rate.

In the obverse, it is highly improbable that U.S. citizens would be required, or even well-advised, under almost any conceivable circumstance, to reduce the quality of their diets. I found ridiculous the pleadings of only a few years ago for Americans to go without foods, such as "meatless Thursdays," in order to increase supplies available to needy peoples in developing countries. Such pleas made no sense at a time when many were concerned that hunger and starvation were accelerating, and they make no sense in the present context where more and more people recognize that recurring cycles in production and demand do not automatically mandate famine for millions. Americans should not and will not be called upon to reduce their own grain consumption as other than a token of concern for their fellow inhabitants. Such a step would do nothing to increase nutritional well-being in other nations around the world and would be counter-productive to the principal world food need of maximizing production.

This position differs from that of some eminent authorities who conclude, quite correctly, from a cursory examination of world grain supply and demand numbers, that there is more than enough grain now available to provide a "good" diet for everyone, if only problems of distribution and ability to pay could be resolved. That kind of conclusion usually is tied to "improving" the diets of the developed nations, such as the U.S., by cutting back on consumption of beef, pork, and other foods. Since I do not see any possibility of the establishment of a worldwide authority rationing food to each one of us, and thus mandating what we shall eat, it is my continuing expectation that people will want to eat better, and that "better," no matter how that may be defined nutritionally, is largely equated with choosing a diet not too unlike what we now have in the U.S.

Of course, there are also many possibilities for change in the U.S. diet. Recent world history is replete with examples of foods newly accepted in the U.S. gaining broad acceptance overseas, and foods not commonly sold in the U.S. having difficulty gaining a foothold in developing markets, no matter how "good" they may be from a nutritional viewpoint. Again, as a grain person, I strongly sense that recent nutritional findings may lead to a reduction in consumption of grain in some forms (the high fat/cholesterol items largely in the meat category) accompanied by a corresponding increase in consumption of grain foods like variety, "whole grain" breads. Dietary changes occurring as a result of development programs overseas,

or in response to changes in the U.S. diet, can greatly
affect the type of grains purchased from the U.S. As I
have noted previously, developmental progress overseas
will be most keenly felt in the U.S. agricultural
economy by way of demand for grains from abroad,
whether for consumption directly as grain foods or
through the livestock-poultry feeding process.

How these trends can vary in relation to individual grains is best illustrated in the rapidly rising
demand for U.S. corn as compared to wheat. At the same
time clearances of both grains are establishing records
in exports almost on a year-to-year basis. For their
respective 1979-80 crop years, the current forecast of
the U.S. Department of Agriculture is that exports of
wheat will amount to 1.4 billion bushels and clearances
of corn will be 2.5 billion bushels. It was only at
the start of the 1970s that exports of corn first
exceeded shipments of wheat. Two decades ago exports
of wheat usually were at least 50 percent larger than
shipments of corn; whereas, in the current season corn
exports will be nearly 80 percent greater than clearances of wheat. Exports of corn first passed the
billion-bushel mark in 1972-73 and passed two billion
bushels for the first time only in 1978-79.

Another basic consideration is that in the 1979-80
season, world trade in wheat and coarse grains combined
is expected to total a record 179 million tons, with
wheat accounting for 44 percent and coarse grains,
mainly corn, for 56 percent. At the start of the
decade of the 1970s, when total world trade was only a
trifle above one hundred million tons, wheat was 51
percent and coarse grains 49 percent. That seven-
percentage-point change in favor of corn is very significant, particularly in light of the massive over-all
trade expansion.

Corn, of course, is desired by overseas buyers for
use as livestock feed; whereas, a very large percentage
of the wheat moving in world trade is ground into flour
for baking into bread, rolls, noodles, and all the
other flour products. Thus, it is easy to perceive how
trends in development overseas, and how their outcomes
when divided between a livestock/poultry economy and a
grain-foods economy, can be of major consequence for
U.S. agriculture.

The one additional point deserving stress in these
preliminary remarks is: if the U.S. is successful as a
producer of grain, and our achievements in this regard
are among our most precious accomplishments, then this
nation's ability to grow corn is at the very pinnacle

of a list of national assets. There are few nations in
the world that do not raise either wheat or rice in a
volume that can provide at least a subsistence diet and
often quite a good diet. On the other hand, only the
U.S., and a very few other countries, are able to raise
corn in anywhere near a commensurate volume.

Two successive seven-billion-plus-bushel corn
crops in the U.S., in 1978 and again in 1979, stand as
stark evidence of this country's success as a corn
producer. The U.S., rather consistently year after
year, has accounted for at least 50 percent of the
world's corn crop, with no other country accounting for
as much as 10 percent. Our share of coarse grain
production, where other countries do compete, has been
rather consistently at 29 percent, rising to 30 percent
this year.

From the mid-1970s to 1978-79, the U.S. share of
world trade in coarse grains increased each year, from
61 percent to 64 percent, and the current expectation
is that the U.S. in 1979-80 will ship just about 70
percent of the record exports of 101 million metric
tons forecast for the season that began on October 1,
1979. Saying it another way, U.S. coarse grain exports
in the 1975-76 season were 46.3 million tons, while
non-U.S. sources accounted for thirty million tons;
four years later, the U.S. outgo is expected to rise to
71.1 million tons, while non-U.S. shippers will not
even hold their own at 29.8 million tons.

Defining the accomplishments of the U.S. as a
grain producer and a grain exporter is important to
providing understanding of this nation's capacity to
serve as an ever expanding source of grain moving to
developed and developing nations. My thesis is one of
confidence in the ability of this country's farmers and
of the marketing system itself to provide an ever-
increasing volume of grain for export to meet needs
overseas. Full-scale success on that score depends on
a number of trends, none of which is more important
than sufficient investment in research to allow continuation of the upward trend in crop production.
Others have stressed the technology needed in the
developing lands to raise their crop production.
Recognition also must be given to the importance of
research in maintaining this country's potential and
responsibility as a supplier of grain to world markets
in developed and developing lands.

In view of the recent surge in demand for grain
from abroad, particularly from the U.S.S.R. in the wake
of that country's disappointing 1979 harvest, a lot of

questions have been raised concerning the ability of the U.S. to meet prospective demand. After all, three of our principal competitors in the grain markets, Argentina, Australia, and Canada, have not only been unable at varying times in recent years to fill actual sales commitments, but have been foreclosed from making sales because of structural problems within their nations in moving grain from producing areas to ports for loading on ships. Many of the problems faced by those nations have not only involved labor strife, which closed major ports for lengthy periods, but also have included overwhelming difficulties related to internal rail and truck transportation, marketing, and even inadequate ship-loading capacity.

Even though the U.S. system is far from perfect, and there are numerous instances of delays due to rail car shortages, barge-traffic problems, shortages of equipment and even strikes, the American performance in moving grain overseas fully matches the accomplishments of producers in boosting crop outturns. In 1972, when a lot of people thought the grain-handling system was going to fall into complete disarray in response to the "surprise" U.S.S.R. purchases that year, total grain exports for the first time exceeded two billion bushels, reaching 2.3 billion. In 1978, the outgo passed four billion bushels for the first time, totaling 4.2 billion. These numbers define with crystal clarity the amazing dimensions of the American grain-marketing system. This leads me to assert that no structural problems exist within the U.S. economy that would serve as a limit on the abilities of this nation to supply the world with expanding quantities of grain. Indeed, the export-marketing system, in which private companies and farmer cooperatives have invested hundreds of millions of dollars, practically assures that this nation's exporting capacity will measure up to any potential demand, even a repeat in the 1980s of the phenomenal growth witnessed during this last decade.

Will agriculture and food-production development abroad impact positively or negatively on U.S. farming? In 1954 the Agricultural Trade and Development Act, or P.L. 480, was passed. I concluded then that U.S. agriculture stands to benefit greatly from the successful encouragement of expanded food production in the developing nations where hunger, malnutrition, and all the other ailments and horrors associated with an inadequate food supply loom as significant problems. I do not agree with the simplistic notion that successful food production development in needy nations will automatically mean the closing of important markets for American agricultural exports.

A great deal of debating occurred twenty-five years ago as to whether P.L. 480 should be structured solely as a device to dispose of "surplus" grain from the U.S., which was a primary impetus for the program. Another concern was whether it was appropriate or not to use the foreign currencies received in payment for the huge grain shipments to encourage increased production in those lands. The matter of using the foreign currencies for these purposes as compared to financing promotions of the sales of U.S.-produced commodities has been a sensitive one for the past quarter of a century. It is apparent that the decision has been in favor of supporting agricultural production increases in the developing nations. While this means funds for cultivation of overseas markets may not have been as abundant as many would have wanted, the net result, measured by almost any criteria, has been highly positive for American agriculture.

Many grain farm leaders in the late 1950s voiced deep concern over U.S. support for agricultural production increases in developing countries. They warned that P.L. 480 could become a Frankenstein's monster, with the eventual emergence not only of self-sufficiency in grain in many nations that were then recipients of large amounts of food aid, but also that some of those countries could become exporters and compete with U.S. farmers. Any number of restraints were put into the original P.L. 480 program and into the recurring extensions to prevent this from happening. One sorry consequence of all this is that it has become popular in some circles to cite those restraints as evidence that P.L. 480, rather than being an important humanitarian effort on the part of the U.S., which it has been, is nothing more than an effort to keep developing nations perpetually dependent upon U.S. grain shipments.

You can sense that I do not agree with that thesis; to the same extent I vigorously disagree with those grain-farming and -exporting interests who voiced concern at the program's start that it eventually could turn upon the U.S. and be counterproductive to the cause of American agriculture. The basic numbers attest to the foolishness of those early worries about the counterproductive possibilities from P.L. 480. In the 1954 crop season, U.S. exports of wheat were 254 million bushels and exports of corn were only 93 million bushels. As noted earlier, the prospective shipments for the 1979-80 crop year will be 1.4 billion bushels of wheat and 2.5 billion of corn. Further, practically all of this year's exports are being made for dollars; whereas, in 1954-55, when the new assistance program was just beginning, the major share of the shipments was financed through agreements that resulted in payment to the U.S. government in foreign currencies, most of which were inconvertible.

It is truly amazing that the role of P.L. 480 in helping to develop markets for U.S. agriculture has so neatly paralleled the scenario spelled out by those who felt this would be the result. In its basic form, establishment of functioning food production and marketing structures in developing countries has served to bolster demands of people in those nations for a rising standard of living, or better said, of eating. Critics of U.S. foreign aid would perhaps cite this as further evidence of the "evil" of the American system, but here again I vehemently disagree. The course of development in country after country in all parts of the world has simply been to allow for a gradual, but perceptible, rise in the standard of eating. Gains in crop production have served as building blocks to strengthen the hands of those nations' governments to the point that they have comfort in allowing for incremental improvements in the quality of the diet.

The basic format in nations like Taiwan and South Korea has been to build year after year and decade after decade upon successes in raising domestic food production. Only in those countries where development programs have failed for a variety of reasons, including the outstripping of gains in food production by unrestrained population expansion, has the development of food production not followed the scenario that was expected.

Another exception to success involves an American agriculture-related industry with which I have considerable familiarity -- flour milling. U.S. exports of flour, which not only serve to export American wheat but also to provide employment for American labor and capital, have not at all benefitted from the P.L. 480 program. Even though fairly large amounts of flour have been, and continue to be, shipped under this program, the impetus of development efforts has largely followed the path of encouraging the construction of foreign flour mills, which has resulted in a cessation of U.S. flour exports to once important markets. Millers in the U.S. expended considerable effort over a period of time a decade or more ago to convince foreign countries that it was economically sound to buy flour from the U.S. rather than wheat. But the trend in the opposite direction has been so powerful that U.S. millers' efforts in this regard are at best half-hearted, except in those instances where the P.L. 480

recipient nation is a major taker of flour and U.S. mills have to compete for a share of the trade. Prime examples in the latter case are Egypt and Sri Lanka.

Another exception to the success of P.L. 480 but of more current vintage is India which has gone in this decade from being a country frightfully dependent on U.S. grain shipments, primarily of wheat, to self-sufficiency and to possibly becoming a competitor of the U.S. in foreign markets. It is startling that India, where per capita grain consumption continues at a low level, has emerged as the nation which threatens the basic thesis that progress in development should be a net plus for American agriculture. If India's problems caused by inadequate monsoon rains lead to a setback in crop production, and if this results in that nation being a sizable buyer, such an occurrence would not prove the basic thesis.

On the other hand, if the government in New Delhi should decide to embark upon a sizable grain export program, as some reports would indicate, then one may have to rethink the entire development process as it relates to American agriculture. The problem, it appears to me, is not so much with the basic concept as it is with a government that sees greater advantage in the export of grains than in encouraging, or even permitting, an increase in the standard of eating within its own borders. The latter, it seems, is the proper and usual reponse to domestic successes in stimulating crop production. I am not sure that India's reaction justifies a reversal of thinking, but it might well be cause for questioning my postulations about the fully positive results for American agriculture in agricultural development abroad.

The third P.L. 480 exception is Iran. While the developed world in the past twenty-five years has seen plenty of revolutions, events in Iran last winter stand as the first upheaval of consequence to be directly related to the energy situation, and thus merits special examination for the purpose of understanding how energy-related occurrences can impact upon U.S. agriculture. Iran, under the Shah, was a very large importer of U.S. grains, particularly of wheat. At one time a P.L. 480 recipient, Iran became an ever-increasing commercial buyer as the country's wealth escalated in the wake of oil price advances of this decade. When the Ayatollah Khomeini took over, his first actions included a brake, and even a ban, on purchases of wheat from the U.S. as a sign of his country's independence of America. This naturally infuriated a number of U.S. wheat-growing interests, particularly those farmers in the Pacific Northwest who grow the white wheat preferred in the Iranian market. When it was disclosed some weeks ago that Iran was going to buy sizable quantities of kerosene and heating oil from the U.S., in one of those bits of madness surfacing with increasing frequency in the world situation, concerned wheat growers saw in this the opportunity to demand that Iran buy U.S. wheat in return for our agreeing to ship that country oil. I mention this as a delicious respite from that other debate suggesting that the U.S. somehow must use its grain-producing and -exporting capacity to demand barters of wheat for oil from OPEC members. In the Iranian instance, the ability of the U.S. to export kerosene was seen as providing the U.S. with the strength to demand that one of the OPEC nations buy our wheat. At least that puts in proper perspective the ridiculousness of the "bushel for barrel" advocates.

If U.S. agriculture is a net gainer from development overseas, it becomes quite important to understand how those gains are actually reflected back to the benefit of U.S. agriculture. Such understanding includes the mechanics by which increased foreign demand for American grain is felt throughout the U.S. agricultural system. The effect primarily concerns the market pricing mechanism, including both the futures markets and cash prices. Changes in foreign demand for U.S. grains are just as quickly reflected through this sensitive system as any other of a myriad of important influences, ranging from droughts that threaten crops to expanded production in a competing exporting country.

The American marketing system provides a unique process for transmitting to farmers swiftly and exactly the benefits (as well as penalties) of the ups and downs of foreign demand. In my judgment this pricing mechanism is a superb device, all too often unappreciated, and in more cases than not, badly misunderstood.

One of the greatest misunderstandings of the pricing mechanism stems from not appreciating how the magnitude of price fluctuations themselves strengthens the ability of the system to broadcast signals to American farmers. There could have been no more powerful signal of the need for an increase in crop production than was provided by the tripling and even quadrupling of prices occurring in the wake of the Soviet Union's purchases in 1972. Similarly, the setback for U.S. grain markets in the 1975-76 period served as a convincing word of caution to farmers that the upward trend in grain demand around the world is not straight up, primarily because of the inability to fine-tune production on either a country or a global basis.

Of great concern are the cries heard frequently that these fluctuations need to be restrained, thus muting the signals sent by the price changes. All too often sponsors of any number of domestic and foreign programs, including proponents of various multilateral and bilateral commodity agreements, present their plans as ways of achieving price stability. All kinds of activities and ideas are proposed with the sole purpose of holding prices within narrow bounds. It is my belief that this is a very undesirable target, if for no other reason than that its achievement would somehow diminish the signaling ability of prices, not just for American farmers but for the entire economy.

Instead of such a disastrous approach to world trade, I would prefer that the U.S. bend every effort in undertaking aid abroad to encourage developing nations to move toward a price-responsive system as quickly as possible. The suggestion is for a system that responds to changes in supply and demand, domestically or worldwide, through price movements in the marketplace. Countries which seek to conduct their affairs with the goal of minimizing price changes for consumers, which is a political decision, work great damage on the ability of their production systems to prosper and to be in tune with ever-rising demands for food.

When a price-responsive system is in place, the signals that go out to that nation's farmers to invest in production practices to increase yields are loud and clear. Similarly, in the event prices in that nation weaken under the weight of competitive offerings from suppliers abroad, that must not necessarily be viewed as discouraging for the ultimate well-being of local farmers, or for the economy of the country in question. In many ways price can serve as a stimulus to future demand, as well as being an influence on production. That effect is not only in the interest of foreign suppliers, such as the U.S., but also of domestic producers. One tremendously important aspect of agriculture, which all too few people have learned, is that the rewards are not the same from year to year, and that certain crop seasons can produce tremendous producer incomes while other seasons might even be characterized by losses. In the long run, though, based on observations not only in the U.S. but in a number of countries overseas, an agricultural system that is price responsive on a world basis serves the best possible objectives of both consumers and producers, and this is a point that cannot be emphasized too often or too forcibly.

Contrary to those who are wont to criticize P.L. 480 dispositions as being counterproductive to development programs overseas, it is my judgment that grain from the U.S. has played and should continue to play a very essential role in the development process. As noted at the start, the availability of grain imports, either on concessional terms or in regular commercial markets, allows a country that important extra margin of safety that is so essential. Better said, export availability helps to build an economy that has the basic elasticity to make room for rising dietary expectations. In Eastern Europe and the Soviet Union the Communist countries, largely encouraged by the availability of grain from the U.S., have adopted internal programs aimed at enhancing diets, primarily by making available greatly increased quantities of livestock and poultry. The result for these nations, which have been hesitant to go the full measure of adopting a price-responsive system, is to accompany the rising standard of living with nearly corresponding increases in domestic production of grain. At the same time, these nations have not hesitated in turning to the U.S. and other foreign sources for supplies in order to lessen deviations from the upward consumption trends which might result from a crop shortfall.

True enough, that crop-shortfall situation has created what some regard as "problems" for the U.S., but which others, including myself, look upon as providing wonderful sales opportunities. Those who equate such programs with "problems" largely consider erratic purchasing of grain by nations like the U.S.S.R. as being undesirable for the U.S. I do not agree, believing that one of the greatest economic strengths of the U.S. is its ability to accommodate to the rapid and major changes from year to year in demand for grains from overseas (with amazingly little stress). This accommodation is largely through stock increase and disposition, accompanied by changes primarily in consumption of grain as livestock and poultry feed. The opportunities American producers have to enjoy extremely profitable years would not be exchanged by most farmers for a system under which year-to-year-income fluctuations are small and where the "guarantee," usually accompanied by achievement of price stability, is one of only "enough income" on an annual basis.

Even though increased reliance on grain imports by emerging nations probably has not been part of a conscious policy, it is important to understand how trade has maintained a key role and has seen its share of consumption rise in the decade of the 1970s. This has occurred in the face of massive increases in production of grains in most parts of the world. For instance, in 1970-71, world trade in wheat and coarse grains combined accounted for 12 percent of total consumption; in 1979-80, the estimate is that world trade will account for more than 16 percent of total use. This means that the increase in trade, from 103.3 million tons to 187.8 million, has been at a faster rate than the amazing climb in consumption from 931.2 million to 1,145.2 million tons in the same period. The share of consumption originating with trade could continue to rise, based on current prospects for production and demand. The U.S. should bend every effort in the cultivation of commercial markets and in the assistance extended to developing countries to make sure this continues to be the case.

The aim must not be to make developing countries dependent on imports of grain. It is to urge a differentiation between the concepts of self-sufficiency and self-reliance. In many cases, striving for self-sufficiency is in error, based on the over-all unfavorable crop production environment in so many populous nations. To come up with the proper balance between domestic production and import supplies requires not only exquisite economic planning but great political skill. Most of the major grain-importing countries take no more than 10 percent to 15 percent of their total use in the form of imports, and in many cases this is a margin of excellence in eating, not the difference between enough and famine. It represents an ideal in attaining self-reliance as distinct from the American dependence on imports for half of our oil consumption.

In order to cultivate world recognition of the importance of trade to future gains in the standard of eating, it is essential to examine the restraints which might rise up to prevent that from happening. The main barrier, as noted earlier, would be the widespread adoption of internal pricing programs which do everything to protect a country from the impact of supply-demand changes reflected in world prices. A prime example is the Common Agricultural Policy of the European Community which shelters the 260 million people in the community, as well as their economies, from changes in world prices for grain. This is accomplished through a variable levy system which raises the import duties on grain as world prices fall and operates in the opposite way when world prices advance. It was only in the few years following 1972, when world prices actually rose above the Common Agricultural Policy targets and the European Community looked to subsidize imports, that European farmers became aware of how their income could be affected by fluctuations in the world markets. One interesting aside here is that sales of the Paris-published Herald Tribune in rural France escalated in this period. A survey done by that newspaper showed that the increase in paper sales was due almost entirely to the eagerness of French farmers to follow the daily changes in grain-futures prices in the U.S.

If an economy is sheltered from world price changes, such as the variable-levy system accomplishes for the European Community, it means consumers of grain, primarily in the form of livestock feed, have no incentive to increase utilization when grain prices are low or to cut back feeding when world prices are rising. The disturbing consequences of such a policy were particularly evident in the post-1972 period when Europe continued to use large quantities of grain for feed and paid minimal attention to the perceived short- ages that were causing price escalations in other parts of the world. As a result, European feed use of grain showed little or no change during this period; whereas American feeders had to face up to the necessity of sharply curtailing use of grain.

In the 1975-76 period, when world prices experi- enced sharp downward pressure, U.S. feeders expanded livestock numbers in response to attractive feeding ratios; whereas, Europe showed little or no response. In other words, U.S. grain farmers and grain merchants are, for all practical purposes, foreclosed from bene- fitting from the economic opportunities that should rule in such a large, developed area like Europe. It is fascinating to note that both the U.S.S.R. and China are quite responsive to prices, preferring to hold back purchases when markets are regarded as too high and engaging in large-scale-buying programs when prices are viewed as attractive.

There are other ways to diminish the role of grain price and to prevent this mechanism from serving a very useful purpose in both the development and cultivation of export markets for American grain. Trade restric- tions of all kinds are powerful disincentives to busi- ness in any number of countries, including the devel- oped nations.

Bilateral and/or multilateral agreements whose purpose is to divide up markets with little or no regard for price or economic forces are also a problem in this context. Last winter a number of countries, including the U.S., carried out long and hard negotiations in Geneva to try to write a new International Wheat Agreement that would have had what is not so euphemistically called "substantive" price and reserve stocking provisions. If implemented in the form proposed at various steps in the negotiating process, the new agreement could have seriously curtailed the effectiveness of market forces. It would be impossible to pinpoint with exactitude what caused the negotiation of that agreement to fall apart, but it is apparent the U.S. representatives saw the way the talks were going as contrary to this country's interests. Specifically objectionable were the efforts of the developing countries to impose upon the U.S. the entire burden of stocks accumulation, whether the stocks were held in the U.S. or in other countries. In the latter case, the U.S. would have had to pay the cost of accumulating and holding stocks, which would have been counter to our own interests.

Even bilateral agreements, such as the one now in place between the U.S. and the U.S.S.R., pose a number of questions as to their effect on the pace of world trade. That agreement, with its minimum U.S.S.R. purchase of six million tons of wheat and corn and its maximum, before consultation, of eight million, probably should be revised to raise the minimum purchase by four million and the maximum by seven million that can be bought prior to consultation with the U.S. government. To a degree, the agreement has discouraged maximization of U.S.S.R. grain purchases. It is true that the existence of the agreement has diffused pressures for export controls, which are much more distasteful than having the agreement. The point that needs to be made in the context of these remarks is that the U.S.-U.S.S.R. agreement is no model for dealing with other countries, particularly those in the developing world where the encouragement of a market economy stands to serve mutual goals.

Even though there is considerable dissatisfaction with the progress of agricultural development, as it relates to upgrading food supplies in the poorest nations of the world and particularly in the wake of accelerated oil price advances, it is my perspective that great strides have been made. The measurement here, as before, is in consumption of grains. If one looks at the rising consumption, for the world as a whole or by economic blocs, it is easy to see how rapid the growth in consumption has been. In my judgment consumption growth, more than production and more than trade, signifies how well the world is getting on in solving global food problems. In the past five years, the annual average increase in world use of wheat and coarse grains has been thirty million tons, compared with about twenty million in the preceding five years. In all but three years of this decade, a period of memorable stresses and extremes, production of wheat and coarse grains combined exceeded consumption in the previous year. This is a measure looked upon by experts within the trade as one of the most telling ways of assessing the real state of the world's supply-demand condition.

It is also my judgment that progress in consumption, which we would expect to continue to grow at a rising rate, is derived substantially from the availability of export grain from the U.S. Investing billions of dollars in developmental programs has undoubtedly played an important role in helping to rationalize the world food situation. Yet, too little attention has been paid to the great developmental strength directly derived from America's ability to produce large quantities of grain in excess of domestic needs. Having that margin of production, available in either the commercial market or through special aid programs, serves to encourage national leaders in developing lands to undertake forward-looking efforts aimed at expanding grain production.

The picture that emerges should be a very satisfying one to Americans. On the one hand, American grain has stimulated developing countries to undertake programs in their own best interests. On the other hand, American farmers have benefitted from rising demand. The combination could hardly be a happier one, far exceeding the highest degrees of optimism voiced twenty-five years ago when P.L. 480 began as an extremely ambitious way of using this country's grain in development channels. Many fits and starts have occurred in the intervening years with many instances of doubt whether development programs were not counter to the interests of American farmers and the American economy. The past twenty-five years, with few exceptions, have proved this is not the case, that the U.S. economy and this country's farmers stand to benefit in a measurable and important way from rising food production abroad, just as the countries which embark on ambitious programs to increase grain production are helped and stimulated in ways not often appreciated by the availability of large quantities of U.S. grain.

A major modification from the past course would be new emphasis on efforts aimed at creating a full measure of price responsiveness in all grain-consuming and -producing nations. As Pollyanna-like as such a platform may be, it is worth striving for, simply because even minimal progress promises important dividends for America and for the world. The American marketing system has proved itself in the last twenty-five years as a highly effective and efficient force. The 1970s has been a decade of phenomenal progress; the system's finest hours lie ahead in a future marked by rapidly rising standards of eating around the world.

13
MEETING ENERGY REQUIREMENTS
IN THE FOOD SYSTEM

Emery N. Castle

In their letter inviting me to prepare this paper Senator Bellmon and Congressman Simon said,

We believe it would be particularly helpful if you would provide us with an analysis of food-system energy requirements from the vantage point of an independent observer familiar with the current and projected demands on the system. Those of us in Congress have heard from energy producers, energy consumers, farm groups, and government agencies such as the Department of Energy and the Department of Agriculture, but we have not had much input from independent sources which could focus on the food system and its requirements.

Such a charge is both a great responsibility and a distinct honor. I am very pleased that my organization, Resources for the Future, has a reputation for impartiality and independence. Perhaps the principal reason we have this reputation is that we strive mightily to obtain competent researchers and then do not interfere with their freedom to draw conclusions from the facts they amass and the analyses they perform. Thus, Resources for the Future takes full responsibility for the competence of its scientists, but it does not take positions on natural-resource-policy issues. This applies to me, president, just as it does to every other person in the organization, so I want to emphasize that I speak only for myself and not for my institution. I am an agricultural economist who has specialized in resource economics for some years at Resources for the Future and, before 1976, at Oregon State University.

My remarks are presented under three main headings -- a perspective on energy use in the food system; possible adjustments to a more expensive and more

constrained energy supply; and broad and general policy
options.

THE FOOD SYSTEM AND ENERGY USE IN PERSPECTIVE

There can be little doubt that the remarkable
productivity of the food system of America has been
associated with low-cost and abundant energy. This is
evident in the extensive use of fertilizer and mechanical power in production, and packaging and refrigeration in marketing. The technology that has been
adopted during the past three decades on farms has
generally involved substituting for land or labor
purchased inputs which usually embody a lot of energy.
Irrigation, field machinery, and such petrochemicals as
pesticides and fertilizers are all examples of production technologies that rely heavily on purchased
energy. Other parts of the food system -- processing,
marketing and distribution, and commercial eating
establishments, as well as home preparation -- also use
a good deal of energy.

This energy-intensive food system has yielded
impressive results. Only about 16 percent of disposable personal income in the United States is spent
on food. West Germany spends a slightly lower percentage, but most industrialized countries spend more.
More than three-fourths of U.S. households exceed
recommended daily allowances of protein, food energy,
calcium, and ascorbic acid, and our marketing system
provides a variety as well as an abundance of foodstuffs, with many conveniences available to the consumer in the form of prior preparation. The system
also produces substantial foodstuffs for people outside
the United States. About one-third of agricultural
output is devoted to export production and this has
been growing for several years.

Yet it cannot be said the food system is without
problems. Agricultural producers are price takers and
they experience considerable instability in income.
There are environmental problems associated with boosting food production through greater use of petrochemical pesticides and fertilizers. There is evidence that
soil erosion has accelerated. While many people consume more than the recommended daily allowances of
certain essential nutrients and protective foods, about
a fourth of the people do not, and the United States is
not without other nutritional problems. Conversely,
many people overeat and the food system produces prepared foods that are often too high in certain elements, such as sugars and sodium.

The United States Department of Agriculture estimates that direct and indirect use of energy by the
food system amounts to about 17 percent of all energy
used by the economy. Of this 17 percent, about half or
8 percent is used directly by the food system.(2)

Interestingly, about twice as much energy is used
in the home preparation of food as is used for farm
production. The amount of energy used in food processing and commercial eating establishments is also
greater than that used in farm production. The amount
of energy used in the marketing and distribution of
food is less than that used on farms, but by only a
small amount.

In production, the greatest energy users are
irrigation, field machinery, and petrochemical pesticides and fertilizers, with the petrochemicals accounting for about a third of all energy used on the
farm. Much more is known about energy use on the farm
than in the remainder of the food system, which perhaps
explains the tendency to concentrate discussion on farm
production even though it accounts for only 15 percent
of the energy used by the total system.

Low-cost, abundant energy has much influenced
agriculture, just as it has the rest of the economy.
Large-scale mechanization, crop specialization and the
expansion of irrigation in the southern high plains
with associated cattle-feeding provide examples of this
influence, as does the extensive use of petrochemical
fertilizers and pesticides. All of this has done a
great deal to enhance the productivity of American
agriculture. However, there is no evidence that American agriculture is significantly more energy intensive
than the remainder of the economy. Sixteen percent of
our disposable income goes for the purchase of food; 17
percent of our total energy supply goes to produce,
market, and prepare that same food.

THE FOOD SYSTEM AND MORE EXPENSIVE ENERGY

During the past six years energy has become more
expensive than have other items used in placing food
before the consumer. From 1972 to 1979 energy prices
to farmers increased 2.6 times as compared to an increase of about 1.8 times in prices paid by farmers for
all other items. Thus, an analysis of the response
that has been made by the food system during the recent
past should provide an indication of future adjustments
to even more expensive energy.

In the material which follows, the assumption is made that scarcity will be reflected in higher prices rather than rationing. It is believed that higher prices are the more probable adjustments; furthermore, it would be exceedingly difficult to predict response to physically curtailed supplies since the type of rationing scheme would indicate the response. Nevertheless, there has been at least one study which has compared the impacts of rationing and higher energy prices on farm production. In that study, Dvoskin and Heady (3) estimated that doubled energy prices would result in a 13 percent increase in farm commodity prices, but that an energy shortage of 5 percent in each region of the country would result in a 26 percent increase in farm commodity prices.

In assessing the probable effects of increased energy costs it is necessary to distinguish between the immediate or short-run effect and the longer-run consequences. It is also necessary to recognize that different industry structures prevail in various parts of the food system. In farm production and in retailing in restaurants and cafeterias, there are numerous producers and a competitive atmosphere. In the case of food processing and in marketing and distribution, a somewhat different situation prevails: there is considerable competition, but the number of firms is smaller, their size is larger, and cost increases are passed on to the consumer more rapidly.

In the short run, producers may be made worse off by increased energy costs because farmers have limited ability to quickly pass energy price increases on to consumers. So long as the value of the commodity exceeds the remaining cost of production, it will pay farmers to produce; thus the closer one gets to harvest, the greater will be the tendency to continue production in the face of cost increases. Since energy costs are not a high percentage of total variable costs, production is not likely to be curtailed by increased energy prices once the production cycle has been started. For example, fuel costs represent 16, 11, and 9 percent of the variable costs for wheat, corn, and cotton, respectively.

In the long run, increased energy costs will be passed on to the consumer, because prospective food prices will have to be high enough to attract and hold resources in the food system. In the absence of careful studies it is not possible to predict the effect of higher energy prices on producers within the food system over the long run. About the only thing certain is that many farmers will be made worse off in the short run.

Because so much more is known about the way resources are used in agricultural production than in
processing, distribution and marketing, and in food
preparation -- either in the home or in commercial
eating establishments -- it is much easier to speculate
concerning farming adjustments to higher energy prices
than it is about adjustments elsewhere in the system.
This is a severe handicap because, as noted, most of
the energy used by the food system is off the farm.
Nevertheless, adjustments made on the farm are important because they usually trigger consequences elsewhere in the system.

A most energy-intensive, on-farm practice is the
use of petrochemicals, especially pesticides and fertilizers. The most rapid increase in fertilizer nutrients per acre occurred during the 1960s; the rate of
the increase had begun to decline before energy prices
began their rapid climb in the early 1970s. In contrast, the use of pesticides, especially herbicides,
has increased rapidly during the 1970s and in the face
of increased-energy costs. While increased-energy
costs can be expected to result in more efficient use
of the petrochemicals, prices would have to increase
substantially before farmers would discontinue using
such materials. It can be said, however, that if these
materials were not available because of extremely high
prices or disrupted lines of supply, there would be an
adverse effect on farm output and surely the comparative advantage of U.S. agricultural production would
suffer.

Field machinery is also a substantial user of
energy in agricultural production. As noted earlier,
1979 fuel costs were about 16 percent of the variable
cost per acre in the production of wheat and about 11
and 9 percent for corn and cotton, respectively.
Continued increases in fuel costs will no doubt continue to stimulate interest in more energy efficient
practices and machines. Even so, the use of mechanical
power and machinery continued to increase throughout
the 1970s at a rate of more than 2 percent per year.
There has been increased interest in minimum tillage
(although this interest began prior to the onset of
higher energy prices), and this can be expected to
continue. Estimates are, however, that the rate of
increase declined somewhat in the late 1970s as compared with the mid-70s. At the present time about 80
to 85 percent of the cropland has conventional tillage.

Higher energy prices can be expected to have a
significant effect on irrigation. Although total
acreage under irrigation continues to increase, there

has been a significant decrease in irrigated acreage in some areas. There are many factors at work in these areas, but declining ground water levels and higher pumping costs, which reflect increased energy prices, are undoubtedly important.

There has been renewed interest in organic farming in recent years, perhaps due in part to higher energy prices. Enough people have actually engaged in organic farming to permit some studies to be made which compare this system of farming to what has become conventional agriculture. These studies and other information indicate that, if adopted, such farming systems may save some energy, but at the cost of a substantial decline in agricultural output from the land now in production. Animal manures and legumes cannot provide sufficient nutrients to serve as an equivalent substitute for the purchased fertilizer which is now used. Only a most significant breakthrough in some new technology, such as the development of biological nitrogen fixation in numerous plant species, could avoid a precipitous drop in production. While such a breakthrough is certainly possible, it is not likely it will occur in the near future.

Any balanced discussion of energy in the food system must consider livestock, for they represent a most significant part of farm output. Indeed, livestock account for nearly one-half of farm receipts from farm marketing. It certainly is true that, on the basis of net energy alone, livestock are inefficient enterprises. The question which arises, however, is whether we would be better served by becoming vegetarians. Ideological issues aside, national vegetarianism would have numerous implications. Granted that excess consumption of animal products can be detrimental to health, there are nutritional advantages to the use of certain of them, primarily because of the quality of the amino acids they provide. Further, because of palatability, animal products are preferred by most consumers, for a part of their diet, to plant sources of food. An even stronger argument is that livestock utilize feedstuffs that would otherwise go to waste. Livestock consumption is one of the few uses of many roughages and forages produced on the land. Should techniques be developed for efficient biomass conversion to energy this situation might change, but it is apparent that we now are far from being able to do this for many roughages and forages.

Livestock species vary greatly in the efficiency in which they process feedstuffs. Furthermore, the efficiency of some species has been improved significantly while it has been relatively stable for others. Higher energy prices can be expected to provide an incentive for improved efficiency for all species. Consider, for example, the attractiveness of multiple births in beef cattle. Unless efficiency can be improved, high consumer prices for many animal products may well result in a shift away from their consumption, and we may be starting to see the development of such a trend: from 1976 to 1979 meat consumption in the United States declined from approximately 195 pounds per capita to about 182.

It is somewhat surprising that greater attention has not been given to the energy wastefulness of meat with high fat content. The product of feedlot methods which use a great deal of grain, fatty meats are undesirable both from the standpoint of energy efficiency and human nutrition. These meats undeniably are popular, however. Apparently, this consumption pattern is continued by virtue of consumer preferences for the tenderness and taste of meat with high fat content as well as grading standards which give the highest grade to meat with the greatest "finish."

While we know a fair amount about farm production, there is less basis for speculation concerning the response of the remainder of the food system to higher energy costs or curtailed energy supplies. The liquid fuels tend to be more heavily used in agricultural production and transportation; electricity, coal, and natural gas use is concentrated most heavily in processing and preparation of foods. Consumer preferences and incomes will undoubtedly have a great deal to do with the way resources are used in the processing, marketing, and distribution as well as in the preparation portion of the food system. The present pattern of resource allocation reflects the fact that food purchases require only about 16 percent of disposable personal income and that per capita personal income has been rising for many years. The percentage of disposable income spent on food is not likely to increase significantly in the near future, and per capita personal income probably will continue to improve, if at a less rapid rate than in the past. Consumers will not surrender readily their consumption of convenience items, nor are they likely to decrease their consumption of food prepared outside the home. Nevertheless, there are undoubtedly energy-intensive practices that will be examined carefully and probably reduced as a result of higher energy prices. Home appliances already are becoming more energy efficient, and it is hoped that our packaging will become less wasteful in the use of energy-using materials.

The concern about the supply of fossil fuels that has become so commonplace during the 1970s has triggered considerable speculation about the greater use of renewable resources for energy production. These include direct solar energy, wind energy, and different forms of energy available from plant materials or biomass, including direct combustion, fermentation, anaerobic digestion, and pyrolysis. All of these processes are technically feasible, but it is not clear just when they will become economical and practical. The most probable development for the remainder of the century is for biomass conversion to be used increasingly, but not as a major source of energy. This could be changed, of course, by a breakthrough in technology, by massive government encouragement of biomass conversion by direct investment or subsidy, or by a change in the relative prices of alternative sources of energy.

Will it be feasible to use renewable resources, via agriculture, to substitute for fossil fuels? This is the basis for the great current interest in the mixture of alcohol and gasoline that in typical U.S. fashion has been christened gasohol. Gasohol is now being produced and used commercially in the United States and elsewhere in the world, but it is important to recognize that U.S. production of gasohol is subsidized substantially. The exemption of ethanol from the federal automobile fuel tax is one form of subsidy, amounting to $16.80 per barrel, and several states provide additional subsidies ranging from five to seven cents per gallon. Ethanol currently is priced in the $1.30 to $1.60 per gallon range, but this may not be an accurate indicator of its cost if it were to be produced extensively.(4) An expansion of alcohol production could be expected to increase the price of agricultural products which serve as raw material. This will result in less productive resources being brought into production, and other resources will need to be attracted away from other activities.

However, there is another, more complicated and dynamic, relationship involved. As has been noted, much agricultural production is now heavily dependent on purchased inputs that are energy intensive. When the price of fuel increases to the point where it would appear to make alcohol economic, the cost of production will automatically rise because of the cost of the energy used in production. This brings into question the energy efficiency of the process itself. It is clear that alcohol production is not highly energy efficient, although the degree of efficiency rests on the relative efficiency of gasohol and other forms of fuel in the end use, the technology of agricultural production, and the use which is made of by-products. Furthermore, the use of coal in alcohol production might permit a reduction in use of other liquid fuels.

There is still another issue that would need to be faced if alcohol production from grains were to be expanded significantly: users would need to be assured of a dependable supply at reasonably stable prices, yet grain prices are notoriously unstable and this instability would be reflected in the cost of producing alcohol.

Given the above, and given the total resources required for alcohol production, the contribution of gasohol to our total energy needs will likely be marginal. It has been estimated that one-sixth of the U.S. corn crop devoted to alcohol production would substitute for only about 2.3 percent of the gasoline used by surface vehicles.(5) Devoting as much as one-sixth of the corn crop to alcohol production would certainly increase corn prices, which in turn would stimulate production and increase the demand for land, fertilizer, and pesticides. Dollar and environmental costs would increase in the agricultural system.

If this nation were to embark on an extensive gasohol program, sources of alcohol other than grains should be investigated. In particular, crops that yield sugar as well as cellulose should be considered. The sugars are easier to process than the grains and woody fibers, but there are economic and technical problems that would have to be faced with both.

The need to develop additional energy sources also will put competitive pressure on resources used by the food system. The competition for water between agriculture and energy development in parts of the West provides a dramatic example. Inasmuch as water used for energy development usually will return much more economically than will the same water used in agriculture, the case for retaining the resource in agriculture usually will have to be made on grounds such as community stability or environmental protection.

PUBLIC POLICY AND ENERGY USE IN THE FOOD SYSTEM

Public policy concerning energy use and the food system should reflect the role it is hoped the food system will play in the larger economy.

One such role is a continuation of the food system as it has evolved to the present, a system highly dependent on energy-intensive technology and reflecting the high incomes and affluence of American consumers. The productivity of the system is such that it produces substantially in excess of domestic consumption and huge quantities of foodstuffs are exported. Not only is the system energy intensive, but also it depends heavily on the fossil fuels. It would be difficult in the short run to substitute other forms of energy for that supplied by the fossil fuels.

Another model for the food system would be, as deliberate public policy, to move it away from heavy dependence on purchased energy generally and the fossil fuels in particular and toward greater energy self-sufficiency. Such a shift, of course, might well result from scarce fossil fuels and very high prices quite apart from deliberate policy. Nevertheless, there are those who believe the present system is basically unstable and that wise policy would be to hasten a change which is believed to be inevitable. There is little doubt, however, that an abrupt shift toward greater energy self-sufficiency would decrease substantially the productivity of the food system. Agricultural product prices would increase and agricultural exports would decline.

There are obviously numerous positions between these two extreme models. One strategy might involve an acceptance and hopeful continuation of the role currently being played by the food system, while recognizing that the present heavy dependence on purchased energy and fossil fuels constitutes a hazard. Attempts would be made to develop alternative energy sources and otherwise provide the emergencies.

To provide focus for policy discussion several propositions are advanced and discussed. In no sense is this list of propositions exhaustive; many others could be advanced, but this exploration permits several important issues to be brought into the open.

Proposition #1. The use of energy in the food system should receive a high priority whether allocation is by price or by quota.

Comment. It is clear that the food system makes an enormous contribution to the economy. It supplies consumers with an abundance of food at low cost when compared to other countries of the world. The system is efficient internationally and much food is exported, which contributes to our balance of payments as well as to the diets of people in the importing countries. Thus, a claim can be made that energy should be made

available to the food system, and if energy is allocated by some means other than price, the food system
should be given high priority. In most cases the
system probably can adjust better to higher energy
prices than it can to absolute shortages of energy. In
the longer run, consumers will pay most of the cost of
higher energy prices in agriculture, although in the
shorter run, some producers, especially farmers, within
the food system will bear more of the burden.

Because of the substantial use of petrochemicals
in farm production and because the liquid fuels are so
important in transporting food products, it may appear
that the food system would be highly vulnerable to
disruption or temporary interruptions in the supply of
liquid fuels. However, some of the dependence on
liquid fuels may be more apparent than real. In the
first place, the use of the petrochemicals for fertilizers and pesticides constitutes a premium market, and
if supplies were available the food system would be
highly competitive. Secondly, natural gas is now used
in the manufacture of nitrogenous fertilizers and
natural gas is now largely produced and used on this
continent rather than coming from overseas. Thirdly,
coal could be substituted for natural gas in the production of nitrogenous fertilizers if this were necessary. In fact, coal probably will become more competitive with natural gas in this use before the end of
the decade.

Nevertheless, the liquid fuels are still necessary
for tractor power and for much of the transportation
needs of the food system; they should have a high
priority if liquid fuels were limited in supply. For
the time being, because we import less than half of our
liquid fuels, there does not appear to be any overriding need to take special measures to provide for the
emergency needs of the food system as distinct from the
economy as a whole.

Proposition #2. Since the number of people living on
farms has been reduced greatly because of farm to city
migration, and because consumers spend only about 16
percent of their disposable income on food, the potential of research to make farming more efficient has
been largely realized. Furthermore, energy use on the
farm is only about 15 percent of total energy used by
the food system. Therefore, there is little need for
public policy to be concerned about additional research
on the food system.

Comment. Careful studies of the rate of return societies receive from investment in agricultural research

suggest that the return is quite high, exceeding 20 percent in the United States at the present time. Such studies also estimate that a one percent increase in research expenditures will yield a 3.7 percent increase in productivity over a thirteen year period.(6) Thus, additional investment in agricultural research can be justified on grounds other than improved energy efficiency in the food system.

Agricultural research also has potential for improving energy efficiency in the food system as well as reducing food-system dependence on imported liquid fuels. Although traditional agricultural research has emphasized on-farm production, increasing attention is being paid within the research establishment to the performance of the entire food system.

There are numerous possibilities for reducing the dependence of the food system on fossil fuels through research. Some of the possibilities which would affect on-farm energy use include:

1. Enhancement of photosynthesis efficiency.
2. Water and fertilizer management.
3. Crop pest-control strategies.
4. Reduced tillage.
5. Development of plants to withstand drought and salinity.
6. Selection for multiple births in beef cattle.

I come to the conclusion that the agricultural research enterprise in this country is highly productive and that increased investment in this activity is in the public interest.

Proposition #3. Regional production of foodstuffs varies considerably in energy use because of differences in the availability of different forms of energy and because of differences in the prevalence of certain practices such as irrigation. Wise public policy will reflect awareness of the differential regional impacts of higher energy costs.

Comment. The existing pattern of agricultural production reflects many influences, including energy availability and the abundance of other natural resources such as ground water. Any attempt to maintain the present geographic pattern of production is bound to be very expensive as the conditions affecting the location of production change. Therefore, any attempt to cushion the regional shock of high-energy costs or declining resource availability probably should heavily emphasize heavily measures which will assist producers to adjust to more realistic patterns of production.

Publicly subsidized water importation to permit a continuation of irrigated agriculture in the face of higher energy costs and declining groundwater levels would be an example of a policy which attempts to maintain present regional patterns of production. Research and educational programs which will assist farmers to adjust to a smaller water supply, and a community development program which would permit accommodation of the indirect impacts of changes in primary production, provide examples of public programs which recognize that regional production patterns will inevitably change in response to external forces.

Proposition #4. Linkage should be developed between the international exchange rate of exported food grains from the United States and imported fossil fuel.

Comment. International trade involves a web of relationships, and it is doubtful that a two-commodity exchange rate could be established. Even if it were possible to do so, it would be unwise to freeze any historical relationship between the two. In a long-run setting, there must be substitution away from the fossil fuels, and attempting to establish and hold a particular relationship would discourage such substitution.

Even so, it is only logical to recognize that the United States is a very efficient producer of food and that this capacity can well become a tool in international relations. The full treatment of such a proposition would take one well beyond the purposes of this paper. The main thrust of this comment is that the "bushel for a barrel" concept probably does not suggest a feasible course of action.

Proposition #5. Grading standards for meats are wasteful of energy as well as questionable on nutritional grounds. While the total amount of energy saved would not be great, the revision of standards would be consistent with more efficient use of energy.

Comment. There is little question but that the meat-grading standards do encourage overfeeding of animals, overeating of fat by consumers, and social waste. The defense of the practice rests on the increased flavor and tenderness of meat, and U.S. consumers are accustomed to such meats. However, the aging and preparation of meats can much influence tenderness. A move toward leaner meat which would substitute roughages and grass for grain would certainly improve energy efficiency and might well improve nutrition and health.

There may well be other examples of government policies and procedures that tend to discourage efficient energy use; government policies and procedures have been developed during periods of abundant and low-cost energy and, as a consequence, their social value may now have changed. It should be emphasized that, even if policies and practices are not changed, the economy will adjust to reality. The possible trend toward reduced lower per capita consumption of meat is an example of such an adjustment.

In summary, the objective of public policy should be to make the food system less vulnerable to severe supply disruptions and heavy dependence on fossil fuels. Fortunately, there are some things that can be done through public policy, and there is evidence that the food system itself is adjusting to a less abundant and more expensive energy supply. Disruptions in supplies pose a more serious threat to the performance of the food system than do higher prices, but rapidly rising energy prices may reduce sharply the income of some producers within the food system for a period. As far as can be foreseen, there does not appear to be great danger that Americans will have inadequate food because of energy shortages. However, if energy use by food producers were sharply curtailed -- for whatever reason -- the variety of foodstuffs available would be reduced, higher consumer prices would result, and agricultural exports would suffer.

FOOTNOTES

(1)   I am indebted to my colleagues Pierre Crosson, Kenneth Frederick, Hans Landsberg, and Harry Perry for comments on an earlier version of this paper. Anne Price was most helpful as a research assistant and Kent Price provided valuable editorial help. I am also in the debt of my administrative assistant, Blossum Carlton, for typing several preliminary drafts of this paper.

(2)   The home preparation of food can be used to illustrate direct and indirect energy consumption. The fuel used to prepare food is considered to be direct energy while the energy to produce the stove on which food is cooked is an example of indirect energy.

(3) Dvoskin, P., and E. O. Heady, "U.S. Agricultural Production Using Limited Energy Supplies, High Energy Prices and Expanding Exports." (Ames, Iowa: Center for Agriculture and Rural Development, Iowa State University, 1976).

(4) Chambers, R. S., R. A. Herendeen, J. J. Joyce, and P. S. Penner, "Gasohol: Does it or Doesn't it Produce Net Energy?" Science, November 1979.

(5) Ibid.

(6) Yu Yao-Chi, and Leroy Quance, Agricultural Productivity: Expanding the Limits, U.S. Department of Agriculture, Economics, Statistics and Cooperative Service, Bulletin 431, (1979).

14
WORLD FOOD AND AMERICAN NATIONAL SECURITY

Steven Muller

Late in 1978, Senator Henry Bellmon and Representative Paul Simon advised me of the Congressional
Roundtable on World Food and Population and asked me to
prepare a concluding paper. Please allow me to pay a
deeply deserved tribute to Senator Bellmon and Representative Simon for their leadership; to their staffs
and others who worked on this project; to the Conference of Southwest Foundations for financial support;
and to the members of the Congress who have participated. This tribute goes far beyond courtesy. One of
the weaknesses of government is a common failure to
aniticipate problems -- to confront them only when they
have ripened to full crisis. All of the democratic
legislatures of the world suffer the experience that
the urgent tends to take precedence over the important.
To concentrate on the problem of world food and population before the actual advent of catastrophe is
leadership of the highest order, and I applaud and
salute those responsible.

I want to share with you a few reflections of my
own on the world food problem that have been prompted
by a year of intensive exposure to the Congressional
Roundtable on World Food and Population as well as by
my experience with the Presidential Commission on World
Hunger. In order to be useful, I will concentrate only
on two points. These two points seem in my own mind to
be of such significance for the future that they merit
priority over all else -- a judgment for which I must
hold myself accountable.

My first point is that the world food problem
constitutes a vital dimension of the national security
of the United States and must be faced in that light.
I do not mean to slight the moral and economic aspects
of the world food problem, but I do choose to view it
as a priority aspect of our national security. By the

same token, I know that any reference to national
security calls instantly to mind our defense establishment and nuclear weapons, and I submit that such a view
of national security is too narrow. To make this whole
point concisely I paraphrase Abraham Lincoln -- not for
effect, but because no other words will do so well:
the world will not long endure half hungry and half
fed.

The time seems to me to have come when we can take
the following assumptions as fact. First, world population will increase over the next century no matter
what is done to retard the rate of growth. Second,
grain foods in particular will play a dominant role in
world food consumption no matter how much is done to
provide alternatives. Third, grain foods will be in
short supply for the forseeable future when measured
against worldwide food needs. Fourth, the United
States is to grain foods what the Middle East is to
petroleum, i.e., so dominant a source that -- whether
by action or inaction -- the United States is inescapably the major actor on the world food scene.

The only circumstance I can conceive that would
negate these facts is worldwide genocide, perhaps as a
consequence of nuclear warfare. No one would make
policy based on such a prospect. However, it opens a
back door to the subject of national security, defined
only as most of us have become accustomed to defining
it. But what is the object of national-security
policy? It is to preserve and promote the peace and
prosperity of our people, in some large part by avoiding or countering any threat to that peace and prosperity from the outside. There is no argument here
concerning the existence of the threat of armed aggression, and there is no denial that strong armed defense
may still be the only effective deterrent to such a
threat. But is armed aggression the only threat?

The peace and prosperity of any people -- in this
case ourselves -- require more than protection against
external aggression. Apart from internal circumstances, which also are vital, the peace and prosperity
of any one nation today are inextricably dependent on a
high degree of international stability. We are aware
of many things that threaten international stability.
Each of them poses a threat to our own peace and prosperity. Lack of food elsewhere in the world is only
one of these threats. It is my belief that in the
future -- the near future -- a world food crisis could
rapidly become a major threat to American peace and
prosperity.

What does that mean? Does one envision hordes of the starving invading our shores to wrest from us the staff of life? That is not what I envision, though for those mindful of the explosive extent of illegal immigration into this country, cognizant of the desperate valor of the so-called boat people of Vietnam, and aware of the unforgivable flood of starving humanity spilling from Cambodia into Thailand, such a vision may no longer seem wholly lunatic. Nor do I assume that starving people elsewhere will terrorize us with missiles or other weapons, although the rise of terrorism recently cannot wholly preclude such a possibility either. All I mean to say when speaking of the world food situation as a potential major threat to the peace and prosperity of the United States is -- calmly and soberly -- that the sudden presence of mass hunger and malnutrition on the globe would create a situation adverse to our national interest, and that such sudden mass hunger and malnutrition is not only possible but increasingly possible.

Human beings have for long endured poverty, disease, ignorance, and maltreatment. They do not tolerate hunger and extreme malnutrition as well. Starvation kills quickly. Extreme malnutrition may cripple first, and kills more slowly. The prospect of starvation obviously does not encourage social stability. In today's world, a significant degree of social instability is inevitably risky to worldwide stability. Hungry people are bound to envy those who are better off. Envy breeds hostility. Hostility on the part of much of the world will not enhance peace and prosperity in the United States. It is clear that the world food situation is fragile. Demand already outstrips supply. Any extended succession of bad harvests could produce crisis. Too little is being done to increase world food production. Is it really difficult to accept the fact that the national security of the United States involves an inescapable concern with the adequacy of the world food supply?

So little has been done to date in consequence of such an accepted fact. If it is indeed true that world food needs represent a valid and potentially major aspect of the American national interest -- of our national security, if you will -- what priority do they have? Some, I acknowledge; too little, I submit. A lot is being done, a lot more is being said, and money is being spent. What is lacking above all is priority, coherence, and consistency of purpose.

Many of our national priorities are reflected in the cabinet departments of the executive branch. Several of the executive departments and agencies have been and are concerned with aspects of the world food problem, but for none of them does it have first priority. This is obviously true of defense, state, agriculture -- whose first priority remains domestic -- commerce, treasury, the national intelligence community, and others. What about foreign aid? International agricultural development has been and is a vital part of our foreign aid program, but it has never been proclaimed as its first, let alone sole, priority. Nor is it the priority concern of any committees of the Congress. That raises questions to which the Presidential Commission on World Hunger has responded. I hope that its report will find an attentive audience. Let me say only that if there is agreement that the world food problem relates to our national security, then the evidence is that we are not yet reacting accordingly, either as a people or as a government. I would, in addition, note the happy coincidence that the very relationship between the threat of hunger and the threat to peace has been recognized explicitly in this year's award of the Nobel Prize in Economics to Dr. Theodore Schultz.

The second point I wish to make supports the first but addresses the problem from a different direction. If we were now to make a durable solution to the problem of world hunger a first priority, what actually could and should we do? My second point is that there is no quick fix, and that the single most important investment to be made is in more research and more training of people. You will be tempted to discount such a statement from me immediately as no more than the reflexive response of a university president to any problem. I hope and believe that it can be argued on more viable grounds.

A useful beginning again may be to state some underlying assumptions that can be treated as factual: First, emergency food aid in cases of critical need plays an indispensable role, and the creation in full of an adequate emergency world grain-food reserve is essential; but the long-term answers to the world food problem do not lie in emergency measures alone. Second, long-term stability with respect to world food depends above all on the ability of the world's peoples to meet their own food needs more adequately, and not to be dependent forever either exclusively or primarily on imported foodstuffs. Third, contemporary technology has the potential to make a significant degree of self-reliance with respect to food available to every one of the world's peoples, even those in the climates and geographies most difficult for this purpose. Fourth, the greatest immediate obstacle to an increase in long-term world food production is the scarcity of relevant research and of persons trained in the skilled application of research and technology. Fifth, greater food productivity among more of the world's peoples requires relevant research and training among those peoples themselves, so with respect to training in particular we are not speaking exclusively or even primarily of Americans.

It is in the context of these assumptions that one can address -- albeit briefly -- the question of how much waste there has been in that part of United States foreign assistance which has tried over the years to address the world food problem. My personal answer is that while substantive results in terms of genuine relief of the long-term problem are disappointingly small and slow, this is not due primarily to the waste of money or effort in the process. The explanation for what might seem a paradoxical conclusion is that so much has been learned that could only be learned by doing. It is not possible here to summarize all of the lessons taught by our experience in international agricultural development, nor would I even be qualified to make the attempt. However, a few major conclusions appear to me to stand out in bold relief: We know more now than we did about the inadequacies of the so-called trickle-down approach which assumes that greater agricultural productivity will result from initial investment in industrial and technological national development. We know more about the consequences of introducing sophisticated technology in circumstances that aggravate rural unemployment on the one hand, and that on the other hand are beyond the trained capacities of indigenous persons to sustain. We have made notable contributions, in cases such as Taiwan and India, where our efforts have been most intensively sustained, where applied research yielded high dividends, and where our programs of assistance were massively comprehensive. The greatest reverses have been experienced in countries where internal strife has eliminated trained talent and has otherwise disrupted agricultural development. Above all, we have learned that immediate results are not attainable, and that a great obstacle presents itself in the absence of trained people and relevant research.

It will demand a substantial and enduring commitment of national funds to enhance our national security by giving appropriate priority to the achievement of worldwide food stability. Even a surge of new appropriations in the immediate future would yield only marginal returns. It is now obvious, for example, that

major investments in irrigation will vastly increase
the agricultural fertility of a great deal of land.
But it is equally true that even one single and relatively modest irrigation project requires skilled,
experienced, extensive conceptualization and design,
comparable quality of maintenance, and informed, knowledgeable agricultural exploitation. The really startling discovery to be made from recent decades of varied
experience is that the indispensable ingredient in
shortest supply is not money but knowledge and trained
human talent.

This discovery may amaze people who have accustomed themselves to think of the United States as an
inexhaustible reservoir of knowledge and trained talent. Nevertheless, even within our own society, viewed
purely from a domestic perspective, a sound case can be
made that agriculture continues to be accorded relatively low stature, both as a profession and as an
academic field of training. An even stronger case can
be made that as a nation our investment in research
devoted to agriculture lags far behind not only our
total investment in military defense, but also in such
concerns as health, communications, and energy as well.
The mention of energy, in fact, invokes an irresistible
analogy. Confronted with a crisis involving reliance
on petroleum products, the United States is embarking
on a crash program of research and training that will
produce acceptable alternative energy sources. Every
citizen might wish that such a national effort had been
launched years ago. One may wonder what dimension of
crisis is necessary to produce an equivalent effort in
pursuit of greater agricultural productivity. Where
the analogy breaks down is that a potential world food
crisis will be global rather than national. The dearth
of research and training is not only domestic but
international. A meaningful American approach must
therefore have a worldwide reach.

One does not need to be a Malthusian to reach
these conclusions. We have learned it is not true that
our globe's resources are not adequate to feed future
populations. There is cutting irony in the recognition
that worldwide food stability is attainable in fact,
and that failure to attain it -- with all the human
misery involved and adverse consequences to the peace
and prosperity of the American people -- would result
primarily from the mistaken belief that the problem
cannot be solved, or by lack of concentrated effort
undertaken in time to solve it. Here, too, the Presidential Commission on World Hunger has made cogent
recommendations. It is beyond my personal capacity to
chart the full course of needed future action. My

purpose is only to point to the need to give due priority to a situation that will unavoidably affect our own national future, and to point out the evident need for programs of research and training that must not only come first, but that also require precious time before there is a long-term effect.

I do not intend to function as a Cassandra. It is not at all my purpose to warn of doom. What I most wish to emphasize, after but one brief though intensive year of this Roundtable and complementary experience, is that the problem of food for the world's people is solvable. The opportunity to solve it is surely more significant than the companion knowledge that a failure to deal with it in time invites disaster whose future advent would be certain even if its dimensions and consequences are unpredictable.

The situation of the United States in relation to the world food problem as we enter the final decades of the twentieth century is extraordinary. It may indeed be extraordinary in such magnitude as to elude full comprehension. Nature has blessed this continent with the land and climate that make it the world's preeminent granary and grazing prairie. Our own efforts have hugely increased our productivity in foodgrains and livestock. The extent to which additional research and training can further increase that productivity is as yet incalculable. Modern technology has provided the means to store, package, and transport American agricultural produce all over the world. The level of expectations with respect to food and diet has been rising rapidly in other lands. Demand for corn and wheat has been created in lands barely acquainted in the past with one, or occasionally, either one. The future prosperity of American agriculture is already linked to a developing international market because its capacity exceeds even the high level of domestic food consumption which marks the American society. It is neither possible, nor would it be desirable, for the United States to become the sole source of corn and wheat products for many of the world's peoples. It is, however, both possible and desirable for the United States to become the balancing factor in achieving world food stability. We have the capacity to supplement and enrich the agricultural output of other peoples, precisely as we are now doing for the Soviet Union.

Unfortunately, there is as much threat as there is opportunity in this extraordinary situation. America can supplement increasing agricultural productivity elsewhere and store and sell enough from storage to

counter a bad harvest, but our own future prosperity in
turn is dependent on increasing agricultural self-reliance on the part of other nations. To encourage demand
for our wheat and corn beyond our capacity to meet that
demand would only breed hatred in disappointment. To
eat in abundance while growing numbers of others starve
would provoke hostility. The blatant manipulation of
food supplies for political gain would result in reciprocal retaliation by the most telling means available.
In an ever more economically interdependent world, the
best means available need not be military but are more
likely to be economic. OPEC should by now have demonstrated the effect on national security of economic
rather than military measures. Fair market prices for
American agricultural products can be justifiably
expected for exports that supplement and enrich the
productivity of others. Fair market prices are not
likely to result from exports to relieve starvation. A
monopoly cartel to enforce market pricing in such a
situation would endanger national security.

To me it seems that the vital ingredient to maximize opportunity and minimize threat in this extraordinary American situation is to give full and effective
priority to the achievement of world food stability.
That requires a level of increased productivity abroad
that limits the American role as supplier to the functions of supplement and enrichment. And that, in turn,
requires a new investment in agriculturally related
research at home -- in large part to reduce the dependence of our present agricultural productivity on
petroleum products -- and abroad. It also demands a
vast new program of training of human talent to prepare
and deploy the skills which modern agriculture employs.
There are high costs involved. The American consumer
will pay more for food as market-priced demand rises
for the products of American agriculture. However,
under the best of circumstances such costs will be a
good investment not only morally, not only for national
security, but also economically. A hugely positive
balance of payments for United States agriculture may
alone have the capacity to balance for ourselves the
negative balance of payments we will run up in other
aspects of consumption.

My ultimate conclusion with respect to world food
is that -- whether we know it or not -- we as a people
and government are now choosing the destiny of the
United States for the next century. To do nothing is
for us today as effective a choice -- although in my
judgment a wholly negative one -- as to make a decision. We are cast in a giant role. We can succeed or
fail, but we cannot hide. How we behave will so dras-

tically affect the future of the world food situation that we will inescapably be assigned the principal responsibility for its evolution. I think we face a promising choice. I hope we make it in time. And I hope that we can make it boldly. We must not endanger our future, and so the right choice must come first. If we can make that choice, I hope that we need then not merely back into the future, but step ahead with purpose.

Printed and bound by CPI Group (UK) Ltd, Croydon, CR0 4YY

23/10/2024

01778239-0001